D1367333

PUBLISHED IN CO-EDITION
WITH SNTL – PUBLISHERS OF TECHNICAL LITERATURE, PRAGUE

DISTRIBUTION OF THIS BOOK IS BEING HANDLED BY THE FOLLOWING PUBLISHERS
FOR THE U.S.A. AND CANADA
ELSEVIER/NORTH HOLLAND, INC.
52, VANDERBILT AVENUE
NEW YORK, NEW YORK 10017

FOR THE EAST EUROPEAN COUNTRIES, CHINA, NORTHERN KOREA, CUBA, VIETNAM AND MONGOLIA
SNTL – PUBLISHERS OF TECHNICAL LITERATURE, PRAGUE

FOR ALL REMAINING AREAS
ELSEVIER SCIENTIFIC PUBLISHING COMPANY
1, MOLENWERF
P.O. BOX 211
1000 AE AMSTERDAM, THE NETHERLANDS

LIBRARY OF CONGRESS CATALOGUE NUMBER:

ISBN 0-444-99736-9 (Vol. 13A)
ISBN 0-444-99734-2 (Set)
ISBN 0-444-41735-4 (Serie)

PRINTED IN CZECHOSLOVAKIA

COMPREHENSIVE ANALYTICAL CHEMISTRY

ADVISORY BOARD

VOLUME XIII

ANALYSIS OF COMPLEX HYDROCARBON MIXTURES

by
Slavoj Hála, Mečislav Kuraš *and* Milan Popl
Institute of Chemical Technology
Prague, Czechoslovakia

CONTENTS PART A

CONTENTS PART B

Wilson and Wilson's
COMPREHENSIVE ANALYTICAL CHEMISTRY

Edited by
G. SVEHLA, PH.D., D.SC., F.R.S.C.
Reader in analytical Chemistry
The Queen's University of Belfast

VOLUME XIII

ANALYSIS OF COMPLEX HYDROCARBON MIXTURES

Part A
SEPARATION METHODS

by
SLAVOJ HÁLA
MEČISLAV KURAŠ
MILAN POPL

Institute of Chemical Technology, Prague, Czechoslovakia

ELSEVIER SCIENTIFIC PUBLISHING COMPANY
AMSTERDAM OXFORD NEW YORK
1981

WILSON AND WILSON'S

COMPREHENSIVE ANALYTICAL CHEMISTRY

VOLUMES IN THE SERIES

Vol. IA	Analytical Processes Gas Analysis Inorganic Qualitative Analysis Organic Qualitative Analysis Inorganic Gravimetric Analysis
Vol. IB	Inorganic Titrimetric Analysis Organic Quantitative Analysis
Vol. IC	Analytical Chemistry of the Elements
Vol. IIA	Electrochemical Analysis Electrodeposition Potentiometric Titrations Conductometric Titrations High-frequency Titrations
Vol. IIB	Liquid Chromatography in Columns Gas Chromatography Ion Exchangers Distillation
Vol. IIC	Paper and Thin-Layer Chromatography Radiochemical Methods Nuclear Magnetic Resonance and Electron Spin Resonance Methods X-Ray Spectrometry
Vol. IID	Coulometric Analysis
Vol. III	Elemental Analysis with Minute Samples Standards and Standardization Separations by Liquid Amalgams Vacuum Fusion Analysis of Gases in Metals Electroanalysis in Molten Salts

Vol. IV	Instrumentation for Spectroscopy Atomic Absorption and Fluorescence Spectroscopy Diffuse Reflectance Spectroscopy
Vol. V	Emission Spectroscopy Analytical Microwave Spectroscopy Analytical Applications of Electron Microscopy
Vol. VI	Analytical Infrared Spectroscopy
Vol. VII	Thermal Methods in Analytical Chemistry Substoichiometric Analytical Methods
Vol. VIII	Enzyme Electrodes in Analytical Chemistry Molecular Fluorescence Spectroscopy Photometric Titrations Analytical Applications of Interferometry
Vol. IX	Ultraviolet Photoelectron and Photoion Spectroscopy Auger Electron Spectroscopy Plasma Excitation in Spectrochemical Analysis
Vol. X	Organic Spot Test Analysis The History of Analytical Chemistry
Vol. XI	The Application of Mathematical Statistics in Analytical Chemistry Mass Spectrometry Ion Selective Electrodes
Vol. XII	Thermal Analysis Part A. Simultaneous Thermoanalytical Examinations by Means of the Derivatograph

Preface

In Comprehensive Analytical Chemistry, the aim is to provide a work which, in many instances, should be a self-sufficient reference work; but where this is not possible, it should at least be a starting point for any analytical investigation.

It is hoped to include the widest selection of analytical topics that is possible within the compass of the work, and to give material in sufficient detail to allow it to be used directly, not only by professional analytical chemists, but also by those workers whose use of analytical methods is incidental to their work rather than continual. Where it is not possible to give details of methods, full reference to the pertinent original literature is made.

From the very beginning the aim of the Editors was to include contributions which specialise in analytical methods of certain key industries. Volume XIII is the first of these "applied" or "industrial" volumes. It covers the analysis of hydrocarbon mixtures, from gases up to complex liquid fractions and solid waxes, used as fuels, lubricants or starting materials for chemical industries. The text is an updated translation of a successful Czech handbook. The authors' *Foreword* throws more light on the backgrounds of the work and the arrangement of material.

Further "applied" or "industrial" volumes are in preparation; soon we hope to publish a longer text on pharmaceutical analysis.

It is my pleasant duty to welcome Mr. Alan Robinson B. A., as the Associate Editor of the Series. In preparing this volume Mr. Robinson was most helpful. Also, Dr. C. L. Graham of the University of Birmingham, England, assisted in the production of the present volume; his contribution is acknowledged with many thanks.

April, 1981 G. Svehla

Foreword

The hydrocarbon materials obtained from petroleum, natural gas, coal, tar sands, oil-containing shales and other petrobitumens are the real driving force of the development of present-day technical civilization. In the form of motor fuels, lubricants, sources of heat and other power, as well as raw material for the chemical industry, more than 4 billion tons of hydrocarbon are consumed annually in the whole world and the demand for these materials is still increasing. However, parallel with the continuing exhaustion of natural resources and the escalating prices of hydrocarbon raw materials, the necessity of using these materials more economically, both in production and utilization, is becoming ever more apparent.

Analysis is one of the important ways of increasing efficiency in this field. Modern analytical techniques permit the observation of the changes in the chemical composition of hydrocarbon fractions very rapidly and with a minimum of demand on the size of the sample. Even in complex high-boiling fractions the content of important hydrocarbon classes and groups can be determined very accurately. In light fractions, hundreds of individual compounds can be determined with high sensitivity and reliability.

The chemical composition of hydrocarbon materials is a measure of quality, based on easily controllable and exact principles. It is therefore widely used today for the control of production, transport, refining and technology, as well as for the purposes of trade and the various forms of consumption and utilization of these materials.

This book, consisting of two parts, describes the techniques for the analysis of hydrocarbon mixtures, from gases up to highly complex liquid fractions and solid waxes.

The first part describes the methods of separation of hydrocarbons, used in the initial steps of the analytical procedures. They comprise distillation techniques, preparative variants of gas and liquid chromatography, methods with shape selectivity, and proved chemical separation methods.

In the second part spectroscopic, chromatographic and chemical techniques are described which are used for group analysis and for detailed analysis of hydrocarbon materials. For the sake of easy orientation the analyses of important industrial hydrocarbon mixtures are summarized in a special review. The concluding chapter of the second part is devoted to the trace analysis of hydrocarbons.

Individual chapters may be read in any order. The presentation is chosen so that all introductory information is available to the reader concerning the respective analytical techniques, including the essential minimum of basic theory. The main emphasis, however, is on practical hints and instructions, detailed descriptions and experimental details which enable determinations to be made without having to consult other sources. Important spectroscopic, chromatographic and other data regarding pure hydrocarbons that are indispensable for routine work are collected in a number of tables.

The most widely used ASTM methods are given in most cases as examples of standard analytical methods. For analysts studying the composition of a given hydrocarbon mixture by non-standard procedures, entailing the selection of suitable combinations of methods, the appropriate parts of the book will provide an excellent guide in affording an overall survey of all techniques used and their efficiency. Readers with a deeper interest in these matters can also find discussion of new trends and developments in the parts dealing with individual analytical techniques, and more than 1700 references to the primary literature in the text.

The various chapters of the book were written by the authors as follows:

1, 2.1, 2.3 – 2.7, 2.10 and 4.1 – S.H.; 2.8, 2.9, 3.1, 3.6, 3.7, 4.5, 5.1 – 5.3, 6.1 and 6.2 – M. K.; 2.2, 2.3, 3.2 – 3.5, 4.2 – 4.4 and 6.3 – M. P.

S. Hála, M. Kuraš, M. Popl

Part A
Separation Methods

Contents

Chapter 1

Introduction

The main source of hydrocarbons for producing refined fuels and oils and for the petro-chemical industry are fossile fuels – mainly natural gas, petroleum, coal and various petrobitumens. In spite of enormous reserves of fossile fuels, the steadily increasing consumption and the steeply increasing cost of hydrocarbon materials is turning attention generally to a reasonable and economical use of all forms of this irreplaceable natural wealth.

Rapid and accurate analytical control of feed stocks and products is an important and often decisive factor in technological progress in the processing of hydrocarbon materials. The contribution of modern methods of instrumental analysis in this field is best seen from the fact that the current methods of checking the properties of various fractions are being increasingly replaced and completed by methods permitting the determination of their chemical composition. Thus the characterization of the fractions is of a more fundamental nature; it is more accurate, better controllable, and mainly it permits a deeper insight into the changes in chemical composition of a fraction in the course of its industrial processing and utilization.

Analyses aimed at determining the chemical structure of the hydrocarbon components and their quantitative representation in a mixture can be classified into routine and research ones. In the first the composition of the mixture is roughly known and mainly the quantitative ratio of the components is checked by a well-tried technique. The most often used methods of routine analysis are standardized. Research analyses are more complex, because only little initial information on the composition of the sample is available. The chemical structures of the components present are determined only in the course of analysis. Research analyses are time-consuming and they often require combinations of various separation and identification techniques.

Nowadays the three following types of analysis are most widely used:

– *Determination of the content of all structural classes present.* In this manner all current fractions, up to the heavy ends, can be determined. Thus the ratio of saturated, unsaturated and aromatic hydrocarbons is obtained as the simplest result. Using more complex separation schemes various structural groups and types (group or type analysis) can be specified in detail within each particular structural class. The results afford a good knowledge of the chemical composition of the whole fraction and therefore this type of analysis is most often used in the evaluation of important starting materials and products.

– *Determination of the quantitative representation of individual hydrocarbons within one structural class.* The difficulty of a detailed analysis increases rapidly with increasing boiling point of the fractions. However, in some structural classes, if these are isolated beforehand, individual hydrocarbons can be determined quantitatively even in heavy fractions (for example *n*-alkanes up to C_{50}, polynuclear aromatic hydrocarbons up to C_{26}, etc.). The analyses of this type are often used for various correlation purposes and in geochemical studies. In petroleum over 450 individual hydrocarbons have been already identified and determined by these procedures.

– *Determination of all the individual hydrocarbons in a fraction.* This detailed analysis affords a maximum of information. However, the efficiency of current analytical techniques permits detailed analysis only of gases and light petroleum distillates. For example, in non-olefinic gasoline about 200 alkanes, cycloalkanes and aromatic hydrocarbons can be determined in a single run.

The presentation of analytical results must be based on an unambiguous classification of hydrocarbon structures. An example of a simple classification is shown in Table 1.1. The basis for the classification is the following hierarchy of the structures: alkanes – cycloalkanes – alkenes – (alkynes) – cycloalkenes – aromatic hydrocarbons. The alkane C – C bond is the lowest structural element, while the aromatic ring is the highest in order. Hydrocarbons with various types of structural elements are classified into groups according to the highest possible order. If the hydrocarbon contains a single double bond in the aliphatic chain it is classified as alkene

TABLE 1.1

Classification of hydrocarbons according to chemical structure

Class	Group	Types (Examples)	z from C_nH_{2n+z}
Alkanes	*n*-Alkanes	even, odd	+2
	Isoalkanes	2-Methyl- (iso-) 3-Methyl- (anteiso-) Isoprenoids	+2
Cycloalkanes	Monocycloalkanes	Alkylcyclopentanes Alkylcyclohexanes	0
	Dicycloalkanes	Alkylhydrindanes Alkyldecalins Alkyldicyclohexyls	—2
	Tricycloalkanes	Alkylperhydroanthracenes Alkyladamantanes	—4
Alkenes (Olefins)	Alkenes	*n*-Alkenes Branched alkenes Terminal alkenes	0
	Alkadienes	Conjugated Isolated	—2
Cycloalkenes	Cycloalkenes	Alkylcyclopentenes Alkylcyclohexenes Alkyloctahydronaphthalenes	—2
	Cycloalkadienes	Alkylcyclohexadienes	—4
Aromatic hydrocarbons	Monoaromatic hydrocarbons	Alkylbenzenes Alkyltetralins	—6 —8
	Diaromatic hydrocarbons	Alkylnaphthalenes Alkyldiphenyls Alkylacenaphthenes Alkylfluorenes	—12 —14 —16
	Triaromatic hydrocarbons	Alkylanthracenes Alkylphenanthrenes	—18

even when the chain is attached to a naphthenic (cycloalkanic) ring. Similarly, a hydrocarbon with two aromatic nuclei, either condensed or separated by a bond, an alkane chain or naphthenic nuclei, belongs into the group of diaromatic hydrocarbons regardless of how many naphthenic rings the molecule contains.

Chapter 2

Separation of hydrocarbon mixtures

Hydrocarbon mixtures are the principal constituents of various natural *petrobitumens,* and they are also produced in large amounts industrially. They occur in various forms—as gases, liquids and solids, and mixtures containing hydrocarbons of all three states simultaneously are no exception. Their complexity also varies within broad limits—from the simplest binary mixtures, up to mixtures containing hundreds and thousands of individual hydrocarbons. Heavy hydrocarbon fractions (oils) belong among the most complex mixtures met with in industry, both in the number of their components and mainly the negligible differences in the properties of the components.

Parallelly with the increasing economic importance of hydrocarbon fractions for the petro-chemical industry, transport and as an energy source, special techniques have also developed gradually which permitted the analysis of such complex mixtures. They are characterized by sequences of various separation and analytical methods which are often combined into complex, multistep schemes. A general procedure for the analysis of hydrocarbon mixtures can be represented in a simplified manner by the scheme in Fig. 2.1. The first step in the scheme is the separation of the fractions, under which is understood the separation of the starting mixture into several fractions with a more limited number of components, or some other simplification of the sample. The next step—the separation of groups—has as its aim the separation of hydrocarbons with differing chemical structure, i.e. obtaining separated structural classes of hydrocarbons (alkanes, cycloalkanes, aromatic hydrocarbons, and others). A sufficiently sharp separation in the first two steps is usually favourable for a successful separation of individual hydrocarbons in the last separation step. The identification of the separated hydrocarbons and their determination is the last and final aim of an ideal analysis. In the case of simpler mixtures a complete quantitative analysis is possible, of course, even when some separation step is omitted – as indicated in the scheme by the arrow on the left. On the other hand, the complexity of some hydrocarbon mixtures is so great that it is impossible or useless to determine

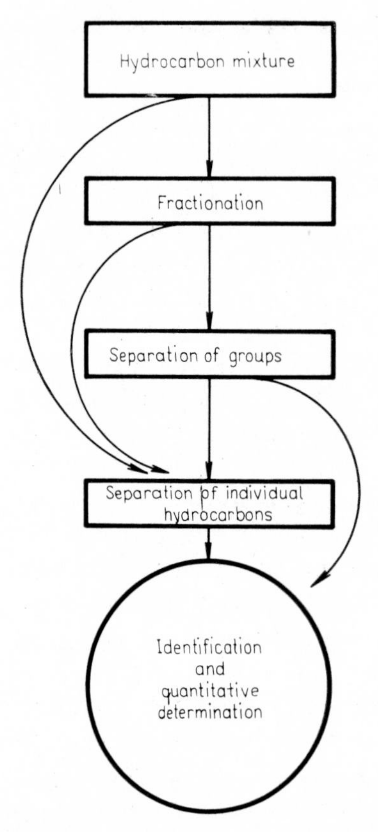

Fig. 2.1 General procedure of the analysis of hydrocarbon mixtures

the content of all hydrocarbons present, and the analysis is limited to the determination of one or several structural groups only (arrow on the right).

As evident from Fig. 2.1 separation methods play an important role in the analysis of hydrocarbons. The separation of substances on the basis of various physical, physico-chemical and chemical principles is today a relatively distinct scientific area (*separation science*) with extensive practical importance for industry and laboratory technique. The theoretical foundations of the majority of separation methods are understood to a considerable depth and they are presented in detail in modern monographs[1-7].

In this chapter separation methods are reviewed which are used in the analysis of hydrocarbons, mainly from the practical point of view. Theoretical foundations of the methods are discussed only to the extent necessary for the understanding of basic relationships. Only those methods are discussed which are of preparative importance, i.e. those permitting the isolation of at least milligram fractions of compositions different from that of the starting mixture. The methods such as *mass spectrometry* and *gas chromatography* in capillary columns, which—in principle—also separate the components of a mixture, but in such minute quantities that they can only be registered with a detector but not isolated, are discussed in subsequent chapters.

Among the physical and physico-chemical separation methods used in the analysis of hydrocarbon mixtures those are mainly used which are based on the differences in the distribution of components between two phases at equilibrium. These include *distillation, extraction,* various *chromatographic* methods, formation of *inclusion* compounds, etc. *Gel chromatography* and *thermodiffusion* operate on a different principle; in these methods the components are separated on the basis of different rates of movement in one phase. The most frequently employed separation methods, described in detail in subsequent chapters of this book, are reviewed in Table 2.1. The methods are arranged according to the system of phases in which separation occurs. Each method in the table is characterized by *selectivity,* range of *utilizability, productivity* and *fractionation capacity*.

The *selectivity* of the methods is compared qualitatively only, according to whether the components are separated on the basis of differences in the structures of the molecules, or on the basis of the differences in the geometry and shape of the molecules. Those methods which separate the components predominantly according to their *volatility, boiling point* and other properties closely connected with the *molecular mass* are designated as *non-selective*. Some separation methods display several types of selectivity simultaneously, but one usually prevails. For example, *thermodiffusion*

TABLE 2.1

List of methods for the separation of hydrocarbons

Method	Phase system	Selectivity of the method	Range of utilizability	Productivity of the method	Fractionation capacity
Distillation	G—L	non-selective	liquids	g/h — kg/d	10
Gas-liquid chromatography	G—L	non-selective or structurally selective	gases, volatile liquids and solids	mg — g/h	20
Sublimation	G—S	non-selective	some volatile solids	g/h	2
Gas-solid chromatography	G—S	non-selective or structurally selective	gases, volatile liquids and solids	mg — g/h	10
Extraction	L—L	structurally selective	liquid and solid substances	g/h — kg/d	2
Liquid-liquid chromatography	L—L	structurally selective	liquids and solids	mg — g/h	10
Crystallization	L—S	non-selective or shape selective	solids	g/h — kg/d	2
Zone melting	L—S	non-selective or structurally selective	solids	g/h — kg/d	2
Liquid-solid chromatography	L—S	structurally selective	liquids and solids	mg — g/h	5—10
Inclusion compounds	L—S	selective by shape	liquids and solids	g/h — kg/d	2
Molecular sieves	L—S	selective by shape	liquids and solids	g/h	2
Gel chromatography	L	non-selective or selective by shape	liquids and solids	g/d	5
Thermal diffusion	L	selective by shape	liquids	g/d	5

can separate a mixture of *n*-alkanes and cycloalkanes according to their molecular shape to a fraction of cyclic molecules and another with linear molecules. If, however, the separated mixture contains *n*-alkanes only, the ability of thermodiffusion to separate according to molecular weight arises, and a separation of lower-boiling alkanes from higher-boiling ones takes place.

The extent of *utilization* indicates which types of hydrocarbon mixtures can be separated by the given method. The mentioned ranges of utilization apply for a common arrangement of methods. In some methods the mentioned ranges can be extended by special measures. For example, the heating of the column head and the receiving system enables the separation even of low-melting solid substances by distillation; on the contrary, gaseous hydrocarbon mixtures can also be distilled using cryostated columns.

The *productivity* of the method is expressed as the amount of the sample which can be worked up by the respective method in a unit of time (at an average *separation efficiency*). Similarly as in the preceding case the values given correspond to the usual arrangement of the methods. A miniature version of the instruments, for example miniature distillation or thermodiffusion columns and other laboratory microtechniques enable in some instances the separation of even lower amounts of samples than indicated in the table. On the other hand some separation methods are adapted in laboratories for large-scale, often continuous processes, the productivity of which greatly exceeds the indicated upper limits.

Fractionation capacity, indicated in the last column of Table 2.1, represents – according to Karger et al.[1] – the maximum number of components which can be separated by a given method in a single operation. Such methods as *sublimation,* simple *extraction, crystallization, inclusion* compounds formation, and separation using *molecular sieves,* can separate the mixture at the best to two parts, and their fractionating capacity is equal to 2. The fractionating capacity of chromatographic methods is much higher and it corresponds to the maximum number of completely separated peaks which can be separated on a given column; this is the so-called *peak capacity,* defined by Giddings[8]. For example, analytical gas chromatography on a capillary column has a peak capacity of up to several hundreds. A chromatographic column with an efficiency of 10 000 *theoretical plates* has in gel chromatography a peak capacity of about 20, in liquid chromatography about 60 and in gas chromatography more than 100. Preparative chromatographic methods, employed in practice for the separation of hydrocarbon mixtures, have a fractionating capacity within the 5 – 20 range, owing to the lower separation efficiency of the columns and other limitations.

In addition to the physical and physico-chemical procedures listed in Table 2.1 the separations of hydrocarbon mixtures based on *chemical reactions* are also discussed in this chapter. The chemical reactions are described which are used for the conversion of reactive hydrocarbons to chemical derivatives which can be easily separated from the remaining unreactive hydrocarbons. These procedures are frequently used, because the elimination of a certain class of hydrocarbons usually simplifies the analysed mixture considerably, so that the separation and the analysis of the remaining (*refined*) mixture is much easier. In some chemical separation processes the original hydrocarbons are regenerated from the separated derivatives, and then analysed by other methods. Direct exploitation of chemical reactions for analytical determinations of the content of some hydrocarbon groups is discussed in Chapter 3.6.

In the terminal section of Chapter 2 the analytical procedures of very complex samples are discussed, which require combinations of various separation methods. The principles of the choice of suitable methods and the standpoint for the selection of optimum sequence of these methods are demonstrated on some examples of proved separation schemes.

2.1 Distillation

Distillation – a method known for thousands of years – has not lost much of its importance even at the present time when *chromatography* and other laboratory high efficiency separation techniques have revolutionized the analysis of complex mixtures.

In the analysis of mixtures of hydrocarbons distillation is used mainly for the preparation of narrower fractions, which are then analysed by combinations of other separation and identification methods. The narrowing of the boiling range or the preparation of several narrower cuts, analysed separately, usually substantially increases the efficiency of subsequent separation methods and facilitates the identification and determination of the constituents. High efficiency distillation is often preferred over *preparative gas chromatography* mainly when higher outputs are required and when the starting material has a broad boiling range. Sometimes distillation is also used as the last step of a separation scheme for the isolation of small samples of pure hydrocarbons.

2.1.1 BOILING POINT AND THE DISTILLATION CURVE

The *boiling point* is defined as the temperature at which – at a given pressure – thermodynamic equilibrium is attained between the liquid and the gaseous phase of the substance. The boiling point and the *boiling range* should therefore be indicated together with the data on pressure at which they were measured. The "*normal boiling point*" refers to a pressure of 10.325 kPa.

The boiling points of a great number of hydrocarbons are measured and several comprehensive collections of these data[7,9–12] exist in literature. Normal boiling points of hydrocarbon *gases* are given in Table 2.1.13, while the boiling points of *gasoline* range hydrocarbons are listed together with *retention indices* in Table 4.1.19. The dependence of boiling points of hydrocarbons on pressure is discussed in section 2.1.4. The boiling points of *n*-alkanes at various pressures are given in Tables 2.1.11 and 2.1.12.

The majority of difficulties during the separation and the analysis of complex mixtures of hydrocarbons are caused by the very small differences in the properties of the hydrocarbons present. The monotonous course of the change of physical properties in *homologous series* is the reason why the differences between the properties of some members of various homologous series are much smaller than the differences between the neighbouring members of the same homologous series. The boiling points represent a typical example of these relationships; in Fig. 2.1.1 the intervals are represented within which the normal boiling points of *gasoline* hydrocarbons of three structural classes lie in dependence on the number of carbon atoms. From the figure it is evident that among the hydrocarbons with an identical number of carbon atoms *aromatic hydrocarbons* have the highest boiling points. The boiling points of *alkanes* and *cycloalkanes* are lower, and above C_7 they overlap. The theoretical distribution of hydrocarbons in various distillation fractions of gasoline can be seen

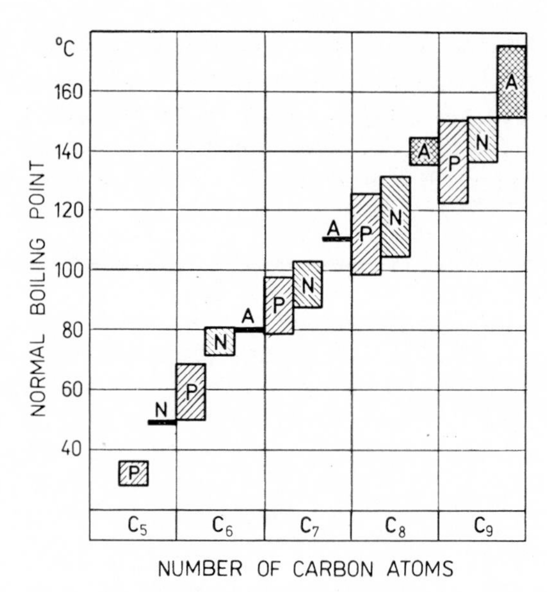

Fig. 2.1.1 Distribution of the boiling points of gasoline hydrocarbons C_5—C_9
P — alkanes, *N* — cycloalkanes, *A* — aromatic hydrocarbons

from Fig. 2.1.1. For example in the fraction 100 – 130 °C of a nonolefinic gasoline the $C_8 - C_9$ alkanes, the $C_7 - C_8$ cycloalkanes, and the C_7 aromatic hydrocarbon are present.

The dependences of the boiling points on the number of C-atoms in the molecule (or on the molecular weight) are different for individual *structural classes* of hydrocarbons. In Fig. 2.1.2 the curves are plotted expressing these dependences for *n-alkanes, isoalkanes* of an *isoprenoid* type, and unsubstituted *polynuclear aromatic hydrocarbons.* From the figure it is evident that the differences between the boiling points of hydrocarbons with the same number of carbon atoms, but with differing structure increase in higher members. This fact complicates the analysis of high-boiling mixtures of hydrocarbons of various structures, because even in fractions with a very narrow boiling range hydrocarbons can be present that differ substantially in their molecular weight. For example the normal boiling point of the C_{25} *n*-alkane is approximately identical with that of the C_{16} aromatic hydrocarbon.

When a certain hydrocarbon fraction is to be analysed, it is important to have an idea of the boiling points of the components present. The boiling point values of the constituents, their interval and mutual differences, are usually decisive for the choice of further separation methods. Therefore the complex hydrocarbon mixtures should be characterized by their *distillation curve,* or at least by the boiling range,

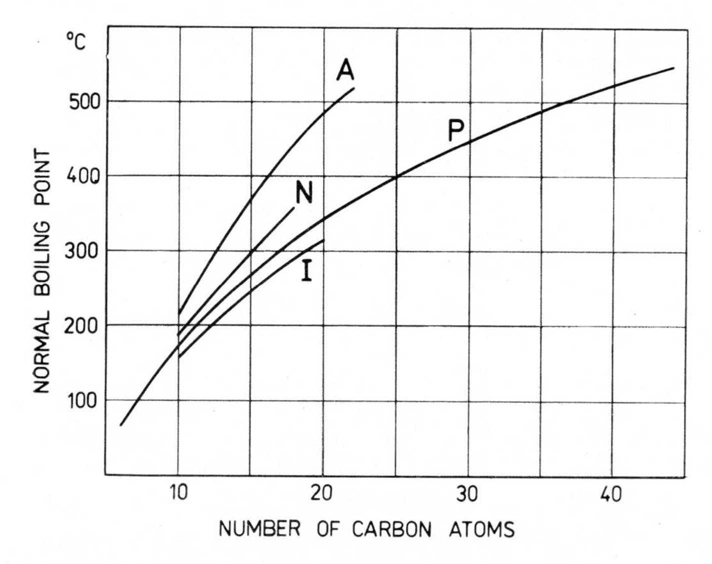

Fig. 2.1.2 Dependence of the boiling points of hydrocarbons on the number of carbon atoms
I — isoalkanes (isoprenoids), *P* — *n*-alkanes, *N* — polycycloalkanes (perhydroaromatic hydrocarbons), *A* — polynuclear aromatic hydrocarbons

before a detailed analysis. These data give an idea of the complexity of the fraction and on the interval within which the molecular weights of the present hydrocarbons vary.

The boiling point curve and the boiling range are determined using a standard method[13,14] in a simple distillation apparatus consisting of a glass flask of 200 ml volume and an oblique condenser. A 100-ml graduated cylinder serves for the collection of the distillate. The flask is heated with a gas burner or electrically. The temperature of the vapours going out through the side-tube into the condenser is measured with a glass thermometer with subdivisions 0.2 K.

Determination of the boiling point curve

The fraction (100 ± 0.5 ml) is measured with a cylinder into a distillation flask and this closed with a stopper through which the thermometer is inserted. The outlet of the condenser tube should extend at least 25 mm into the receiving cylinder, but it should not be below the 100-ml mark. The flask is heated so that the first drop of the distillate should fall into the receiver within 5—10 minutes for fractions boiling up to 150 °C, and within 10—15 minutes for liquids boiling above 150 °C. The temperature at which the first drop fell into the receiver is recorded as the *initial boiling point* (IBP). The receiving cylinder is placed so that the tip of the condenser tube touches the wall of the cylinder and the distillate flows down along it. The distillation rate is kept at 4—5 ml/min and the temperature is recorded at which the level of the distillate in the receiver reaches 5, 10, 20, 30, 40, 50, 60, 70, 80, 90 and 95 ml. Distillation is continued until the last drop of the liquid in the flask is evaporated, and this temperature is recorded as the *end point* of the distillation (EP). If any residue remains in the flask, the maximum temperature attained during the distillation, after 95 ml of the distillate have been collected, is recorded.

Each recorded temperature is corrected for normal pressure by adding the correction value k, calculated according to equation:

$$k = 0.00012\,(760 - b)\,(273 + T) \tag{2.1.1}$$

where b is the barometric pressure at the time of the distillation test in Torr and T is the temperature in °C registered with the thermometer. The results are presented as the corrected initial boiling point temperature, and the corrected temperatures of the distillation, corresponding to the measured volume % of the fraction. The volume % of the distillate, the residue and the losses are also noted.

It is typical of petroleum fractions that the wider the distillation interval and the higher the mean boiling point, the more hydrocarbons will be contained in the fraction. This is a consequence of the fact that with increasing molecular weight the differences between the boiling points of neighbouring members of homologous series (Fig. 2.1.2) steadily diminish, and that simultaneously with increasing molecular weight the number of existing *isomers* also rapidly increases (Table 2.1.1).

The boiling range determined by the standard procedure on a simple distillation apparatus is well reproducible and convenient for the common characterization of fractions. The measured values of the initial and the end boiling point do not however correspond to the *true boiling points* of the lightest and the heaviest components present in the fraction. The course of the *ASTM boiling point curve* is far from the true distribution of the boiling points. The approximation to the true values is the higher the more efficient the distillation apparatus employed. According to the definition of Greene et al.[15] the distillation plot measured on a column

TABLE 2.1.1

Number of hydrocarbon isomers

Class of hydrocarbons	Number of isomers								
	C_5	C_6	C_7	C_8	C_9	C_{10}	C_{11}	C_{12}	C_{13}
Isoalkanes[a]	3	5	9	18	35	75	159	335	802
Alkylcyclopentanes	1	1	6	15	55	150[b]			
Alkylcyclohexanes	—	1	1	8	40	225[b]	900[b]		
Alkylbenzenes	—	1	1	4	8	22	51		

[a] The number of isoalkane isomers is: C_{15} — 4.3×10^3; C_{18} — 6.0×10^4 and C_{20} — 3.6×10^5.

[b] Extrapolated values.

with 100 *theoretical plates* (see Section 2.1.2) and registered for fractions of 1 % volume is considered as the curve of *true boiling points* (TBP). Such a TBP curve gives more accurate information on the boiling range, on the complexity of the fraction and on the boiling points distribution. However, the time necessary for such distillation is about 100 h, and therefore the TBP curve is commonly determined on less efficient columns.

ASTM distillation record usually agrees with the TBP curve in the 50 % point. A correlation exists between the *slope* of the ASTM and TBP curves at the 50 % point, expressed by the equation[16,17]:

$$\text{TBP slope} = 1.15\,(\text{ASTM slope}) + 0.22 \qquad (2.1.2)$$

The slopes of the distillation curves are expressed in K per 1 % of distillate, and they can be calculated from the linear segments in the proximity of the 50 % point; for the ASTM distillation the slope is usually measured between the 20–80 % points, while for the TBP curve the interval is narrower, for example between the 40–60 % points. On the basis of the correlation between the slopes of the curves and the correlation between initial boiling points a calculation procedure has been developed for the transformation of the whole ASTM curve to the TBP curve and *vice versa*[16]. A good agreement between the calculated TBP curve and the experimental curve is illustrated by Fig. 2.1.3 in which the distillation curves of a gasoline fraction are shown.

The analytical exploitation of the distillation plot for the determination of the number of components and their content in the fraction has lost its original importance, because gas chromatography affords in a shorter time more detailed information on the composition; for example by gas chromatography even dissolved hydrocarbon gases can be determined

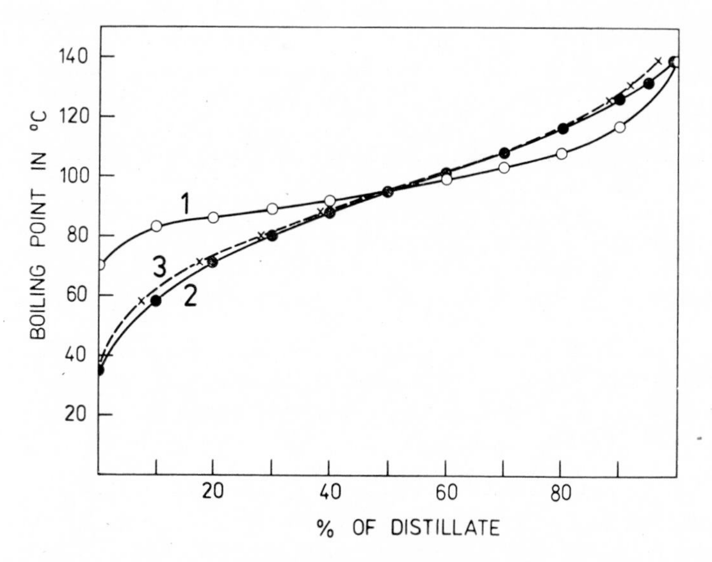

Fig. 2.1.3 Comparison of the experimental TBP curve (2) of the gasoline fraction with the TBP curve calculated (3) from ASTM distillation (1) (E. J. Hoffman[16])

in light fractions, which are usually included in the distillation losses of current distillation. On the other hand, on analysis by gas chromatography a possible content of non-volatile (high-boiling) components in the fraction can be overlooked if the relationships between the total signal and the volume of the injected sample is not being observed.

Using a gas chromatograph provided with an *integrator,* distillation characteristics of hydrocarbon fractions can be obtained with great accuracy and much faster than on distillation columns. The method called *simulated distillation*[15] enables the determination of the boiling range distribution on a gas chromatograph, which is equivalent to the TBP curve obtained by means of a distillation column with 100 theoretical plates. The method is also suitable for high-boiling hydrocarbon fractions, with the final boiling point about 500 °C. A few milliliters of sample suffice for the determination which takes only about one hour.

Determination of the boiling range distribution by gas chromatography (simulated distillation)[18]

A sample of a hydrocarbon fraction is analsyed by gas chromatography on a column with a stationary phase which separates hydrocarbons in order of their boiling point (for example the silicone phase SE-30, OV-1 or OV-101). During the analysis the column temperature is programmed and the area under the chromatogram is recorded throughout the run with an integrator. The cumulative area is registered on the chromatogram at short intervals (for example 10 s).

Under the same conditions a calibration mixture of known hydrocarbons (for example *n*-alkanes) covering the boiling range expected in the sample is analysed on the same column. From the retention times of the hydrocarbons and their boiling points a calibration curve is constructed, according to which a corresponding boiling temperature is assigned on the sample chromatogram to each interval. When the analysed sample displays distinct peaks of *n*-alkanes on the chromatogram, they may be used as boiling point calibration marks (see Table 2.1.12), and the calibration mixture need not be run.

On the evaluated chromatogram of the sample the point at which the cumulative area count is equal to 99.5 % of the total area corresponds to the temperature of the final boiling point. At the start of the chromatogram the temperature is found that corresponds to the first 0.5 % of the total area, and this is marked as the initial boiling point. The area values of each interval are divided by the total area and thus the percentages of the sample recovered at each interval are obtained. From the boiling temperatures assigned on the basis of the calibration curve to each interval, and from the percentual representation of each interval the boiling range distribution of the sample is tabulated or plotted in the usual manner. The determination can be made faster if a short capillary (10 m) coated with OV-101 phase[18a] is used as a column.

The distillation curve simulated by gas chromatography coincides well with the curve of true boiling points determined with an efficient distillation column. Certain differences may be met in samples containing a large proportion of high-boiling polycyclic hydrocarbons, because in the interval of high temperatures the dependence of retention times of these hydrocarbons on normal boiling points deviates from the calibration curve of *n*-alkanes. At reduced pressure these deviations are much smaller, so that significant differences from laboratory distillation practically do not occur even in the case of high-boiling aromatic fractions, because these fractions are also distilled at reduced pressure in order to prevent cracking of the sample. Worman and Green[19] give as an example the comparison of a simulated distillation data of a *heavy gas oil* with a high content of polycyclic aromatic hydrocarbons, with the true boiling point curve, measured on a 100 theoretical plates spinning band column, at a 133 Pa pressure (Table 2.1.2). The agreement is very good; the checking of the method on an artificial mixture of pure hydrocarbons confirmed that the simulated distillation data practically coincide with the theoretical boiling point curve.

TABLE 2.1.2

Boiling point curve of heavy gas oil[19]

Distillate in weight %	Normal boiling point in °C		Distillate in weight %	Normal boiling point in °C	
	TBP[a]	SD[b]		TBP[a]	SD[b]
IBP[c]	190	176	60	410	406
10	318	302	70	425	425
20	341	338	80	445	443
30	357	358	90	...	469
40	377	375	95	...	492
50	390	391	100	...	542

[a] TBP = true boiling points curve
[b] SD = boiling point curve simulated by gas chromatography
[c] IBP = initial boiling point

Simulated distillation is especially suitable for routine analyses of high-boiling fractions. It can also be applied to *crude oils* and other hydrocarbon fractions with a higher content of constituents boiling above 540 °C. In such a case the sample is rechromatographed with addition of 10 % of an *internal standard*[19]. The fraction of the sample boiling below 540 °C is calculated from the equation:

$$\text{weight \% below } 540\,^{\circ}C = \frac{100 \text{ weight standard}}{\text{weight sample}} \cdot \frac{\text{area counts sample}}{\text{area counts standard}} \tag{2.1.3}$$

Simulated distillation gives best results with samples distilling within the interval of several hundreds of K. The standard procedure[18] is therefore limited to fractions having a boiling range greater than 38 K. However, by adjusting the procedure even narrower fractions can be measured. In order to obtain accurate data on the boiling range of very narrow high-boiling fractions it is recommended to evaluate the simulated distillation by a simple *graphical method*[20]. From the chromatogram – which in the case of narrow fractions usually has the form of a simple peak of area A – the boiling range is determined from two points corresponding to A/2 – – 0.3413 A and A/2 + 0.3413 A, i.e. instead of 99 % of the area the distillation limits are restricted to 68.3 % of the area around the 50 % point. In the case of a chromatographic peak with a normal *Gaussian* distribution this corresponds to the area between the positive and the negative *standard deviation* σ.

Gouw et al.[21] have described a computer program (*Fortran*) for the processing data of simulated distillation. The results from a gas chromatograph, processed by a computer, are presented as a volume and weight percent TBP curve and as a machine drawn graph. The simulated distillation data can be converted to weight or volume percent data by use of response factors[21a].

The method of simulated distillation has been further elaborated by O'Donnell[22] for the characterization of the *fractionation ability* of column used for the distillation of multicomponent mixtures with a wide-boiling range (crude oils) and for the calculation of the distillation curve of any finite cut of the analysed fraction.

For aromatic concentrates the distillation curve can also be derived from the results of *low-voltage mass-spectrometric* analysis[22a]; the calculated curve agrees well with the simulated distillation.

2.1.2 DISTILLATION COLUMNS

Simple mixtures of hydrocarbons with sufficiently distant boiling points can be separated to pure components by the distillation itself. In the case of close-boiling mixtures the degree of separation depends on the *relative volatility* of the components and on the *separation efficiency* of the distillation process.

For an ideal system the relative volatility α of a binary mixture of the components 1 and 2 is expressed by the equation

$$\alpha = \frac{y_1 x_2}{y_2 x_1} = \frac{P_1^0}{P_2^0} \qquad (2.1.4)$$

where x is the mole fraction of the component in liquid phase, y is the mole fraction of the component in the gas phase, and P^0 is the vapour pressure of the pure component under the given temperature. Equation (2.1.4) shows that the relative volatility in *ideal* systems is a constant at a given temperature, which is determined by the ratio of vapour pressures of the pure components, and independent of the ratio of components in the mixture. The value of the relative volatility cannot be affected too much by a change of temperature, because in the case of hydrocarbon mixtures the ratio of vapour pressures changes only little with temperature. Generally the value α decreases a little with increasing temperature.

The dependence of the vapour pressure of pure hydrocarbons on temperature is most often expressed by *Antoine's equation*:

$$\log P^0 = A - \frac{B}{T + C} \qquad (2.1.5)$$

which for the boiling point calculation at a certain pressure can be written in the reversed form:

$$T = \frac{B}{A - \log P^0} - C \qquad (2.1.6)$$

where A, B and C are constants which are tabulated for a large number of hydrocarbons[7, 9, 11]. In Table 2.1.3 the constants of Antoine's equation are listed (for P° in kPa and T in °C) for several hydrocarbons used for the testing of distillation columns (see below). For binary mixtures of hydro-

TABLE 2.1.3

Constants of Antoine's equation (2.1.5), (2.1.6) for some hydrocarbons (for P° in kPa and T in °C)

Hydrocarbon	Norm. b.p. in °C	A	B	C
n-Heptane	98.427	6.025 17	1 266.871	216.757
2,2,4-Trimethylpentane	99.238	5.936 79	1 257.840	220.735
Methylcyclohexane	100.934	5.951 79	1 272.864	221.630
Benzene	80.099	6.037 00	1 214.645	221.205
Toluene	110.626	6.079 98	1 345.087	219.516

carbons that form *ideal mixtures* α is considered to be a constant within the interval of the boiling points of the components. For practical calculations the mean value of the relative volatility values at boiling points of both pure components is taken in such cases.

If the vapour tensions of pure components are not available, the relative volatility of ideal mixtures of hydrocarbons can be determined approximately from the *normal boiling points* of pure components (T_1, T_2) and the boiling point of the mixture T_M (in K), according to the equation[23]:

$$\log \alpha = \frac{T_2 - T_1}{T_M}(3.99 + 0.001939 T_M) \tag{2.1.7}$$

In the case of *non-ideal* mixtures of hydrocarbons the relative volatility can change with the composition of the mixture; then only the equality between the first two terms in the relation (2.1.4) is valid, for example, in the mixture of toluene and methylcyclohexane the relative volatility has a distinctly lower value for a mixture rich in methylcyclohexane, than for a mixture rich in toluene. In non-ideal mixtures the relative volatilities are determined experimentally from the composition of the liquid and the gaseous phase. For this purpose one-step – or in the case of mixtures with low relative volatility – multistep apparatus for the measurement of the liquid-gas equilibrium[1] are used.

Rectification. During distillation the separation effect is obtained by a process in which the liquid mixture is partly evaporated and the vapour is condensed separately. The mutual dependence of the composition of the liquid and the gas phase at equilibrium is expressed by the equation:

$$y_1 = \frac{\alpha \dfrac{x_1}{x_2}}{1 + \left(\alpha \dfrac{x_1}{x_2}\right)} \tag{2.1.8}$$

For the more volatile component 1 the equation (2.1.8) can be simplified:

$$y_1 = \frac{\alpha x_1}{1 + (\alpha - 1)\, x_1} \tag{2.1.9}$$

From the equation (2.1.9) it follows that the higher the concentration of the more volatile component in the vapour phase, the higher its concentration in the liquid phase and the higher the value of relative volatility. The composition of the vapour and the liquid phase in dependence on the relative volatility of the components can be easily read from the graph of *ideal equilibrium curves* calculated from the equation (2.1.9). The curves for six different α values are shown in Fig. 2.1.4.

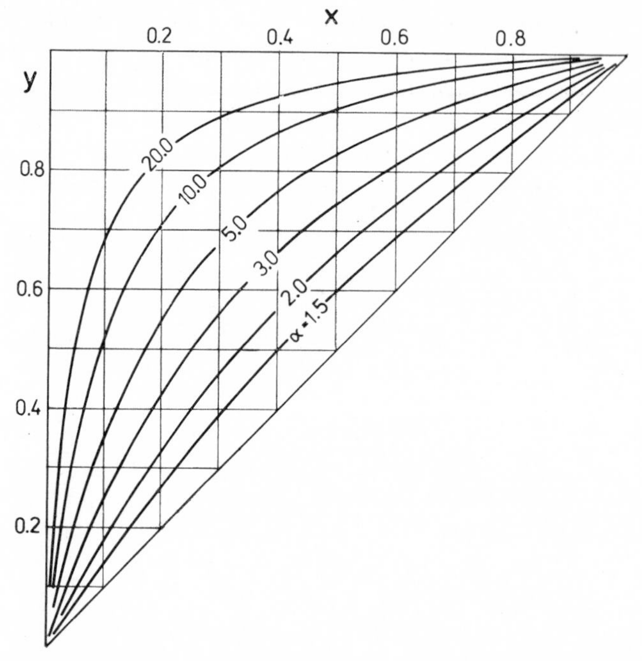

Fig. 2.1.4 Graph of equilibrium curves of ideal binary mixtures with relative volatility (α) 1.5—20
x — mole fraction of the more volatile component in the liquid phase, y — mole fraction of the more volatile component in the gas phase

At values of α higher than 1 the condensate obtained by the condensation of the vapour phase is enriched by a more volatile component and in the remaining liquid mixture the content of the less volatile component is increased. Repeated evaporation of a part of the condensate and the separate condensation of the vapours leads to further concentration of the more volatile component in the distillate. Practically, this effect is made use of in *rectification sections* of distillation columns, where the ascending vapour comes into contact with the liquid distillate, which partly returns into the column. In the rectification section of the column equilibrium is attained repeatedly between the ascending vapour and a countercurrent of the condensate, while less volatile components from the vapour are condensed and a vapour considerably enriched in more volatile components leaves the column. The *separation efficiency* of a distillation column is expressed by the number of consecutive attainments of equilibrium, taking place during the passage of the vapour through the column, and the number of these steps is called the *number of theoretical plates* (n). In addition to the way in which the equilibrium attainment between both phases is accelerated, the number of theoretical plates is also dependent – *inter alia* – on the height of the rectification section (L). Therefore, for quantitative evaluation of the distillation columns the *height equivalent to a theoretical plate* ($H = L/n$) is a more suitable measure for their separation efficiency. H is a height segment of the rectification section which causes the same change in the composition of a fractionated mixture as equilibrium attainment between the liquid and the vapour phase.

Factors affecting the operation of the distillation column. Various factors influence the result of the separation by distillation, i.e. the purity of the separated components and the time necessary for the achievement of the separation; some of these factors depend on the construction of the column and others can be affected partly or completely by the operating conditions.

One of the constructional factors that most strongly affect the result of separation is the efficiency of the rectification section (expressed by the height equivalent to a theoretical plate H) and its length. Although a high number of theoretical plates can be achieved by prolongation of the column – for example 150-plate columns used by Rossini and co-workers[7] were 4.5 m long – the column height is limited in the case of today's commercial laboratory columns to 1 – 2 m for practical reasons.

Another important factor is the *throughput* of the column, i.e. the amount of vapour passing through the column in a time unit; it is given as the volume of liquid (in ml/min) formed on condensation of this vapour and it is measured by the number of drops falling from the condenser drip-tip in the column head. The throughput can be controlled by the intensity of the boil in the pot, and its upper limit is given by the construction, and especially by the resistance (back pressure) of the rectification section. When the upper limit of the throughtput is attained, the velocity of the flowing vapours in the column is so high that it even prevents the free downward flow of the condensate through the rectification section and the column is *flooded*.

An important characteristic of the rectification section is the *operating hold-up* defined as the amount of vapour and liquid (expressed in ml of liquid) which is present in the rectification section when the column is operating in a steady state. According to the construction of the rectification section the operating hold-up depends more or less on the throughput and the amount of condensate which returns from the column head into the rectification section. Columns with a high hold-up are not suitable for the separation of mixtures containing some components at low concentration. During distillation single components of the separated mixture penetrate gradually into the rectification section and form its operating hold-up. In the case when the amount of some component is lower than the operating hold-up, this component does not fill the rectification section completely, and in consequence it is separated by a smaller number of plates than the column can afford. For efficient analytical distillations it is recommended that the amount of the separated component in the distillation flask should exceed at least 5 – 10 times the operating hold-up of the distillation column selected.

The rectification section of the column offers a certain resistance to vapour flow, resulting in a *pressure drop* between the pot and the column head. This is a further important factor affecting the operation of the column. The higher the resistance of the rectification section, the greater the pressure drop necessary for the maintenance of the required throughput. When the column is flooded the pressure drop increases steeply. A high pressure drop is an unfavourable factor, restricting the boil-up rate and causing an increase in pressure in the pot, and thus the boiling temperature increase of the pot charge. During work under reduced pressure there is a greater tendency of the column to flooding. In addition to this the utilization of low pressures for a decrease of boiling temperature in the distillation flask is limited.

At a steady operation of a distillation column all vapours passing through the rectification section are condensed in the column head and the condensate is divided into two parts with a valve; one part returns to the rectification section as *reflux,* and the second part is collected as distillate (product). The amount of reflux divided by the amount of distillate collected at the same time intervals is called *reflux ratio*. For calculations the amount of both liquids is usually expressed in moles. For practical control of the operation of a laboratory column the reflux ratio is determined by counting the number of drops falling from the drip tip over the rectification section, as opposed to the number of drops from the take-off drip-tip over the distillate receiver. Often the distillate is collected so that two time intervals alternate regularly, one during which all the condensate from the column head is returned as reflux (total reflux), and the other when all the condensate is withdrawn as distillate. In this procedure the reflux control value is monitored by an automatic timer and the reflux ratio is given approximately by the ratio of both selected time intervals.

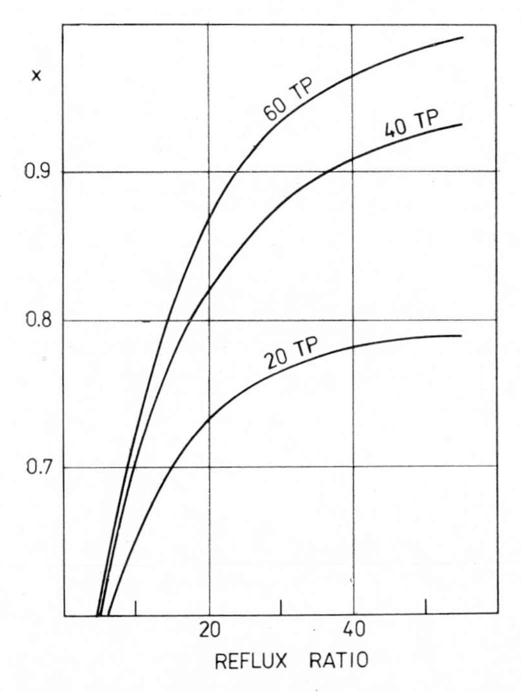

Fig. 2.1.5 Effect of the separation efficiency of the column and of the reflux ratio on the purity of the distillate during the rectification of a mixture of *n*-heptane and methylcyclohexane (50 : 50 mol %)
x — mole fraction of *n*-heptane in the distillate (according to H. Kolling[24])

The reflux ratio and the number of theoretical plates of the column are closely connected. The column has its highest separation efficiency at total reflux, i.e. when no distillate is withdrawn. With increasing product take-off the separation efficiency decreases. The requirements of a high separation efficiency and of a high rate of distillate collection are evidently contradictory, and therefore it is necessary to find a suitable compromise for each distillation. If the column has a higher number of theoretical plates than necessary for the separation of the given mixture, the operation can be carried out at a low reflux ratio, thus shortening the time of distillation. On the other hand the low separation efficiency of the column can be compensated to a certain extent by a high reflux ratio. The limits within which the reflux ratio affects the purity of the distillate obtained using packed columns with various numbers of theoretical plates[24] are evident from Fig. 2.1.5. The dependences show that when a 50 % mixture of *n*-heptane and methyl-cyclohexane is separated on a column of, for example, 20 theoretical plates, the increase in the reflux ratio above 40 has almost no effect on the purity of the collected *n*-heptane, and in fact it means only a useless extension of the time of distillation. On the other hand, the decrease of the reflux ratio below 15 rapidly worsens the separation of both hydrocarbons. For analytical distillations a reflux ratio equal to the number of theoretical plates of the column is chosen as optimum. According to Rose[25] it is not useful to keep the reflux ratio below 2/3 or above 3/2 of the number of theoretical plates.

Kinds of laboratory distillation columns. From the days of the old alchemists, who already knew the favourable effect of *deflegmation* on the result of distillation, progress in the separation efficiency of distillation is for the most part the result of the improvement of mass transfer and heat transfer in the rectification section. An ideal rectification section should mediate perfect contact of the ascending vapour with the refluxing condensate, further permit a high throughput, and at the same time it should have the lowest operating hold-up volume and pressure drop. These conflicting requirements are fulfilled by rectification sections of the distillation columns used to various degrees – dependent on their construction. Detailed descriptions of columns of various types are given in comprehensive monographs[26,27]. According to the arrangement of the rectification section, the distillation columns can be divided into the following types:

Open-tube columns. The rectification section of these columns is an empty tube, along the inner wall of which the reflux flows down as a thin film. The fractionating power of the section is dependent on the tube diameter and the upward velocity of the vapour. In tubes with larger diameter the area of the liquid film available for exchange with the ascending vapour is higher, but simultaneously the average distance between the vapour and the liquid phase is also larger, which worsens the contact between the two phases. Columns with an empty helical tube (*Jantzen column*) are also used, as well as columns with a tube the internal surface of which is increased by sintering or by indentations (*Vigreux column*). Current columns with empty tubes have height equivalents of a theoretical plate $H = 5 - 15$ cm, and they are characterized by a very low operating hold-up and a negligible pressure drop. They are suitable for distillations and microdistillations of low-boiling hydrocarbons and for vacuum distillations of easily separable mixtures. The separation efficiency of empty tube columns can be considerably increased if an extremely low throughput is maintained during the distillation. Such operation conditions can be maintained only with difficulty, and they are possible only when the thermal insulation of the rectification section is perfect. A Vigreux column with a heating jacket is suitable both for atmospheric and vacuum distillations of high-boiling hydrocarbon fractions, especially for a preliminary fractionation of crude materials with a very broad distillation range.

Plate columns. The rectification sections of these columns are partitioned at regular distances by plates of various constructions. The reflux flows from the upper plates to the lower ones and it forms a layer of liquid on each plate through which the ascending vapour must bubble. The columns of this kind are of various constructions, for example all-glass *Bruun bubble-cap column, perforated-plate columns,* etc. During the distillation the plate columns are easily monitored and they can operate at a high boil-up rate. The efficiency of a real plate, for example in the commercial Bruun column, attains 70 – 80 % of a theoretical plate. Owing to their large hold-up volume the plate columns are employed for the distillation of large volumes. The disadvantage of these columns is their high pressure drop; for example the Bruun column with 50 real plates has a pressure drop of about 3.3 kPa at the minimum, and 9.3 kPa at the maximum throughput[26]. For this reason plate columns are not suitable for vacuum distillation.

Packed columns. Such columns can be prepared easily in the laboratory and they are also mass produced in various perfectly equipped and automatized versions. The rectification section consists of a vertical tube filled with fine packing particles with large surface, mostly made of metal, glass or some ceramic material. For the correct function of a packed column it is necessary to flood it at the start of the distillation, in order to wet the packing completely. In the steady state the reflux forms a film on the packing, which has a much larger surface area for contact with the vapour phase than in columns with an empty tube. For the high efficiency of the column it is important that the diameter of the packing particles should be minimally 8 times smaller than the inner diameter of the column; under this condition the packing distributes the reflux uniformly over the whole diameter of the section .

For laboratory distillations of hydrocarbon mixtures a packing is most suitable, consisting of small wire helices of the type HELI-PAK. They are produced in three sizes, for columns with 0.5 – 1.5 cm, 1 – 3 cm, and over 2.5 cm internal diameters. A column with a rectification section of 2.5 cm diamater, packed with HELI-PAK of medium size (1.3 × 2.5 × 2.3 mm), has the height equivalent of a theoretical plate $H = 1.3$ cm, when the throughput is 50 % of the maximum value. The pressure drop of such a column per 1 m of height is about 0.6 kPa, and it permits operation under reduced pressure.

Columns with integral assembly packings. They contain insertions of various forms, in their rectification tube which prolongs the way both for the ascending vapour and the refluxing liquid and thus affords a high surface for the liquid-vapour interaction.

The simplest solution is a rectification section made of *concentric tubes*; in the rectification section a second tube is inserted that leaves only a narrow space of about 1 mm thickness for the contact of the vapour with the liquid film. The operation characteristics are similar to those in empty tube columns, but the separation efficiency is much higher; perfectly constructed columns made of concentric tubes of very accurate dimensions have a $H = 0.5 - 1$ cm.

Another well known type of column with integral packing is *Widmer column,* containing a glass helical insert. Metal packings preformed are also used, for example the column with *Stedman screen*.

Columns with an integral packing of the HELI-GRID type are widely used. They are constructed so, that one continuous piece of wire coil packing is tightly fitted into the rectification tube. The packing consists of a firm rod around which a wire helix is tightly wrapped. When the packing is put into the tube, the annular space between the central rod and the inner wall of the rectification tube is uniformly filled.

Between the windings of the coil a liquid film is maintained which has a large area, but which does not present a large resistance to the ascending vapours. The distillation columns with a rectification section of the HELI GRID type combine a high separation efficiency and a high throughput with a relatively low hold-up volume and a low pressure drop. They are produced in various sizes, including miniature versions for analytical distillations of 10 – 50 ml samples. A miniature column with 8 mm diameter of the rectification section, 90 cm long[29], provides an $H = 0.7$ cm at total reflux and 0.33 ml/min throughput value.

Spinning band columns. The favourable effect of the mechanical mixing of vapours in the rectification section on the efficiency of the distillation column has been known, in principle, for a long time[26,27]. In spite of this the development of the spinning band columns was slow, mainly owing to constructional problems of vacuum-tight driving mechanism. In comparison with the simple and efficient packed columns the original prototypes of spinning band columns were much more complex and clumsy at comparable efficiencies. The increasing requirements put on the separation efficiency of analytical columns led to the renewal of interest in spinning band columns[30] and stimulated an intensive development that brought about remarkable results.

The present types of *spinning band stills* commercially available have excellent operating properties and they represent important progress in the development of the rectification sections. They are produced in two versions: with a rotating *flat band* of metal mesh, or with a rotating *teflon spiral.* Both types are shown in detail in Fig. 2.1.6. A scheme of a spinning band distillation column is shown in Fig. 2.1.7.

The width of the metal band corresponds to the diameter of the rectification tube. Bands are made of various types of inert metallic screen; for hydrocarbon mixtures a band of stainless steel screen with 16 mesh

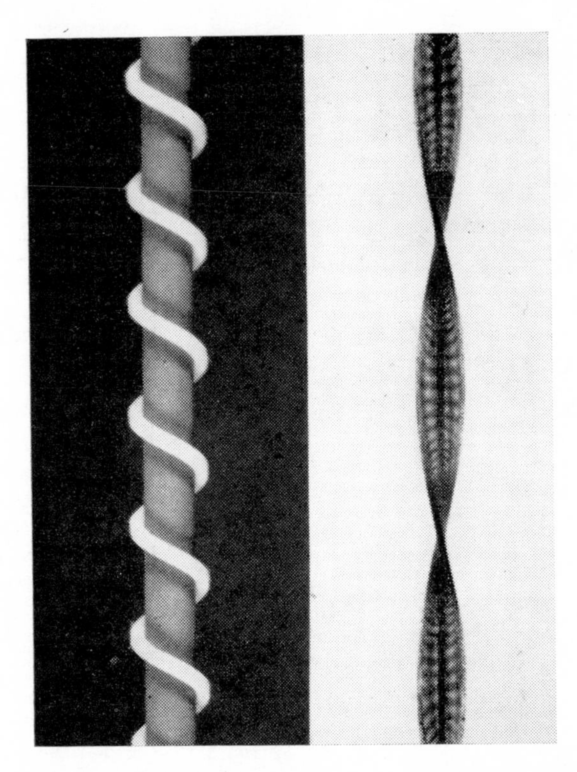

Fig. 2.1.6 Detail of the annular teflon band (on the left) and the stainless steel band (on the right) from spinning band columns

per 1 mm^2 is suitable. The band is constructed of two layers of metal mesh wrapped in a spiral fashion on a flexible drive shaft. The protruding fine wires at the edges of the band fit closely to the rectification tube, and they mix intensively the descending reflux film during the spinning. The band is driven either magnetically, or in newer types of columns, the drive shaft passes through the vacuum septum-seal in the column head, and it is directly connected with the motor. The spinning band columns are produced in various sizes, from a miniature column for the distillation of 0.5 – 10 ml of sample, having an efficiency of 30 theoretical plates, up to laboratory columns with a pot of several litres volume. They operate at one speed of the revolving of the band, i.e. 1 125 rpm. The columns are provided either with electrically heated jackets, or they are adiabatic, which is reflected in a higher efficiency.

Columns with *annular spinning teflon band* are constructed as fully automated analytical units with a separation efficiency of up to 200 theoretical plates. The teflon band is made of a solid teflon rod around which a teflon helix is wound. The diameter of the helix is slightly smaller than that of the rectification tube (Fig. 2.1.6). The teflon band is driven

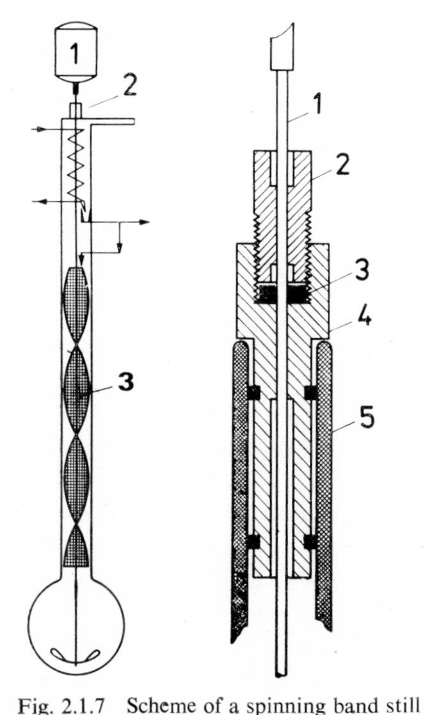

Fig. 2.1.7 Scheme of a spinning band still
1 — motor, 2 — drive shaft seal, 3 — spinning band

Fig. 2.1.8 Drive shaft seal of the rotating column
1 — drive shaft, 2 — nylon threaded septum retainer, 3 — septum seal, 4 — teflon shaft bearing, 5 — top of the column

by a steel flexible shaft passing through the whole length of the band. The upper end of the drive shaft passes through a septum-seal in the column-head and is directly connected with the motor. The speed of revolving is regulatable and it is measured by a *stroboscope*. The highest separation efficiency is achieved at about 7 000 rpm. A minor oscillation of the speed around the set value has no pronounced influence on the separation efficiency. Only when the band speed decreases by 1 000 rpm some loss in efficiency may result. In these columns the height equivalent to a theoretical plate has an enormously low value, $H = 0.3 - 0.5$ cm. At current throughputs of $0.25 - 1$ ml/min the operating hold-up of the column is 0.4 to 0.6 ml.

The columns with spinning metal band and annular spinning teflon band have a great advantage in that the drive shaft passes through the whole length of the rectification section, right to the bottom of the pot flask, where it ends with a teflon stirrer. The intensively stirred content of the pot eliminates overheating of the batch and irregularities in boiling. In combination with the thermostating of the pot flask it enables a sensitive regulation of the column. The stirring of the content of the pot flask suffices for the maintenance of smooth boiling even in *vacuum distillation*.

The sealing of the rotating drive shaft is solved simply and effectively (Fig. 2.1.8). The band drive rod protrudes through the teflon shaft bearing tightly fitted in the column head. In the broadened part of the teflon shaft bearing there is a septum-seal which is perforated with the sharp end of the shaft during the mounting, and the shaft is drawn through and connected with the motor. The septum is greased with vacuum grease and during the operation of the column it is lightly pressed to the rotating shaft with a threaded nylon septum-retainer. The septum may also be greased during the operation and it will hold a medium vacuum for dozens of working hours. For distillations at pressures under 100 Pa two septa may be used.

When the metal band or the teflon spiral is rapidly rotated the ascending vapours are very intensively mixed with the refluxing liquid, which ensures an intimate contact between the two phases. In addition to a rapid equilibrium the rotation produces a pumping action of the helical band. The helix is wrapped so that it forces down the excess liquid from the rectification section into the pot. Therefore the spinning band columns allow high throughputs and they have a very low tendency to flooding.

In separatíons of close boiling complex hydrocarbon fractions by distillation, the spinning band columns give the best results. Work with spinning band columns is no more difficult than distillation on columns of other constructions. However, certain differences should be kept in mind: When the annular teflon spinning band still is to be set in operation first the heating of the pot is switched on and the content is stirred at very low speed of the teflon band until the liquid starts to boil. The teflon band should not be spun at a higher speed until the whole rectification section is fully wetted[28] with the refluxing liquid, or the teflon helix might be damaged. Only when an abundant reflux appears in the column head the revolving of the teflon band may be increased to the operational value, i.e. maximum 7 200 rpm. The heating of the pot flask is then gradually decreased until the number of drops of the reflux in the column head is decreased to the value corresponding to the required throughput (4 – 16 drops per min), and then the still is allowed to equilibrate.

In columns with a metal band no special danger of damage during the spinning in a dry state is eminent, but even here it is better to stir the content of the pot flask during the setting in operation by switching the band on for short periods only until the liquid begins to boil. The heating of the pot flask is slowly increased until liquid appears in the bottom part of the rectification section. Then only the still motor and the band are set in motion. The heating of the pot is no longer increased, but the jacket of the rectification section should be gradually heated, and the liquid ascending in it observed. When the liquid begins to reflux in the column head the heating of the jacket is set so that the required throughput is obtained. *Adiabatic columns* which are vacuum jacketed and silvered are set in operation similarly as described for annular teflon spinning band columns.

The stainless steel band can be rapidly damaged if the distilled mixture contains hydrogen halides as impurity, further by some sulphur containing compounds or other corrosive chemicals. The teflon band is insensitive to corroding mixtures, but the band life will be shortened if it is operated at excessively high temperatures. If it is not exposed to temperatures above 175 °C it can operate for a very long time. In the authors' laboratory hydrocarbon fractions and common organic solvents have been distilled on an annular teflon spinning band column for more than 7 years without any sign of wear being observed on the teflon helix.

Another difference is the control of the separation of the components on an annular teflon spinning band column. As the separation efficiency of a 200 theoretical plates column is so high that it can separate hydrocarbons differing by 0.5 K in their boiling points, it is impossible to follow the changes in the composition of the vapour in the column head on the basis of the negligible changes of its temperature only. An arrangement is practical where a *gas chromatograph* is operating near the still. A drop of the distillate is then analysed by GC at required intervals, and the column operation and the collection of fractions is monitored according to the results. Samples of distillate for chromatographic analysis can be taken directly with a *Hamilton microsyringe* from the drop hanging on the take off drip tip.

If the isolation of smaller amounts of a high-purity product is required the maximum efficiency of the column can be exploited if the distillation is carried out with the so-called *intermittent take off*: the column is kept at total reflux until the chromatographic analysis shows that the purity of the head product no longer increases. Then a small amount of product is taken off and the column is returned to total reflux until a new equilibrium is established. This procedure is repeated until the necessary amount of the product is obtained.

At current distillations of multicomponent mixtures two alternative procedures are used: either the heating of the pot flask is kept constant and the throughput decreases during the distillate withdrawal, or a constant throughput is maintained, requiring a gradual increase in the heat input to the pot mantle in the course of distillation. In the first case it can happen that during the passage from one component to another the head temperature decreases, because the heat input to the pot no longer suffices for the maintenance of the reflux of the higher boiling component in the column head. For such cases the annular teflon spinning band stills are provided with a device which interrupts the collection of the distillate automatically when the temperature in the column head increases or decreases by more than 0.1 K than allowed, and switches the column over to total reflux, signalizing visibly and audibly that the still conditions are changed.

Testing of the distillation columns. Three factors given by the construction of the column have a great influence on its ability to separate a certain mixture: pressure drop, number of theoretical plates, and operating hold-up. Therefore in a new column all three factors are determined experimentally.

The measurement of *pressure drop* is the simplest. To do this a U-manometer suffices. One of its arms is connected with the pot flask and the second with the outlet from the condenser in the column head. The pressure drop is measured at various throughputs, from a minimum one up to the flooding point. The measured values serve for orientation only because the pressure drop can vary on the same column and at equal throughput by up to 100 % in dependence on the structure and the molecular weight of the hydrocarbons distilled. However, even an approximate value of pressure drop permits an evaluation of the suitability of a certain column for distillations at reduced pressures.

The *number of theoretical plates* is measured by some of the standard binary testing mixtures for which the curves of the equilibrium states of the liquid and the gas phase are known. For columns with an expected high number of theoretical plates testing mixtures with a low relative volatility are used – and *vice versa*. For the 10 – 90 theoretical plates range a mixture of *n*-heptane (n_D^{20} 1.3876) and methylcyclohexane (n_D^{20} 1.4231) is used, the relative volatility of which is α = 1.083 (ΔT = 2.5 K). For the 50 – 150 theoretical plates range a mixture of 2,2,4-trimethylpentane (n_D^{20} 1.3915) and methylcyclohexane is recommended with a relative volatility α = 1.049 (ΔT = 1.7 K).

Analytical columns used for batchwise distillations are usually tested at total reflux. The number of plates is determined at a steady state operation of the column, from the difference of the composition of the testing mixture in the pot flask and in the column head. The measured number of plates is dependent, to a certain extent, on the difference in volatility of the separated components and on the composition of the distilled mixture. Therefore a comparison of various distillation columns is more

TABLE 2.1.4
Dependence of the refractive index of *n*-heptane-methylcyclohexane mixture on the molar fraction of *n*-heptane (x)

x	n_D^{20}	x	n_D^{20}
0.00	1.4231	0.60	1.4005
0.10	1.4191	0.70	1.3971
0.20	1.4151	0.80	1.3939
0.30	1.4113	0.90	1.3907
0.40	1.4076	1.00	1.3876
0.50	1.4040		

accurate when the same testing mixture is employed. Columns with higher efficiency are tested with a mixture containing a more volatile component at a lower concentration (for example 20 or 30 mol %), so that the distillate contains a sufficient amount of the less volatile component. The accuracy of the calculation is then higher than when the more volatile component in the distillate greatly predominates. The composition of the samples of the testing mixture withdrawn from the column head and from the pot is determined from their refractive indices, density, or by gas chromatography. The number of theoretical plates n of the rectifying section of the column is given for testing mixtures with ideal behaviour by *Fenske's equation*:

$$n = \frac{1}{\log \alpha} \log \frac{\left(\dfrac{x}{1-x}\right)_{\text{head}}}{\left(\dfrac{x}{1-x}\right)_{\text{pot}}} - 1 \tag{2.1.10}$$

where x is the molar fraction of the more volatile component.

Determination of the number of theoretical plates of a distillation column

The pot flask of a cleaned and dried column is filled with half its volume of a suitable testing mixture. The amount of the mixture should be at least such that the volume of the more volatile component exceeds several times the operating hold-up of the column. The content of the pot is set to boiling at an average throughput and total reflux, and allowed to continue until equilibrium is attained. The course of attaining equilibrium is followed by analyses of small samples collected at regular intervals from the column head. In doing this the traces of humidity or other volatile impurities are eliminated from the column head drop by drop. The column is equilibrated when several consecutive samples of the distillate have the same composition.

At equilibrium 1 ml of distillate is withdrawn from the column head; simultaneously an equal volume of sample is also withdrawn from the pot flask. For this a sampling tube is used, introduced under the liquid surface, or the sample is taken with a syringe, through the septum on the side neck. The composition of the samples withdrawn is determined by refractometry at 20 °C.

When the testing mixture is *n*-heptane-methylcyclohexane[31] the molar fraction of *n*-heptane is computed for both samples with a 0.01 accuracy from the graphical dependence of the refractive index on the composition, plotted according to Table 2.1.4. The mole fraction of *n*-heptane in the samples from the column head and from the pot is put into equation (2.1.10), and for α the relative volatility of the mixture, 1.083, is substituted.

Another, purely algebraic procedure, is mentioned in the monograph ref.[7]. The number of theoretical plates n calculated from equation (2.1.10) refers to the rectifying section only and not to the entire column, because one plate, corresponding to the transfer from the liquid level in the pot to the bottom part of the rectifying section is subtracted from the right side of the equation.

The data necessary for the testing mixture 2,2,4-trimethylpentane–methylcyclohexane have been published by Walsh et al.[32]. At low concentrations of 2,2,4-trimethylpentane the relative volatility of the mixture deviates a little from the constant value according to the equation

$$\alpha = 1.04774 + 0.05126\,(0.003\,13)^x \tag{2.1.11}$$

where x is the molar fraction of 2,2,4-trimethylpentane.

For the dependence of the refractive index (at 20 °C) of the mixture on the composition, the equation is proposed

$$n_D^{20} = 1.4231 - 0.0376x + 0.0061x^2 \tag{2.1.12}$$

where x is the mole fraction of 2,2,4-trimethylpentane in the mixture.

When a larger number of measurements is performed it is advantageous to use Table 2.1.5 and plot the dependence of the refractive index of the mixture on the corresponding number of theoretical plates. This simplifies the calculation of the efficiency of the columns: for samples withdrawn from the pot and the column head refractive indices are measured at 20 °C, and corresponding numbers of theoretical plates are read from the

TABLE 2.1.5

Refractive index versus the number of theoretical plates for the mixture 2,2,4-trimethylpentane—methylcyclohexane[32]

n_D^{20}	Number of TP	n_D^{20}	Number of TP
1.4226	0	1.3978	90
1.4221	10	1.3957	100
1.4209	20	1.3943	110
1.4188	30	1.3933	120
1.4157	40	1.3927	130
1.4120	50	1.3923	140
1.4079	60	1.3920	150
1.4040	70	1.3918	160
1.4005	80		

graph for both values of refractive indices; the difference of these two values gives the number of theoretical plates of the column directly. For example, if the mixture in the pot has $n_D^{20} = 1.4157$ and the sample from the column head $n_D^{20} = 1.3943$, the efficiency of the distillation column is estimated to $110 - 40 = 70$ theoretical plates.

For the testing of columns of low efficiency (up to 25 theoretical plates) the testing mixture benzene-tetrachloromethane is often employed. Both components are easily accessible in sufficient purity and necessary amount; therefore they are also suitable for the testing of stills with a large volume of the pot flask, high hold-up, etc. Even in this case the number of theoretical plates is determined directly from the refractive index (at 25 °C) of the samples withdrawn from the pot and the column head. A graph is used for this purpose, plotted on the basis of the tabulated values of the

TABLE 2.1.6

Refractive index versus the number of theoretical plates for the mixture benzene—tetrachloromethane

n_D^{25}	Number of TP	n_D^{25}	Number of TP
1.4617	37	1.4741	18
1.4621	36	1.4753	17
1.4626	35	1.4767	16
1.4630	34	1.4781	15
1.4634	33	1.4795	14
1.4638	32	1.4810	13
1.4643	31	1.4825	12
1.4648	30	1.4840	11
1.4652	29	1.4855	10
1.4657	28	1.4871	9
1.4663	27	1.4885	8
1.4669	26	1.4899	7
1.4675	25	1.4912	6
1.4682	24	1.4922	5
1.4691	23	1.4932	4
1.4699	22	1.4940	3
1.4708	21	1.4948	2
1.4718	20	1.4954	1
1.4730	19		

refractive indices and corresponding numbers of theoretical plates (see Table 2.1.6), or else, the number of plates can be read directly from a more detailed table, published for example in monograph ref.[26].

From the number of theoretical plates n found and the length of the rectification section L (in cm) the height equivalent to a theoretical plate H is calculated according to the equation.

$$H = L/n \tag{2.1.13}$$

When a column is tested, it is advantageous to measure the number of theoretical plates at various throughputs. A few measurements will show to what extent the separation efficiency of the column will decrease with increasing throughput. The knowledge of this dependence facilitates the choice of the optimum throughput during the distillation.

The operating hold-up volume is determined on a column operating at a steady state, under total reflux, using a testing mixture containing one non-volatile and one volatile component.

Determination of the operating hold-up

The dependence of the refractive index on the weight fraction of the phthalate is determined by using a testing mixture consisting of di-n-butyl phthalate and n-heptane.

The column pot is filled with a weighed amount of the testing mixture of known composition and the column is allowed to run at total reflux until equilibrium is attained. Using the same technique as for the testing of the number of theoretical plates 1 ml of liquid is withdrawn from the pot and the weight fraction of phthalate is determined in this sample. From the increase of phthalate found in the pot the amount of n-heptane is calculated, forming the operating hold-up volume of the column according to equation

$$H_W = W\left(1 - \frac{a}{b}\right) \tag{2.1.14}$$

where H_W is the weight of n-heptane in the operating hold-up of the column, W is the weight of the testing mixture put into the pot, a and b are weight fractions of the phthalate in the original mixture in the pot charge and in the mixture after equilibrium attainment, respectively.

This procedure is used for the determination of the hold-up at several throughputs.

The method cannot be applied to spinning band stills and other types of columns with a very small operating hold-up, in which the changes in concentration of the testing mixture are too small.

The separation of hydrocarbon mixtures by distillation can be improved in some respects if a further suitably chosen component is added to the separated mixture. The added substance is either inert and does not have any effect on the relative volatility of the separated components during destillation, or an agent is sought that suitably affects the volatility of the components.

Distillation with an inert component. Generally separation of two components can be improved if an inert substance is added to them the boiling point of which is between the boiling points of the two components. This method was elaborated by Bratton et al.[33], and it is called *amplified distillation.* The inert substance is selected so that it can be easily separated from the distilled components either chemically or by some other simple procedure. The majority of the examples of distillations carried out in the presence of an inert substance, described in literature, concerns the separation of nitrogen- and oxygen-containing organic compounds (cf.[6,34]). However, the method can also be applied to hydrocarbon mixtures. It was demonstrated by a model experiment[34] that during the distillation of a mixture of benzene (b.p. 80.1 °C) and toluene (b.p. 110.6 °C) in the presence of *n*-propyl acetate (b.p. 101.3 °C) on a 24 theoretical plates column both the purity of the benzene and toluene fractions is increased and the amount of both hydrocarbons in the intermediate fractions is decreased.

The main limitation of the method is that the added inert component can only improve the separation of mixtures which separate well on the column even without any addition. If the efficiency of the distillation column does not suffice for the separation of a mixture, then the addition of an inert substance does not bring about any improvement. A favourable effect can be obtained only if the inert component really distils between the two components so that it fills the inner hold-up of the column for a certain time and thus assumes the role of an intermediate fraction. The amount of the inert component in the separated mixture should be at least a multiple of the operating hold-up of the column. In the case of multicomponent mixtures several inert components can be added simultaneously. In distillations of more complex mixtures of non-hydrocarbon mixtures it is the hydrocarbon fractions which are used as inert components with advantage, because they contain an almost continuous series of components with various boiling points. In these cases the effect of the inert components is combined with the influence on the volatility of some constituents. If the distilled mixture is available in a small amount only a large excess of the inert fractions is added (a 15–20 fold amount), which enables separation by distillation with minimum losses even of microamounts of mixtures.

In spite of some evident advantages this method is not very often used for the separation of hydrocarbon mixtures. The main reason is the relatively restricted choice of inert components with suitable boiling points, and the difficulties with a quantitative separation of the components from hydrocarbons.

Another often used variant of this method is distillation with an addition of a *high-boiling* component. The boiling point of the added component is in this case higher than the boiling point of the highest boiling component of the distilled mixture. Towards the end of distillation the inert component displaces even the last traces of the higher boiling fractions of the distilled mixture from the pot flask into the rectifying section, and it remains alone in the pot as distillation residue. This method is employed most often for the distillation of small samples and very precious liquids, when it is desirable to restrict the amount of substance remaining in the column to a minimum.

In distillations on columns with a small operating hold-up the boiling point of the inert component added may be 50 K and more higher than the boiling point of the last displaced substance. By this the possibility of the contamination of the last fractions of the distillate with the added components is practically excluded. For the distillation of hydrocarbons of the C_8-C_{12} range *n*-hexadecane is suitable for this purpose.

In the case of columns with a high hold-up (plate or packed columns) it is necessary to displace the components of the distilled mixture even from the rectifying section, if a maximum yield of heavy fraction is required. For this purpose an inert component is selected with a boiling point only about 10–20 K higher than the boiling point of the last component of the separated mixture. The added compo nent can then be distilled over into the rectifying section and can fill it without the heat regimen of the column being drastically changed. The possibility of penetration of the added component into the last fractions of the distillate is of course higher than in the preceding case.

Extractive distillation. Among practically occurring hydrocarbon mixtures only the mixtures of stereoisomers, the mixtures of neighbouring members of homologous series and the mixtures of some structural isomers behave ideally. The diagram of the equilibrium composition of the phases of a binary ideal mixture is shown in Fig. 2.1.9-I. Other mixtures, especially if containing hydrocarbons of different polarity, mostly display positive deviations from *Raoult's law*, and marked deviations from ideality. The

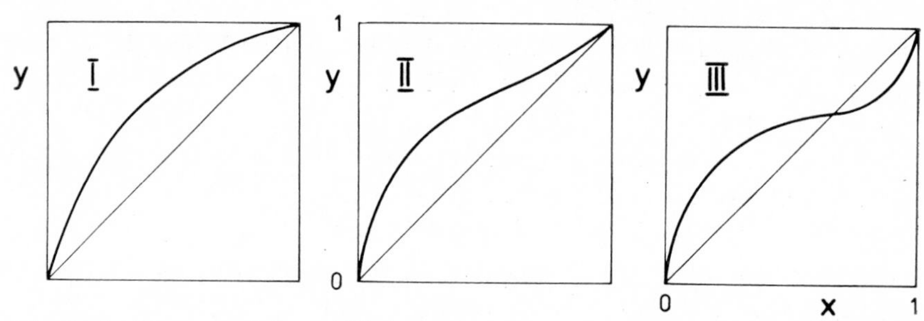

Fig. 2.1.9 Diagrams of the equilibrium composition of the phases of binary mixtures x — mole fraction of the more volatile component in the liquid, y — mole fraction of the more volatile component in the gas phase, *I* — ideal mixture, *II* — non-ideal mixture with a positive deviation from Raoult's law, *III* — azeotropic mixture

equilibrium curves of such mixtures are shown in Figs. 2.1.9-II, and 2.1.9-III. In the first case the equilibrium curve approaches asymptotically to the diagonal ($x = y$), which means that in the range of higher concentrations of the low-boiling component the relative volatility approaches closely to one and the mixture cannot be separated by a regular distillation. An example of such a system is the mixture methylcyclohexane-toluene; the relative volatility of these hydrocarbons drops below 1.07 if the amount of methylcyclohexane increases above 90 mol. %. In the latter case, represented in Fig. 2.1.9-III, the deviation from Raoult's law is so high that the equilibrium curve dissects the diagonal at the so-called *azeotropic point,* and the mixture of this composition boils constantly at temperature, which is lower than the boiling point of the lower-boiling component. For example a mixture of benzene (b.p. 80.1 °C), and cyclohexane (b.p. 80.7 °C) forms an *azeotropic mixture* with a minimum boiling point of 77.7 °C.

The relative volatility (α) of real binary mixtures is expressed by the equation valid for ideal mixtures (2.1.4), in which *activity coefficients* are introduced as a correction for non-ideality:

$$\alpha = \frac{\gamma_1}{\gamma_2} \cdot \frac{P_1^0}{P_2^0} \tag{2.1.15}$$

where P_1^0 and P_2^0 are vapour pressures of the pure components and γ_1 and γ_2 are activity coefficients of the components 1 and 2.

Activity coefficients of the components of ideal mixtures are equal to one within the whole concentration range. In real mixtures they are a quantitative measure for the deviation from the ideal behaviour and their values change with the concentration of the components. Fig. 2.1.10

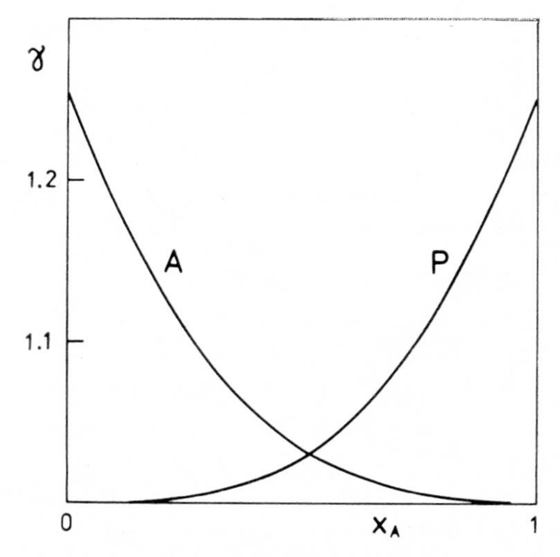

Fig. 2.1.10 Dependence of activity coefficients (γ) of toluene (A) and 2,2,4-trimethylpentane (P) on the mole fraction of toluene in the mixture (x_A)

illustrates this dependence for the typical non-ideal hydrocarbon mixture toluene – 2,2,4-trimethylpentane. From the figure it is evident that the activity coefficient, equal to one for the pure component, increases with decreasing concentration of the component in the mixture. For the calculation of limit values of activity coefficients at infinite dilution (γ° see Fig. 2.1.11) various equations have been deduced, based on the properties of the mixtures[35–38], or on the properties of pure components[39–43].

From the equation (2.1.15) it follows that in real systems the possibility exists of changing the relative volatility of the components by changing the ratio of activity coefficients. In *extractive distillation* this is achieved by introducing into the rectification section of the distillation column a high-boiling non-hydrocarbon solvent that changes the activity coefficients of the separated components differently. Fig. 2.1.11 shows how the increasing concentration of the solvent (expressed by mole fraction x_S) affects the activity coefficients and the relative volatility of the hypothetic mixture of the aromatic and the non-aromatic hydrocarbon. With increasing content of solvent in the mixture the activity coefficient of the non-aromatic hydrocarbon (γ_P) increases more rapidly than the activity coefficient of the aromatic hydrocarbon (γ_A). By this the value of the γ_1/γ_2 ratio in equation (2.1.15) increases and the relative volatility of both hydrocarbon components is higher in the presence of solvent (α_S) than the relative volatility of the original binary mixture: $\alpha < \alpha_S$. The

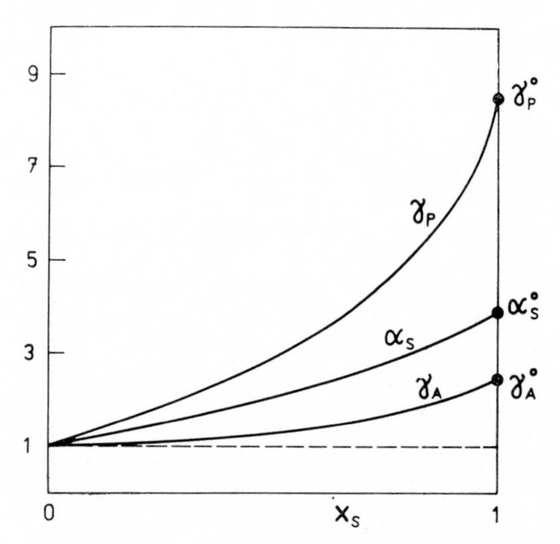

Fig. 2.1.11 Dependence of the activity coefficients of a hypothetic aromatic (γ_A) and non-aromatic (γ_P) hydrocarbon and their relative volatilities (α_S) on the mole fraction of the solvent (x_S) in the mixture; γ° and α_S° are limit values at infinite dilution of the hydrocarbon in the solvent

effect of various solvents on the increase of the relative volatility of the n-heptane – methylcyclohexane mixture[44] is evident from Table 2.1.7; among the solvents mentioned aniline displays the maximum effect. A number of other examples can be found in refs.[26, 45]

TABLE 2.1.7

Relative volatility (α_S) of the n-heptane—methylcyclohexane mixture in the presence of various solvents[44]

Solvent	x_S^a	α_S
None	0	($\alpha = 1.07$)
Aminocyclohexane	0.76	1.16
Nitrobenzene	0.82	1.31
Furfural	0.79	1.35
Aniline	0.78	1.40
Aniline	0.92	1.52

[a] Molar fraction of the solvent in the liquid phase

The value α_S is usually increased with increasing concentration of the solvent in the mixture, and it achieves the limit value (α_S^0; in Fig. 2.1.11) at $x_S = 1$, i.e. at infinite dilution of both hydrocarbons in the solvent. At infinite dilution the mutual interactions of the hydrocarbon molecules are negligible, and the value α_S^0, or the ratio α_S^0/α is a suitable measure for the effectiveness of the solvent added.

The relative volatility at infinite dilution can be easily determined for various systems by *gas chromatography*[46, 47]. For this purpose a chromatographic column should be prepared, containing the tested solvent as the *stationary phase*. From the corrected retention times of both hydrocarbons (t_1 and t_2), measured on the prepared column, the limit relative volatility is calculated according to equation

$$\alpha_S^0 = t_2/t_1 \tag{2.1.16}$$

The method is advantageous for the measurement of the dependence of α_S^0 on the temperature of the system. In Table 2.1.8 the results of such a measurement are presented, for example, for a mixture of methylcyclohexane-toluene on an aniline column, using a circulation gas chromatograph[46].

Table 2.1.8

Dependence of the relative volatility of methylcyclohexane and toluene on temperature at infinite dilution in aniline (determined by gas chromatography[46])

Temperature °C	($\alpha^{\circ}s$)aniline	Temperature °C	($\alpha^{\circ}s$)aniline
25	8.78	100	4.22
50	6.67	125	3.50
75	5.22	150	2.97

For the same pair of hydrocarbons, at 147 °C and 75 % concentration of aniline in the mixture, $\alpha_S = 2.59$ was found from the equilibrium data; this is in good agreement with the value $\alpha_S^0 = 2.97$ at 150 °C, given in the table. From the table it is evident that the relative volatility of the measured pair of hydrocarbons can be increased threefold if the extractive distillation takes place at low temperature (under reduced pressure). Another simple technique of the evaluation of the solvents for extractive distillation by means of gas chromatography has been elaborated by Tassios[48]. A selection of solvents for hydrocarbons with close molar volumes is given by Spelyng and Tassios[49].

A laboratory arrangement of extractive distillation is represented schematically in Fig. 2.1.12. A high-boiling solvent (S) is introduced continuously in to the upper part of the distillation column and allowed to flow down the rectifying section against the ascending vapours of the distilling mixture of hydrocarbons (M). In consequence of the large difference in boiling points the solvent remains liquid and is accummulated in the still pot. After distilling off of the lighter component into the distillate (D) the receiver is exchanged and the higher boiling hydrocarbon component is distilled off. In the still-pot a residue of the solvent is left. The distillation apparatus for laboratory extractive distillation can also be arranged as a two-column unit, with the regeneration and the recycling of the solvent[26, 27, 45, 50].

Extractive distillation offers a broad range of possibilities for the separation of various hydrocarbon mixtures with a very low relative volatility (α = 1.0 to 1.2). In addition to ideally behaving mixtures with close boiling points it also makes the separation of non-ideal mixtures possible, which have an *asymptotic* course of the equilibrium curve (*n*-heptane-toluene), as well as of constant boiling azeotropic mixtures. In the last two cases distillation in the presence of a suitable solvent brings the hydrocarbon mixture back to ideal behaviour, i.e. the equilibrium curves of the type II and III, shown in Fig. 2.1.9, converge in the presence of solvent to the type I.

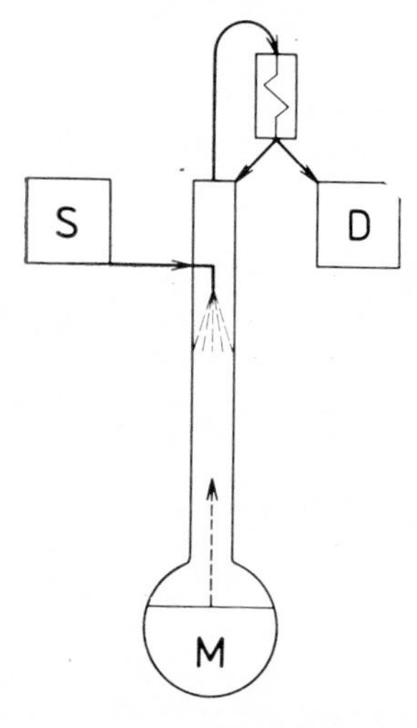

Fig. 2.1.12 Scheme of laboratory discontinuous extractive distillation
S — feed of high-boiling solvent, *D* — withdrawal of distillate, *M* — hydrocarbon mixture distilled

The advantage of extractive distillation is its ability to separate even multicomponent mixtures and hydrocarbon fractions with a not too broad distillation range (10 to 20 K), containing hydrocarbons of various structural classes. In contrast to industrial processes of extractive distillation, using only a few solvents, such as furfural, acetone, phenol or cresol, the choice of solvents for a laboratory apparatus is very broad.

In spite of the mentioned advantages extractive distillation has its main importance in industrial separations, while for the analysis of hydrocarbon mixtures it is used only rarely. The reason is that the method requires a more complex distillation apparatus (a column with a device for the introduction of the solvent into the rectifying section), and, moreover, similar separation effects can be achieved in the laboratory with other simpler laboratory techniques, as for example azeotropic distillation, extraction, chromatography, etc.

Azeotropic distillation. Many liquid mixtures with especially large deviations from ideality form constant boiling mixtures, called *azeotropes*. When distilling such mixtures the composition of the liquid phase changes only up to the azeotropic point (see Fig. 2.1.9-III), at which the vapour has the same composition as the liquid phase. Further fractions of the distillate contain then both components in a constant ratio until one component in the distilled mixture is completely exhausted.

The tendency of hydrocarbons to form azeotropic mixtures is the higher, the closer boiling points both hydrocarbons have and the more they

TABLE 2.1.9

Effect of intermolecular interactions on the properties of liquid mixtures
(~ strong interactions; ~ weak interactions)

	Intermolecular interactions	Type of mixture	Behaviour on distillation
a	○ ● / ~ ~ / ○ ●	→ non-miscible	→ azeotrope with minimum boiling point
a, b		→ partly miscible	→ azeotrope with minimum boiling point
b	○~● / ~ ~ / ○~●	→ solution	→ non-ideal, with a positive deviation; → azeotrope with minimum boiling point
c	○~● / ~ ~ / ○~●	→ solution	→ ideal
d	○~● / ~ ~ / ○~●	→ solution	→ non-ideal, with a negative deviation; → azeotrope with maximum boiling point
e	○~● / ○~●	→ formation of compounds	

differ in *polarity*. Therefore, when analyzing hydrocarbon mixtures, the formation of azeotropes should be kept in mind, especially in the case of narrow fractions that contain simultaneously saturated and unsaturated hydrocarbons. The separation schemes of such fractions must be chosen with regard to the fact that the separation of azeotropic components by regular distillation is impossible.

The liquid mixtures, containing in addition to hydrocarbon an equally volatile non-hydrocarbon polar compound, form azeotropes almost in all instances. Generally the behaviour of the liquid mixtures depends on the difference in the interaction between similar molecules and different molecules. In Table 2.1.9 various possibilities of *intermolecular interactions* are represented schematically for mixtures containing two kinds of molecules (white and black rings). If the *attractive* forces between the molecules of the same species are much stronger than between unlike molecules (cases a and b in Table 2.1.9), then immiscible or partially miscible mixtures or solutions result which form azeotropes with a minimum boiling point, or solutions with a positive deviation from ideality (the vapour pressure of the solution is higher than would correspond to *Raoult's law*). In the case when the interactions between the like and unlike molecules are approximately equalized (case c), the solution behaves ideally. The second extreme is represented by the cases when attractive forces between the unlike molecules predominate (cases d and e). This is manifested by a negative deviation from ideality, which can result in an azeotrope with a maximum boiling point. Extremely strong interactions between unlike molecules already approximate on or can cause the formation of compounds.

Azeotropic mixtures containing only hydrocarbons, or a hydrocarbon with a non-hydrocarbon component, have almost exclusively minimum boiling points. One of the known exceptions is the mixture of 1,3,5-trimethylbenzene with pentachloroethane, that forms an azeotrope with a maximum boiling point[7].

In contrast to azeotropes composed of hydrocarbons only, that complicate the separations, the distillation of constant boiling azeotropic mixtures of hydrocarbons with non-hydrocarbon component is used as a special separation technique. Azeotropic distillation enables, similarly to extractive distillation, a separation of hydrocarbon mixtures boiling within a narrow temperature range and containing various structural types. The laboratory arrangement of azeotropic distillation is simple and any distillation column may be used for this purpose. A non-hydrocarbon compound of suitable volatility – also called *entrainer* – is added into the column pot containing the mixture to be separated. The entrainer forms an azeotrope with a minimum boiling point with one or several components of the mixture. This increases the differences in the boiling points of hydrocarbon components and the first fractions of the distillate contain predominantly those hydrocarbons that have formed an azeotrope with the greatest depression. The entrainer is selected so that it may be eliminated from the distillate easily, for example by separation of phases after cooling, washing out with water, extraction, or some other simple method. It is advantageous that the excess of non-hydrocarbon compound entrains the hydrocarbons forming an azeotrope from the column quantitatively, without regard to the hold-up volume of the column.

When hydrocarbon mixtures are purified and separated, various alcohols, glycols, phenols, organic acids, ketones, ethers and other polar compounds are used as entrainers. The entrainers mentioned are mostly *non-selective,* i.e. they form azeotropes with hydrocarbons of various classes. These azeotropes, however, are not formed equally easily and they differ in their boiling point depression. Of a certain homologous series of hydrocarbons only a few members form azeotropes with a given entrainer; their boiling points lie within a not too broad interval around the boiling point of the entrainer.

The selection of suitable azeotropic entrainers is facilitated through the published sets of data on binary and multicomponent azeotropic mixtures[7, 51, 52]. An entrainer is selected so that the boiling point decrease of a certain hydrocarbon or a certain group of hydrocarbons should suffice for a separation, by distillation, from other components of the mixture. In order to shorten the time of distillation it is advantageous if the concentration of hydrocarbon in the azeotrope is high. On the contrary, in *microdistillations* it is more advantageous when the entrainer predominates in the azeotrope.

General tendencies of how the hydrocarbon component and the entrainer affect the boiling point and the composition of the azeotrope are evident from Figs. 2.1.13 and 2.1.14. For the sake of greater clarity the points indicating the composition and the boiling point of the azeotrope are always connected with the boiling point of one component with a dashed line. Fig. 2.1.13 shows the azeotropes of ethanol with *n*-alkanes. As is evident from the figure those *n*-alkanes enter the azeotrope the boiling points of which are within the interval 42 K below the boiling point and 47 K above the boiling point of ethanol. Alkanes that boil below the boiling point of the entrainer form azeotropes rich in hydrocarbon. With increasing boiling point of the *n*-alkane the boiling point of the azeotrope also increases and the content of hydrocarbon in the azeotrope decreases.

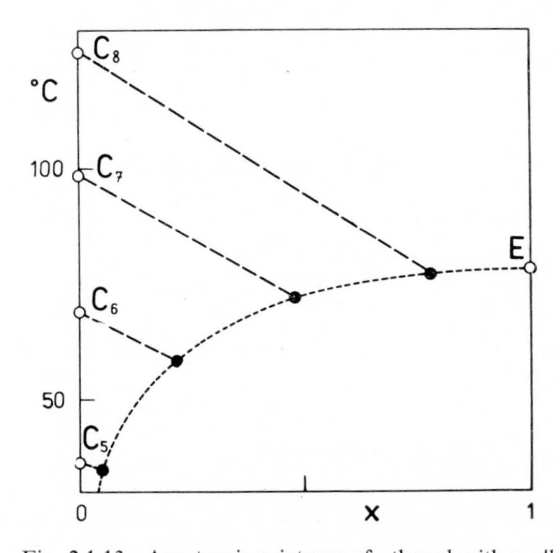

Fig. 2.1.13 Azeotropic mixtures of ethanol with *n*-alkanes
x — weight fraction of ethanol in the azeotrope, C_5—C_8 — normal boiling points of *n*-alkanes, E — normal boiling point of ethanol
Black points on the dotted line indicate the boiling points and the composition of azeotropes

The upper limit for the mentioned example is *n*-octane with an azeotrope boiling 1 K below the boiling point of pure ethanol only. As other alkane hydrocarbons display the same tendency, it is possible to use the dotted line in Fig. 2.1.13 for the determination of approximate boiling points of ethanol azeotropes of branched alkane hydrocarbons, boiling between *n*-pentane and *n*-octane.

Fig. 2.1.14 shows the azeotropes of *n*-hexane with various *n*-alcohols. With *n*-hexane only methanol, ethanol and *n*-propanol form azeotropes.

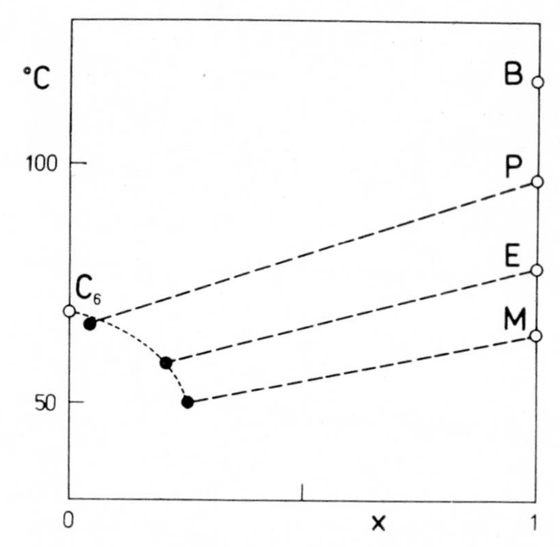

Fig. 2.1.14 Azeotropic mixtures of *n*-hexane with alcohols
x — weight fraction of alcohol in the azeotrope, C_6 — normal boiling point of *n*-hexane, *M*, *E*, *P*, *B* — normal boiling points of methanol, ethanol, *n*-propanol and *n*-butanol
Black points on the dotted line indicate the boiling point and the composition of the azeotrope

n-Butanol boils too high above hexane and does not give an azeotrope. The content of hydrocarbon in the azeotrope is again the higher the higher the boiling point of the alcohol. From Fig. 2.1.14 the generally valid fact is evident, that a maximum depression, relative to the lower boiling component is observed in azeotropes in which the hydrocarbon and the entrainer have approximately equal boiling points.

For the separation of hydrocarbon fractions the difference is important between the azeotropes of hydrocarbons with the same boiling point, but with different structure. Fig. 2.1.15 shows a comparison for benzene (normal boiling 80.1 °C) and 2,4-dimethylpentane (b.p. 80.5 °C). Comparison of the azeotropes with ethanol shows that in the case of alkane hydrocarbon the depression is greater than in the case of the aromatic hydrocarbon (the boiling point of the cyclohexane azeotrope lies between them, close to the alkane azeotrope). Hence, when this mixture is submitted to azeotropic distillation with ethanol, the hydrocarbons will distil in the following order: alkane hydrocarbon, cycloalkane hydrocarbon, and finally aromatic hydrocarbon. The original small differences of 0.4 K

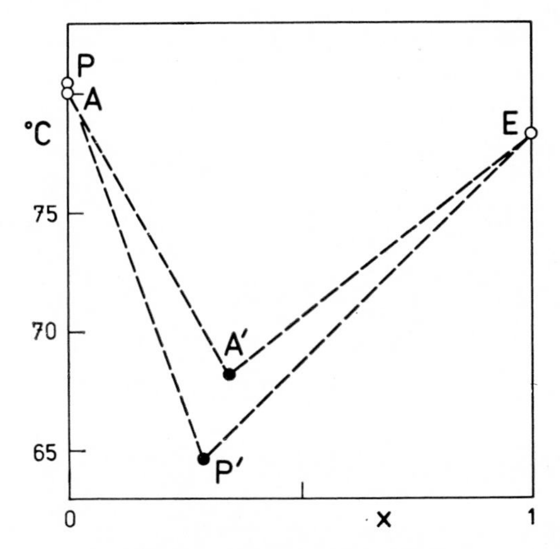

Fig. 2.1.15 Depression of the boiling point of benzene and 2,4-dimethylpentane on azeotropic distillation with ethanol
x — weight fraction of ethanol in the azeotrope, *A*, *A'* — normal and azeotropic boiling point of benzene, *P*, *P'* — normal and azeotropic boiling point of 2,4-dimethylpentane

between the boiling points of the alkane and the aromatic hydrocarbon increase to 3.5 K on azeotropic distillation.

Mair[53] has described *fluorinated entrainers* which give the greatest depression with aromatic hydrocarbons and the smallest one with alkane hydrocarbons. On distillation of a mixture of hydrocarbons having an identical boiling point with the addition of a perfluorinated cyclic ether or perfluorinated tributylamine, the order of the distilling hydrocarbons is reversed, i.e. the aromatic hydrocarbon distils first, followed by the cycloalkane hydrocarbon and finally the alkane hydrocarbon. In Fig. 2.1.16 the depressions of the azeotropes of methylcyclohexane and 2,2,4-trimethylpentane with a perfluorinated cyclic ether $C_8F_{16}O$ are compared. Although the pure methylcyclohexane boils 1.8 K higher than 2,2,4-trimethylpentane, the boiling point of the methylcyclohexane azeotrope is 2.2 K lower[53] in comparison with the isoalkane azeotrope.

Entrainers suitable for laboratory distillation of hydrocarbons with boiling points from 50 °C up to 200 °C are listed in Table 2.1.10. For the successful result of an azeotropic distillation it is necessary to separate the mixture beforehand by a regular distillation to as narrow fractions

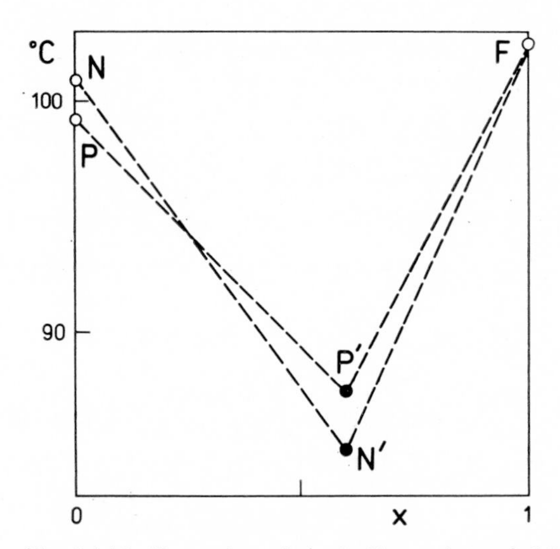

Fig. 2.1.16 Depression of the boiling points of 2,2,4-trimethylpentane and methylcyclohexane on azeotropic distillation with perfluorocyclic oxide $C_8F_{16}O$
x — volume fraction of the agent in the azeotrope, *N, N'* — normal and azeotropic boiling point of methylcyclohexane, *P, P'* — normal and azeotropic boiling point of 2,2,4-trimethylpentane

as possible. A suitable entrainer is selected according to the mean boiling point of the hydrocarbon fraction. Rossini et al.[7] recommend using an entrainer boiling 0 to 30 K below the boiling point of the hydrocarbons to be separated. If the material for separation is scant, it is more advantageous to use a lower-boiling entrainer, about 30–40 K below the boiling point of the hydrocarbons, so that the concentration of the hydrocarbons in the azeotrope is low. In distillations of large amounts of hydrocarbons, when it is important to consider the rational utilization of the pot flask volume, an entrainer is selected the boiling point of which is approximately equal to that of the hydrocarbon fraction.

In separations of hydrocarbons with close boiling points maximum effect can be achieved when the selected entrainer increases the already existing differences between the boiling points of the hydrocarbons of various structures. For example, for a mixture containing a lower-boiling alkane and a higher-boiling cycloalkane hydrocarbon, the entrainers given in Table 2.1.10 in the 1st group are suitable. In contrast to this, for a mixture of a lower-boiling cycloalkane and a higher-boiling alkane

TABLE 2.1.10

Reagents for laboratory azeotropic distillation of hydrocarbons

Reagent	Boiling point (°C)
1. group	
Acetone	56.5
Methanol	64.7
Ethanol	78.4
Acetonitrile	82
Isopropyl alcohol	82.5
Acetic acid	118
Ethylene glycol monomethyl ether (Methylcellosolve)	124.5
Ethylene glycol monoethyl ether (Cellosolve)	135.1
Ethylene glycol monomethyl ether acetate	144
Ethylene glycol monobutyl ether (Butylcellosolve)	171.2
Diethylene glycol monomethyl ether (Carbitol)	194.2
2. group	
Perfluorocyclic ether ($C_8F_{16}O$)	102.5
Heptacosafluorotributylamine $(C_4F_9)_3N$	178.4

hydrocarbon, the use of an entrainer from the 1st group would cause boiling point depressions which would nullify the original difference in volatility. In such a case the separation can be improved by the addition of some fluorinated entrainer from the 2nd group, that decreases the boiling point of the cycloalkane hydrocarbon more than that of the alkane hydrocarbon; this increases the original difference in volatility between the two hydrocarbon components still more.

The composition and the boiling point of the azeotrope changes generally with changing pressure during the distillation. Systems are even known in which the azeotrope can disappear by a change of the distillation pressure. The effect of pressure on the boiling point of the azeotropic mixture can be predicted using a simple graphical method, elaborated by Nutting and Horsley[52,54]. The plot of the vapour pressure against temperature both for an azeotropic mixture and the two pure components gives straight lines in the *Cox diagram* [log P versus $1/(T\,°C + 230)$], which are shown in Fig. 2.1.17 for various types of azeotropic mixtures with a minimum boiling point. Straight lines A and B represent the vapour pressures of pure components and the straight line C is the vapour pressure of the azeotrope. In system I in Fig. 2.1.17 the boiling point of the azeotrope varies in the whole considered range of pressures according to the straight line C. In system II the azeotropic straight line C crosses the straight line of the component A at a certain pressure, which means that

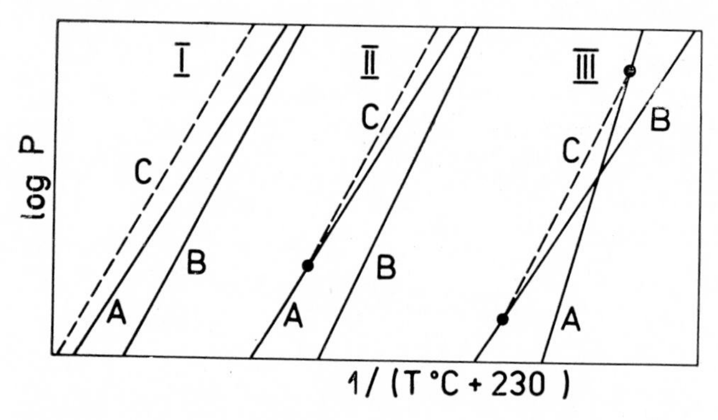

Fig. 2.1.17 Cox's vapour pressure versus temperature plot diagram for binary azeotropes with minimum boiling points
A, B — vapour pressures of pure substances, C — vapour pressure of the azeotrope. In systems *II* and *III* the crossing points of the azeotropic straight line indicate the pressures at which the azeotrope disappears

the vapour pressure of the azeotrope no longer exceeds the vapour pressure of the lower-boiling component, and at this point the system becomes *nonazeotropic*. In system III azeotropy can be suppressed both by decreasing the pressure, or by increasing the distillation pressure (for example the mixture methanol-acetone). The mentioned graphical methods can be used for the prediction of the pressure effect even when only the normal azeotropic boiling point is known. The graph is plotted by constructing a straight line with a slope equal to the mean slope of the straight lines of both pure components through the boiling point of the azeotrope.

Steam distillation. Distillation of two immiscible liquids is a special case of azeotropic distillation. The vapour pressures of immiscible components do not affect each other during simultaneous evaporation, so that the total vapour pressure above the mixture is given by the sum of the vapour pressures of pure components. When such a mixture is heated, the distillation begins at the moment when the sum of the vapour pressures of the components becomes equal to the outer pressure, i.e. at a temperature lower than the boiling point of the lowest-boiling pure component.

For the purification and the separation of hydrocarbon fractions water is used as the immiscible component. Steam distillation is widely utilized both as a purification operation for the separation of hydrocarbons from the admixed nonvolatile substances, and also as a mild method of distillation of high-boiling hydrocarbons.

A binary mixture of a hydrocarbon with water begins to distil at a temperature when the vapour pressure of the pure hydrocarbon is equal

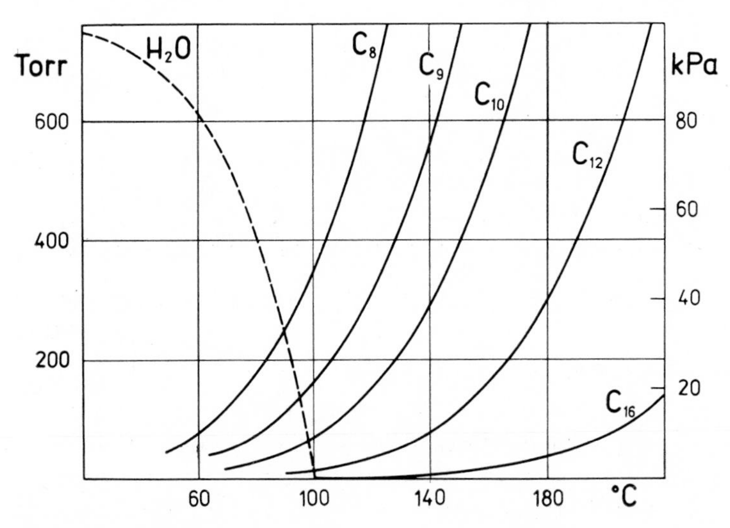

Fig. 2.1.18 Graph for the calculation of the distillation temperature of *n*-alkanes on steam-distillation
C_8—C_{16} — curves of the vapour pressure dependence of *n*-alkanes on temperature, H_2O — dependence of the value (vapour pressure of water minus normal pressure) on temperature

to the difference of the atmospheric pressure and the water vapour pressure. Graphically this relationship is expressed in Fig. 2.1.18 for several *n*-alkanes. In this graph (according to Badger and McCabe[26, 27]) the changes of the hydrocarbon vapour pressures are plotted against temperature, and the curve is shown of the temperature dependence of the value: normal atmospheric pressure minus the water vapour pressure. Every point of intersection of this curve with the curve of the vapour pressure of pure hydrocarbon gives the normal distillation temperature of the hydrocarbon in the presence of steam. From the figure it is evident

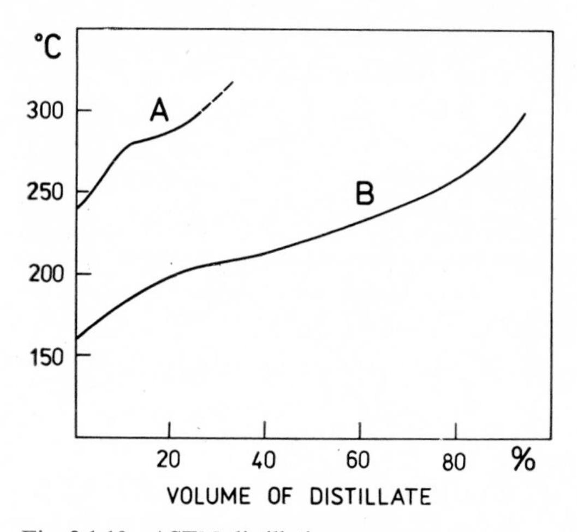

Fig. 2.1.19 ASTM distillation curves
A — naphthenic crude, B — fraction obtained from the naphthenic crude by steam distillation

that, for example, *n*-alkanes C_8-C_{12} — the normal boiling range of which is 125–216 °C — distil in the presence of steam within a much narrower range, 89–99 °C. The decrease of the distillation temperature is distinct and it corresponds roughly to the vacuum distillation at an approximately 1 kPa pressure (cf. Fig. 2.1.30). From Fig. 2.1.18 it is evident that the upper limit of hydrocarbons which can be steam distilled at normal pressure lies somewhere near *n*-hexadecane. Hence, the boiling point decrease for a hydrocarbon may attain up to 200 K. In Fig. 2.1.19 the *ASTM distillation curve* of a naphthenic crude oil is compared with the distillation curve of a fraction obtained from this material by steam distillation. The end point shows that the highest boiling hydrocarbons distilling with steam have their normal boiling points about 300 °C.

The high-boiling hydrocarbons do not distil with steam at the same rate as the low-boiling ones. The number of moles of hydrocarbon which can be distilled over with one mole of steam is given—at a 100 % efficiency—by the ratio of the vapour pressure of the hydrocarbon and the vapour pressure of water at the temperature of distillation. The high-boiling hydrocarbons, the vapour pressure of which at about 100 °C is low, need much more steam for distillation than the lower-boiling hydrocarbons. The ratio of the weight of the hydrocarbon (W_{CH}) and the weight of water (W_{H_2O}) in the distillate can be computed from the equation:

$$\frac{W_{CH}}{W_{H_2O}} = \frac{P_{CH}}{P_{H_2O}} \cdot \frac{M_{CH}}{18} \qquad (2.1.17)$$

where P_{CH} and P_{H_2O} represent the vapour pressures of the hydrocarbon and the water, respectively, at the distillation temperature, and M_{CH} is the molecular weight of the hydrocarbon. For example using the values taken from the graph (Fig. 2.1.18) and introduced into the equation (2.1.17) the differences in the distillation of *n*-octane and *n*-dodecane can be compared. While in the first case 3.1 kg of *n*-octane is distilled over with 1 kg of steam, in the second case only 190 g of *n*-dodecane is distilled over with the same amount of steam.

The content of hydrocarbon in the distillate in mol % can be calculated from the equation:

$$\text{Mol \% of hydrocarbon} = \frac{P_{CH}}{P_{CH} + P_{H_2O}} \cdot 100 \qquad (2.1.18)$$

Although steam is commonly introduced into the pipe stills in industrial distillations of heavy hydrocarbon fractions, this method is not used in laboratory distillations. Steam distillation is carried out in laboratories almost exclusively in simple distillation apparatus, consisting of a pot for the development of steam, a distillation flask containing the hydrocarbon and aqueous phase, a descending condenser, and a receiver. The arrangement of the apparatus is evident from the scheme in Fig. 2.1.20. The steam is led through a tube to the bottom of the distillation flask, where it bubbles through the hydrocarbon layer and leaves the flask together with the hydrocarbon vapours for the condenser. In the receiver the distilled hydrocarbon again separates from the

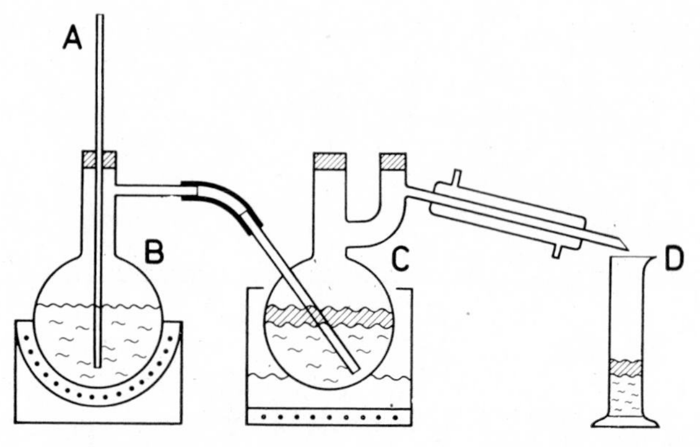

Fig. 2.1.20 Apparatus for laboratory steam distillation
A — vent tube, B — steam developer, C — distillation pot with the hydrocarbon layer, D — distillate receiver

aqueous condensate. If the flask containing the hydrocarbon layer is not heated from outside during the distillation, a part of the steam will condense in the flask and the volume of water in it will rapidly increase. An adequate heating of the distillation flask is best carried out using a steam bath, as shown in the figure.

2.1.4 VACUUM DISTILLATION

On prolonged heating of hydrocarbon fractions above 250 °C thermal decomposition sets in distinctly. Therefore the high-boiling fractions are distilled under reduced pressure, which is selected so that the boiling points of the hydrocarbons present are below the temperature of decomposition. Owing to the elimination of air from the apparatus, the oxidative reactions are also considerably restricted.

The dependence of the boiling point on the pressure is *logarithmic,* i.e., when the pressure is decreased the temperature decreases, slowly at the beginning, and rapidly at low pressures. The boiling points of individual hydrocarbons under reduced pressure can be calculated with great accuracy from equation (2.1.6) if the constants of the equation are known. For *n*-alkanes C_1-C_{40} the calculated boiling points at various pressures are listed in Tables 2.1.11 and 2.1.12.

The boiling range of heavy hydrocarbon fractions is usually determined by the standard testing method[55] at which about 200 ml of sample are distilled in a simple apparatus under reduced pressure, at a 4 – 8 ml/min rate. The values given in Tables 2.1.11 and 2.1.12 can be used for the

TABLE 2.1.11

Boiling points of *n*-alkanes C_1 — C_{12} at various pressures

Number of C atoms	Boiling point in °C at pressure					
	101.325 (760)	26.664 (200)	13.332 (100)	6.666 (50)	2.666 (20)	kPa (Torr)
1	—161.49					
2	—88.63					
3	—42.07					
4	—0.50					
5	36.07	1.9				
6	68.74	31.6	15.8			
7	98.43	58.7	41.8	26.8		
8	125.67	83.6	65.7	49.9	31.5	
9	150.80	106.7	87.9	71.3	52.0	
10	174.12	128.2	108.6	91.2	71.1	
11	195.89	148.3	127.9	109.9	89.0	
12	216.28	167.1	146.1	127.5	106.0	

approximate determination of the boiling range of the hydrocarbon fraction at reduced pressure, or, for a conversion of the distillation range determined by *vacuum distillation* to that at normal pressure. For more accurate calculations detailed tables are used[31] and the converted temperature can be further corrected on the basis of the structural composition of the fraction. In doing this the *characterization factor K*, calculated[17] from the average (by volume) boiling point of the fraction and the density at 15 °C, is taken as a base. For a characterization factor $K = 12$ the corrections are equal to zero; within the 10 – 14 K range and the pressure 1.333 kPa (10 Torr) the corrections added to the converted temperatures range between +5.5 and −5.5 °C. The corrections increase with decreasing pressure.

In addition to the decrease of the boiling points of individual hydrocarbons in the mixture, the decrease of the distillation pressure can also

TABLE 2.1.12

Boiling points of *n*-alkanes C_{12} — C_{40} at various pressures

Number of C atoms	Boiling point in °C at pressure					
	101.325 (760)	13 332 (100)	6.666 (50)	1.333 (10)	0.133 (1)	kPa (Torr)
12	216	146	127	91	51	
13	235	163	144	106	65	
14	253	179	159	121	78	
15	271	194	174	134	91	
16	287	209	189	148	103	
17	302	223	203	162	115	
18	317	237	216	173	126	
19	331	250	228	184	138	
20	344	261	239	196	147	
21	356	273	251	207	157	
22	369	284	263	217	167	
23	380	294	272	227	176	
24	391	305	283	236	185	
25	402	314	292	245	193	
26	412	324	301	254	202	
27	422	334	310	263	210	
28	432	343	319	271	218	
29	441	351	327	279	225	
30	450	359	335	287	232	
31	459	368	344	295	240	
32	468		352	303	247	
33	476		360	311	254	
34	483		366	316	260	
35	491			324	267	
36	498			329	273	
37	505			336	279	
38	512			343	285	
39	518			348	289	
40	525			354	296	

cause a change in the relative volatility. The effect of the distillation pressure P (in the 10 – 1 500 Torr interval) on the relative volatility α of a pair of hydrocarbons of b.p. t_1 and t_2 (in K) at the boiling point of the mixture, T (in K), is expressed by the equation[23]:

$$\log \alpha = \frac{t_2 - t_1}{T}\left(7.30 - 1.15 \log P + \frac{T}{179 \log P}\right) \qquad (2.1.19)$$

In homologous series of hydrocarbons the decrease of the distillation pressure narrows the differences between the boiling points of neighbouring members to a certain extent (see Fig. 2.1.30), but the relative volatility increases slightly. In such instances the use of vacuum distillation can bring about a sharper separation of the components than when atmospheric distillation is applied.

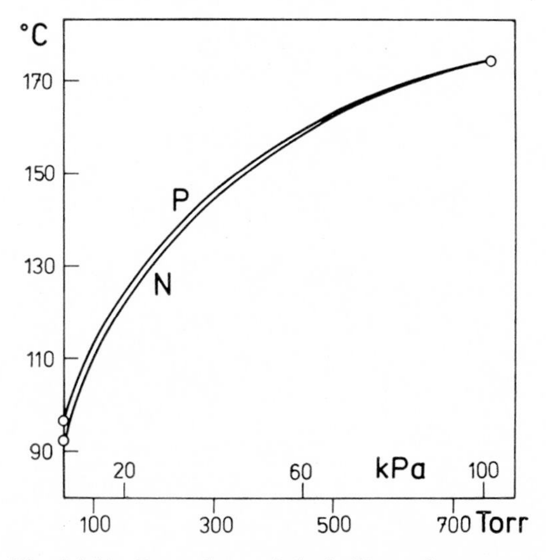

Fig. 2.1.21 Dependence of the boiling point on pressure for sec.-butylcyclohexane (N) and hypothetic n-alkane (P) with identical normal boiling point

For the separations of hydrocarbon mixtures the changes of relative volatility of hydrocarbons of various structures, but with close boiling points at atmospheric pressure, have a greater practical importance. In some such mixtures the relative volatility hardly changes within a broad range of pressures. For example, the relative volatility of the mixture *n*-heptane-methylcyclohexane, which is used for testing the distillation columns, changes only from the value 1.08 to the value 1.05 when the pressure decreases from atmospheric to 40 kPa. However, if two hydrocarbons with identical normal boiling points have sufficiently large differences in heats of evaporation, then the boiling point of one hydrocarbon decreases more on the pressure decreasing than that of the other, and the mixture can be separated by vacuum distillation.

In connection with the *API Research Project 6* it was found[7] that under reduced pressure the majority of cycloalkane hydrocarbons has a larger change in boiling points than the straight-chain hydrocarbons and weakly branched alkanes. Fig. 2.1.21 shows the pressure dependence of the boiling point of a cycloalkane hydrocarbon (*sec.*-butylcyclohexane) and a hypothetic *n*-alkane of the same normal boiling point. The curve of cycloalkane (*N*) deviates distinctly from the curve of alkane (*P*) at low pressures, and the difference between both hydrocarbons attains 4 K at a pressure of 6.666 kPa (50 Torr). These differences are smaller between cycloalkanes and corresponding monomethyl- and dimethylalkanes, but sufficient for the separation of both structural types by a combination of atmospheric and vacuum distillation. The curves of higher-branched isoalkanes (trimethyl and tetramethyl derivatives) do not display any utilizable difference in comparison with the curves of cycloalkanes.

Hydrocarbons of various structures can also have such different dependences of boiling points on pressure, that an originally low-boiling component of the mixture can become a higher-boiling one. For example

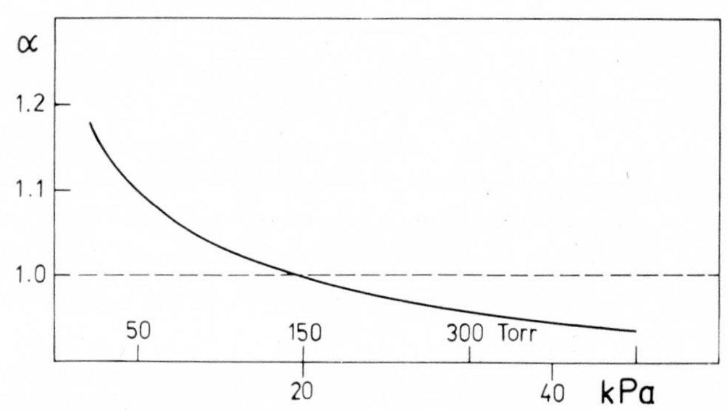

Fig. 2.1.22 Dependence of the relative volatility (α) of an *n*-tridecane-dicyclohexyl mixture on pressure (according to T. Feldman et al.[56])

in the mixture of dicyclohexyl and *n*-tridecane, recommended for the testing of column efficiency in vacuum distillation[56], *n*-tridecane is the lower boiling component at normal pressure; when the pressure decreases the boiling points converge, and eventually, at pressures below 20 kPa (150 Torr) dicyclohexyl becomes the lower-boiling component. This is shown graphically in Fig. 2.1.22. Relative volatility of the mixture at 20 kPa pressure is equal to 1, which means, that the mixture boils constantly within the whole concentration range (but it is not the azeotrope) and has the same composition of the liquid and the vapour phase. A decrease of pressure brings about an increase in relative volatility above 1.1, while an increase in pressure above 20 kPa brings the relative volatility close to 0.9.

In the case of aromatic hydrocarbons the different pressure dependence of boiling points has been found for *mononuclear* and *binuclear* hydrocarbons[7]. For example, alkylnaphthalenes and alkylbenzenes with the same normal boiling point can be separated by vacuum distillation, because alkylbenzenes exhibit a higher boiling point depression.

Laboratory vacuum distillations of hydrocarbon fractions up to pressures about 0.1 kPa are carried out either in a simple distillation apparatuses or with distillation columns. In both cases the maintenance of a smooth and regular boiling is one of the main problems. Disturbances caused by pressure variations during distillation can be eliminated by connecting the apparatus to a *manostated* vacuum system with sufficiently large reservoirs[31]. Another source of irregular boiling is the *overheating* of the distilled liquid, which may be observed especially in common apparatuses where a higher layer of liquid is heated in the pot flask from below. The liquid at the bottom of the pot is under hydrostatic pressure, and at low pressures it has up to several tens of K higher boiling point than the liquid near the surface. The overheated liquid, when moving to the level, evaporates in irregular bumps that make precise distillation impossible and can even endanger the whole apparatus. In simple distillation apparatuses this is prevented by a thin stream of bubbles of an inert gas being introduced to the bottom of the distillation flask by a capillary. Another, better solution, is employed in commonly used laboratory rotary vacuum evaporators (Fig. 2.1.23). In these apparatuses the distillation flask rotates in an inclined position, and it is heated either by an infrared lamp or radiator or a water bath. The liquid in the flask is evaporated rapidly and smoothly from the film spread by rotation over the inner wall of the flask. During the distillation the flask can be refilled continually. Apparatuses of this type are usually used for simple isothermic distillations – mainly for the separation of a larger amount of light, volatile components.

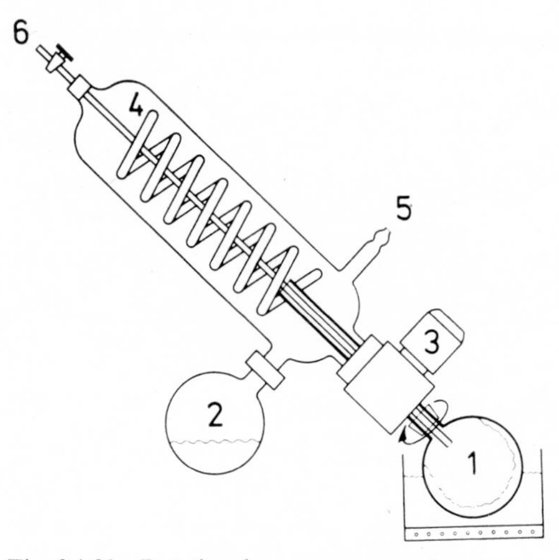

Fig. 2.1.23 Rotational vacuum evaporator
1 — distillation flask, 2 — receiver, 3 — motor rotating the distillation flask, 4 — water condenser, 5 — connection with pump, 6 — feed of the mixture distilled

Vacuum distillation on columns. When mixtures have to be separated by column distillation the requirement of smooth boiling is still more imperative than in simple distillation. The capillary introduction of an inert gas disturbs an accurate distillation, and therefore it is better to avoid it. Special constructions of the column pot, provided with a heating body located so that the liquid in the pot circulates intensively, are more advantageous. New types of *spinning band stills* have an excellent solution: the content of the pot is stirred with a teflon stirrer (see Fig. 2.1.7), so that the pot can be heated with an ordinary heating jacket without the danger of overheating.

The choice of the distillation pressure in vacuum distillation on columns is limited by the *pressure drop* in the rectifying section. The pressure in the column pot is equal to the sum of the pressure in the column head, and the pressure drop. The difference of pressures in the pot and in the column head is accompanied by a temperature gradient which causes, in the case of columns with a high pressure drop, the contents of the pot to be overheated more than in columns with a low back pressure of the rectifying section. Therefore columns with an empty rectifying tube, columns with concentric tubes and spinning band columns are most suitable for vacuum distillations up to a 0.1 kPa pressure, because they have the lowest pressure drops. Packed columns and columns with integral assembly packing can also operate under reduced pressure, but the operation pressure is selected depending on the back pressure of the rectifying section, from 1 – 40 kPa.

The *separation efficiency* of the columns under reduced pressure is determined as under atmospheric pressure, using suitable testing mixtures. For columns that have no more than 20 theoretical plates at 2.6 kPa *n*-dodecane-cyclohexylcyclopentane or *n*-tridecane – dicyclohexyl[56] mixtures are suitable. Hawkins and Brent[57] recommend the mixture ethylbenzene – chlorobenzene as a testing mixture; this mixture at a pressure drop from the normal to 2.6 kPa displays only little change in the relative volatility, i.e. from 1.10 to 1.12.

For testing under reduced pressure the mixture *n*-heptane – methylcyclohexane, employed currently for testing under atmospheric pressure, is also suitable. With this mixture the separation efficiency of distillation columns operating in atmospheric and vacuum distillation can be compared objectively[57]. The results have shown that the maximum efficiency of the distillation column at total reflux is independent of whether the distillation is carried out under reduced or atmospheric pressure.

If the same throughput is maintained in the column under reduced pressure, the vapours flow through the rectifying section at a much higher velocity than under atmospheric pressure, because under reduced pressure they have a larger volume. The higher velocity of the vapours results in a better contact between the gas and the liquid phase and it represents a factor increasing the separation efficiency, but on the other hand, the time of contact of both phases is shortened at a higher velocity, which worsens the separation efficiency. Both these contradictory effects become clearly manifest in packed columns when a decrease in pressure increases the separation effect of the column up to a certain maximum, but at low pressures it decreases again. For various types of packings the maximum efficiency was observed in the 20 – 27 kPa range[58]; in other types of columns this range was 7–13 kPa[59]. In analytical columns with a small back pressure of the rectifying section (spinning band, concentric tubes, *Hyper-Cal*) it was found that a reduction of pressure decreases the separation effect, but only at medium and high throughputs. At low throughputs the separation efficiency remains unchanged[29].

Hence, in vacuum distillation the columns can operate with the same (or higher) separation efficiency than under atmospheric pressure. The maintenance of the separation efficiency requires, as a rule, operation at lower throughput, which results in a prolongation of the distillation time. This fact invalidates the other advantages of vacuum distillation to a certain extent.

Flash distillation. In apparatuses with a differential collection of distillate the heavier fractions of the batch remain in the pot all the time, being thus exposed to heat for a long time. In simple distillations the mixture is usually heated 1–2 hours, in distillations carried out with efficient columns the mixture is exposed to elevated temperatures for 10 to 20 hours, and sometimes even longer. Under these conditions the chemical composition of the distilled material can be changed considerably under the effect of heat.

The risk of thermal decomposition of the distilled substances is practically eliminated by stills for flash distillation. For laboratory fractionation of hydrocarbon mixtures equipments with a disturbed *falling-film* are mainly used. In principle they operate by introducing liquid for distillation into the apparatus where it is mechanically spread over a vertical cylindrical surface heated at the required temperature. The volatile components are evaporated from the falling-film almost instantly and they are then condensed and led into the receiver. The surface of the film freed of the volatile components is incessantly renewed by spreading which enhances distillation considerably. The non-volatile material is conducted from the bottom of the heated zone into another receiver. In construction this apparatus corresponds to the scheme in Fig. 2.1.24. In some types the operating surfaces are arranged so that the hot surface surrounds the condenser passing through the centre of the apparatus.

Flash distillation combines two great advantages: the distilled liquid, occurring in the form of a thin film, is not overheated, and the time of stay in the heated zone is less than one minute. Equipments of this type are used mainly for fractionations of high-boiling hydrocarbon materials with a very broad boiling range. The distillation is isothermal and it can be carried out under atmospheric pressure, in a stream of an inert gas or in a vacuum. The limits of distillation cuts and their yields are controlled by the temperature of the evaporation surface, pressure in the apparatus and the time of contact with the hot evaporation surface.

Flash distillation has proved successful in the fractionation of *crude oil*[60,61]. Fractions can be distilled out from the crude oil up to gas oil with great output and without any thermal damage of heavy fractions. The fractionation scheme used in the *API Research Project 60* for the characterization of the heavy ends of crude oils includes flash distillation on two different apparatuses. The conditions of distillation are the following.

Fractionation of crude oil by flash distillation

From crude oil the lightest fractions are distilled off under atmospheric pressure on a falling-film flash still[62]. The feed rate of the crude oil is 1 000—1 200 ml/min, the temperature of the evaporation surface is 100 °C, sweep rate of helium 9 l/min. The distillate obtained boils within the 38—165 °C interval.

A second distillation on the same apparatus under isothermal conditions, but under a 0.7—0.8 kPa pressure yields the rest of the volatile components boiling up to 200 °C.

The crude oil freed from the light fractions is distilled on an ASCO 4-inch *Rota-film* (wiped-wall) still[60]. At a 1.33 kPa pressure and temperature of the evaporation surface of 180 °C the 225—315 °C fraction is distilled off. On the second passage at 1.33 kPa pressure and 251 °C temperature the gas oil fraction is separated (315—370 °C).

The residue is then separated on the same apparatus under the conditions of molecular distillation. At a 0.5—0.6 Pa pressure, 245°C temperature, and a feed rate of 600 ml/h the fraction 370—535 °C is obtained; on further passage at the same pressure, evaporation surface temperature 370 °C, and a feed rate of 450—500 ml/h, the last fraction is separated, boiling at 535—675 °C.

The boiling range of fractions, derived from thermal and pressure conditions during distillation is in quite good agreement with the data of *simulated distillation* (see Section 2.1.1). The boiling range distribution of the distillates prepared from various crudes under the same conditions agree well. The overlapping of adjacent distilled fractions is minimum; for example, for the fraction 370 – 535 °C simulated distillation shows that 93 – 95 % of the material distils within the mentioned limits.

Molecular distillation

When pressure drops below 10 Pa the rectification sections of the distillation columns rapidly lose their separation efficiency because the very low density of the vapours no longer permits an efficient interphase exchange. The *mean free path* of molecules, which is about 10^{-15} cm at normal pressure, increases at a decreasing of the pressure in the apparatus, and at a 0.1 Pa pressure it already exceeds 5 cm. Distillation in a high vacuum, about 0.1 Pa, is called molecular distillation, and it has a completely different character from the distillation under atmospheric pressure or in a medium vacuum. Apparatuses for molecular distillation have their condensation surface close to the evaporation surface, and they operate under a pressure at which the mean free path of the molecules exceeds this distance. Under the effect of the temperature gradient in the apparatus, the molecules of the volatile components leave the surface of the heated liquid, and the majority of them reaches directly the cooling surface and condenses there. This distillation is a *non-equilibrium* process at which the rate of evaporation is dependent on the vapour pressure of pure components at the temperature of the evaporation surface and on the molecular weight of the component. In order to increase the rate of evaporation and to decrease the possibility of thermal decomposition, it is especially advantageous if the distilled liquid flows down over the evaporation surface in the form of a thin film. A very efficient all-glass still for laboratory molecular distillation is shown schematically in Fig. 2.1.24. After degassing the liquid is introduced to the upper part of a vertical cylindrical evaporator (finger) that is heated with circulating oil from a thermostat. Within a very short time, before the liquid film comes down the finger, the volatile components are evaporated and condensed on the cooled cylindrical surface which surrounds the finger at a distance of a few cm. The condensed fractions are then conducted into the distillate receiver. In comparison with the rate of evaporation of volatile components from the surface of the liquid film, their diffusion from the deeper layers of the film to the surface is relatively slow. Therefore the rate of distillation is very much enhanced by mechanical spreading of the film on the evaporation surface, which incessantly restores the content of the volatile components on the film surface. In the apparatus shown the falling film is stirred with a rotating glass helix that fits on the evaporator walls. The helix is driven magnetically with a motor and it turns against the film flow. Non-volatile components of the distilled mixture flow from the bottom part of the evaporator into the receiver for the residue.

Other types of stills with a stirred falling-film have a reverse arrangement, i.e. the evaporator is on the inner surface of a jacket and the condenser is placed in the axis of the apparatus. The film surface is renewed with rotating wipers, made of teflon or carbon.

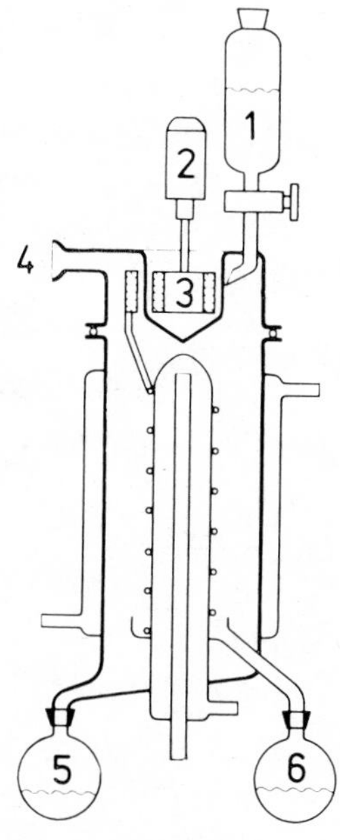

Fig. 2.1.24 Apparatus for molecular distillation with falling stirred film
1 — feed of the degassed mixture, 2 — motor, 3 — magnetic transmission to the glass helix drive, 4 — connection with the vacuum pump, 5 — distillate receiver, 6 — receiver of the residue

Generally the stills with a stirred film are very efficient, and in view of the fact that the substance is in contact with a hot surface for only a few tens of seconds, or less, they considerably decrease the possibility of thermal decomposition. As already said in the preceding section, apparatuses of this type are used in laboratories with advantage for *flash distillations* under normal pressure or in a medium vacuum[60, 61]. From the point of view of the possibility of thermal decomposition, stills with a *rotating* evaporation surface are the mildest[26, 27]. Under the centrifugal force the liquid film on the evaporator is only several hundredths of a mm thick, and the exposure time in the hot zone is of the order of tenths of seconds only.

Using molecular distillation hydrocarbons up to about C_{50} can be distilled without decomposition. The degree of separation of certain components is dependent – equally as in equilibrium distillation – on their relative volatility. In addition to vapour pressure the molecular weights of the components also affect the relative volatility in molecular distillation. Under the supposition of the validity of *Raoult's law* the theoretical relative volatility (α_T) of a binary mixture in molecular distillation is given by the equation:

$$\alpha_T = \frac{P_1^0}{P_2^0}\sqrt{\frac{M_2}{M_1}} \tag{2.1.20}$$

where P^0 is vapour pressure of the pure component and M is the molecular weight of the component. From the equation it follows that molecular distillation can bring about the separation of isomers, if they have differing vapour pressure, or substances with identical vapour pressure, if they differ in their molecular weight. However, the relative volatilities determined experimentally differ, considerably from the theoretical value calculated from the equation (2.1.20). If the factors disturbing the ideal course of distillation are included in the activity coefficient, then the *real relative volatility* (α_R) is expressed by the equation:

$$\alpha_R = \frac{\gamma_1 P_1^0}{\gamma_2 P_2^0}\sqrt{\frac{M_2}{M_1}} \tag{2.1.21}$$

The study of the separation of various testing mixtures has shown that the relative volatility increases with a temperature decrease and a decrease in distillation rate. An increase in temperature and other factors that shorten the mean free path of the molecules worsen the separation.

Under the conditions of molecular distillation the individual components do not have a defined boiling point as in equilibrium distillation. Evaporation from the liquid surface takes place to a different extent at all temperatures, as long as the temperature *gradient* is preserved. However, the temperature at which the rate of distillation of the compound from the non-volatile solvent is maximum is characteristic of a given compound, and it is called *elimination temperature*[26]. Compounds with a very different elimination temperature can be separated by molecular distillation even in a one-stage apparatus, either by a single passage or by a repeated recycling of the residue. However, owing to their low efficiency in the majority of cases single-stage apparatuses are used only for the separation of volatile components of high-boiling mixtures from non-volatile substances, or for the preparation of broader distillation cuts from complex fractions. In the preceding section, an example of the fractionation of crude oil by combined flash and molecular distillation on one apparatus is shown. Cuts can thus be obtained without a measureable decomposition from the heavy ends, up to a fraction with a normal end boiling point of 675 °C[60,61].

Multi-stage molecular stills of cascade type have a higher separation efficiency. Melpolder et al.[63] have described a countercurrent column consisting of one large and 19 small distillation units connected in series, of which each has an efficiency of 0.8 of a theoretical plate. The equilibrium on the column with a 1 500 ml sample batch is attained after 16 hours of distillation. At a 40 ml/h throughput and 30 : 1 reflux ratio the rate of distillate withdrawal is 28 ml/24 h. The tested mixture of *n*-alkanes was separated on this column to pure fractions of *n*-nonadecane, *n*-eicosane and *n*-heneicosane.

The apparatus described by Mair et al.[64] has 50 stages and an effective volume of the pot flask 2.7 litres. The condensate from the still pot flows down into the rectifying tube which is inclined 3.5° from the horizontal position. The rectifying tube consists of two sections of which each contains a system of stainless steel troughs and tubes, forming 25 plates. The bottom part of the rectifying section is heated and the evaporated fractions are condensed on the upper cold wall; from there they flow down to the trough and through the tubes to the next higher plate. In this manner the distillate proceeds upwards through the rectifying section into the still head from where the predominant part of the distillate flows down as reflux along the inclined bottom of the rectifying-tube back into the pot flask. The apparatus works at 4 Pa pressure, 80 ml/h throughput, and 55 : 1 reflux ratio. The equilibrium in the apparatus is stabilized after 6 hours. On this still too pure hydrocarbons C_{18} to C_{21} were obtained from the tested mixture of *n*-alkanes.

A similar countercurrent column with 37 stages was constructed by Malyusov et al.[65,66]. It consists of a glass rectifying tube of 4.5 cm diameter, 89 cm long, inclined 4° from the horizontal. Inside the tube a metallic condenser with oblique metallic leaves is located, which transfer the condensate to plates.

A *semi-micro* rotary still for molecular distillation was constructed by Watt[67]. A rotating glass cylinder, inclined 5° from the horizontal, serves as evaporator. The distilled liquid is spread by rotating over the inner cylinder walls in the form of a film, and it is stirred intensively. The water-cooled metallic condenser is static and consists of 5 conical surfaces over which the condensate flows toward a fraction-cutting device. The still operates at 0.67 Pa pressure and 100 – 120 rpm. The amount of the distilled mixture can vary from 2 to 10 g.

2.1.5 SPECIAL DISTILLATION TECHNIQUES

Low-temperature distillation. For the separation of liquefied hydrocarbon gases, with boiling points below 30 °C (see Table 2.1.13), special columns are used, provided with a condenser for cooling with liquid nitrogen or carbon dioxide. The columns are perfectly insulated with a vacuum jacket with a metallic reflecting sheet so that an exchange of heat with the surroundings is prevented.

As evident from Table 2.1.13 some commonly occurring hydrocarbon gases have very close boiling points and a low relative volatility. For example the pair isobutene and 1-butene has a relative volatility $\alpha = 1.03$, and they require at least 176 theoretical plates for their separation. The rectifying section of the columns for low-temperature distillation have, therefore, high separation efficiencies, up to more than 200 theoretical plates. Columns with integral assembly HELI-GRID packings are most widely used. Packed columns, columns with a helical empty tube, and other types are also used. The constructional details and the operating procedure are described in detail in the literature[26,68]. Commercial apparatuses are highly automated, including an automatic recording of the distillation curve.

Low-temperature distillation of hydrocarbons has only a preparative

Table 2.1.13

Boiling points of hydrocarbon gases

Hydrocarbon	Formula	Boiling point in °C	Relative volatility
Methane	CH_4	—161.49	
Ethylene	C_2H_4	—103.71	
Ethane	C_2H_6	—88.63	
Ethyne	C_2H_2	—84	(subl. point)
Propene	C_3H_6	—47.70	> 1.3
Propane	C_3H_8	—42.07	
Cyclopropene	C_3H_4	—35	
Propadiene	C_3H_4	—34.5	
Cyclopropane	C_3H_6	—32.80	
Propyne	C_3H_4	—23.22	
Isobutane	C_4H_{10}	—11.73	> 1.2
Isobutene	C_4H_8	—6.90	> 1.03
1-Butene	C_4H_8	—6.26	> 1.05
1,3-Butadiene	C_4H_6	—4.41	> azeotrope
n-Butane	C_4H_{10}	—0.50	
Methylcyclopropane	C_4H_8	0.73	> 1.04
trans-2-Butene	C_4H_8	0.88	
Cyclobutene	C_4H_6	2.9	> 1.09
cis-2-Butene	C_4H_8	3.72	
1-Buten-3-yne	C_4H_4	5.1	
1-Butyne	C_4H_6	8.07	
Methylenecyclopropane	C_4H_6	8.8	
2,2-Dimethylpropane	C_5H_{12}	9.50	
1,3-Butadiyne	C_4H_2	10.3	
1,2-Butadiene	C_4H_6	10.85	
Cyclobutane	C_4H_8	12.51	
3-Methyl-1-butene	C_5H_{10}	20.06	
1,1-Dimethylcyclopropane	C_5H_{10}	20.63	
1,4-Pentadiene	C_5H_8	25.97	
3-Methyl-1-butyne	C_5H_8	26.35	
2-Butyne	C_4H_6	26.99	
2-Methylbutane	C_5H_{12}	27.85	
1-t-2-Dimethylcyclopropane	C_5H_{10}	28.21	
1-Pentene	C_5H_{10}	29.97	

importance today. Previously used analytical evaluation of technical hydrocarbon gases by distillation technique has been replaced by gas chromatography.

Thermal rectification. When the pressure drops below 100 Pa the current rectifying sections of the distillation columns rapidly lose their separation efficiency. The low vapour density and their great velocity prevents an effective exchange of matter between the liquid and the vapour, and therefore a countercurrent contact rectification ceases to be effective at low pressures.

Enrichment of vapour in more volatile components can, however, be achieved, even at low pressures and in equilibrium conditions by multiple repeating of partial evaporation of the liquid mixture, or by partial condensation of vapours. This method – called thermal rectification[26] – is based practically on the same principle of enrichment, as the long known *redistillation* techniques. On columns for thermal rectification the access of heat to the condensate and the withdrawal of heat from the vapours are regulated separately so that a large number of simple distillation steps are carried out in the rectifying section.

The simplest column for thermal rectification consists of an evacuated sealed glass tube, containing a small sample of the mixture over which a thermal gradient is allowed to sweep in a horizontal position[26].

Vertical columns of earlier construction have the rectification tube divided into a series of electrically heated and air-cooled zones. The reflux flowing down the inner wall of the tube is partly evaporated in each hot zone, the vapours ascend in the tube and are partly condensed in each cooled zone. The separation efficiency of such a column is about 15 theoretical plates[26].

A rotating column for thermal rectification has been described by Byron et al.[69]. The column consists of two 1 m long concentric tubes; the outer one, of 4.5 cm diameter, is heated and the inner, metallic one, of 2.5 cm diameter is cooled and rotates around its axis. The vapours ascend in the annular space between the tubes and the reflux flows in the form of a film down the inner surface of the outer tube. The components evaporated from the reflux are mixed with the main stream of vapours, from where the heavier fractions again condense on the inner cooled tube. The condensate is sprayed by the centrifugal force from the rotating tube onto the opposite wall where it is combined with the down-flowing reflux film. The enrichment is the result of a partial evaporation and a partial condensation and not the result of the interphase exchange. It is true that the column can operate under atmospheric pressure or under a medium vacuum, but its main advantage is a good separation efficiency at pressures about 1 Pa.

The construction of rotating columns has been further improved by Perry and Cox[70]. Their column, called "*Brush-still*", consists of a vertical heated rectification tube (made of a 9 cm diameter glass tube), in which a metallic, water-cooled condenser rotates at 450 rpm. This condenser

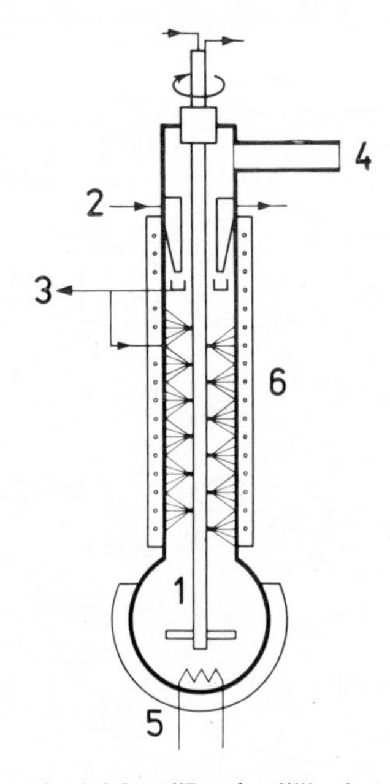

Fig. 2.1.25 "Brush-still" scheme for thermal rectification
1 — metallic, water-cooled rotor provided with bristles and terminated with a stirrer, 2 — column head condenser, 3 — distillate withdrawal and reflux, 4 — connection with the vacuum pump, 5 — heating of the pot, 6 — heating of the rectifying tube

is provided with sweeping bristles (see the scheme in Fig. 2.1.25). The height of the rectification section is 28 cm; working pressure is about 0.1 Pa. During the distillation the condensate is flung by centrifugal force onto the heated wall of the rectification tube where it is distributed with the bristles made of fine wire. The volatile parts are again evaporated, condensed on the rotating brush and are again brought onto the hot wall of the tube. Thus the mixture passing the tube is submitted to a series of repeated evaporations and condensations. The separation efficiency of the column at a 50 ml/h throughput is 9–10 theoretical plates; at much higher throughputs, 700–800 ml/h, it does not drop below 7 theoretical plates. The column of this construction can operate effectively even without reflux. It separates the substances with good effieiency, which are otherwise distillable on molecular stills only.

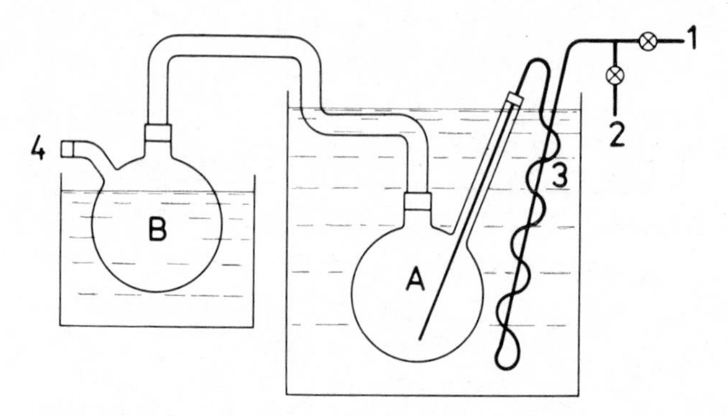

Fig. 2.1.26 Apparatus for low-temperature distillation under carbon dioxide
A — distillation pot, B — receiver, 1 — introduction of carbon dioxide, 2 — connection with the vacuum pump, 3 — heat exchanger, 4 — connection to manometer

Low-temperature vacuum distillation in a closed system. For a mild quantitative isolation of volatile substances from a large volume of non-volatile material a method of low-temperature vacuum distillation in a stream of *carbon dioxide* has been elaborated[71].

The distillation apparatus is shown in Fig. 2.1.26; it consists of two 1 litre flasks connected with a stronger tube, a simple device for the introduction of carbon dioxide, and a connector to the vacuum pump and the manometer. The mixture to be distilled is put into the flask *A*. Before the start of the distillation the flask *A* is cooled in an acetone-carbon dioxide bath and the whole system is evacuated. Then the pump is disconnected and the apparatus is filled with carbon dioxide of high purity ($> 99.9\,\%$). The cooling bath of flask *A* is taken off and the flask *B* is cooled with liquid nitrogen. The resulting pressure in the apparatus is about 13 Pa. A slow stream of carbon dioxide is led over a heat exchanger into the flask *A* slowly heated in a thermostat to the required temperature. Thus the distillate is transported into the container *B*. When the distillation is over, both the cooling and the heating baths are withdrawn and the apparatus is filled with carbon dioxide up to atmospheric pressure. When the container is opened the solidified carbon dioxide is allowed to sublimate and the distillate remains in the flask *B*.

The testing of the distillation system with a synthetic mixture (2.5 g) of *n*-alkanes C_6-C_{28} showed that after 30 minutes distillation at 50 °C, using 100 g of carbon dioxide, hydrocarbons C_6-C_{17} distilled over into the container. The yield of hydrocarbons C_6-C_{13} in the distillate was quantitative; a certain overlapping of the distillate and the residue took place in the interval $C_{14}-C_{17}$ only. This interval can be changed by choosing a suitable distillation temperature, which makes the method sufficiently flexible. In view of the low temperature, the presence of an inert gas, and the closed system in which the losses caused by the suction into the pump are excluded, the method is advantageous mainly for quantitative isolations of less stable components spread at low concentration in a large volume of nonvolatile material. The procedure is milder than steam distillation.

Sublimation. When some crystalline hydrocarbons are heated, their vapour tension sometimes equals the outer pressure before the temperature of melting has been attained. Such hydrocarbons then sublimate, i.e. the crystals are evaporated on heating directly into the gas phase, without previous melting of the substance. Also when the vapours are cooled the hydrocarbon separates from the gas phase directly in crystalline form.

Although sublimation can be considered as a special type of distillation of solid substances, its separatory power cannot be increased on the principle of a countercurrent contact rectification, because no liquid phase is present which would function as reflux. In the analysis of hydrocarbon mixtures sublimation is used mainly for the separation of sublimating hydrocarbons from non-sublimating substances. Sublimation is very advantageous as a purification operation, and if the properties of the hydrocarbons permit, it should be given preference to other methods. The sublimating hydrocarbon can be separated from the non-volatile impurities in a practically quantitative yield, and the purity of the sublimate often exceeds 99.5 % after this single operation. Sublimators are of simple construction and a number of various types and modifications are described in literature[26, 72–74]. Sublimation can be easily improvised in common laboratory glassware, or the small apparatuses for molecular distillation can also be used. In view of the negligible material losses the method can easily be adapted as a *microtechnique*. In

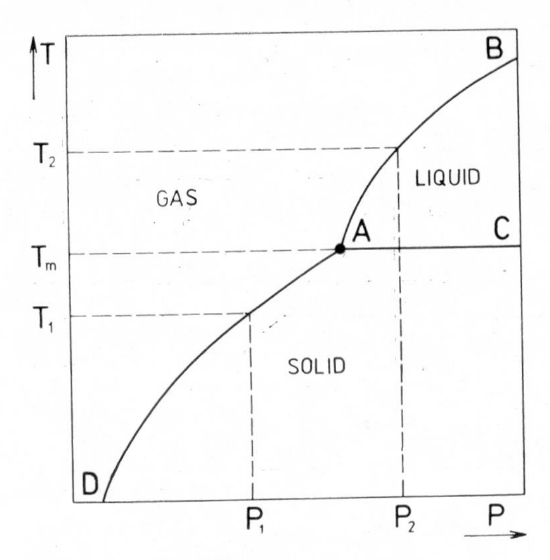

Fig. 2.1.27 Phase diagram
A — triple point, *T* — temperature, *P* — pressure, *AB* — dependence of the boiling point on pressure, *AC* — dependence of the melting point (T_m) on pressure, *AD* — dependence of the sublimation point on pressure

small sublimators milligram amounts of sample can be purified without difficulty. Microgram amounts can be sublimated, for example, between the microscope slides on a *Kofler block*.

A measure of the sublimation ability of a substance is its *sublimation point*, defined as the temperature at which the vapour pressure of the solid substance is equal to the outer pressure. The effect of pressure on the sublimation point is evident from the phase diagram in Fig. 2.1.27. Point *A* is a ternary point at which the solid, liquid and gas phases are at equilibrium and coexist simultaneously. Curve *AD* expresses the dependence of the sublimation point on pressure, similarly as the curve *AB* expresses the dependence of the boiling point on pressure. In analogy to the boiling point the sublimation point also decreases with the decrease of pressure. The melting point is independent of pressure (straight line AC). If the outer pressure is P_2 then the substance will melt on heating at temperature T_m and at temperature T_2 it begins to boil. In the case when the outer pressure is P_1 then the substance attains its sublimation point at temperature T_1 and passes directly into the gas phase. From the diagram two practical important facts are evident: 1. The temperature at which the substance sublimates can be decreased by decreasing pressure. 2. Some substances that do not sublimate at atmospheric pressure can be sublimated at decreased pressure at a temperature below the melting point.

The temperature at which a certain compound is sublimated in the laboratory always lies below its sublimation point. The sublimation begins to proceed at an observable rate at the so-called *sublimation temperature* which is defined as the minimum temperature at which the first observable amount of sublimate appears on the condenser surface. In contrast to the sublimation point which is a thermodynamic constant of a substance – the same as the boiling point –, the sublimation temperature depends on the conditions of the determination, especially on the distance of the condenser surface from the substance, and the time interval during which the substance is kept at a given temperature (usually 10 – 20 minutes). Practical sublimation temperature is mostly selected between the sublimation temperature and the sublimation point. Low-melting substances are usually sublimated at temperatures 10 – 20 K lower than their melting point; substances with high melting points are sublimated at temperatures 50 – 150 K lower.

Substances can be sublimated at atmospheric pressure, in a vacuum, or even in a stream of inert gas. Even some high-molecular substances that do not practically sublimate under usual conditions can give the necessary amount of sublimate if an apparatus is used the condensation surface of which is only a few mm (or tenths of mm) distant from the substance, or if a high vacuum of the order of 10^{-1} to 10^{-3} Pa is employed. Generally the sublimation rate is increased with temperature, pressure decrease, diminishing the distance between the evaporation and the condensation surface, and increasing the surface area of the sublimand. Sublimation rate can be also enhanced by an inert gas streaming from the sublimand to the condensation surface.

In comparison with the use of sublimation as a purification operation it is much less used for the separation of sublimating substances. In view of the low separation efficiency the result depends on the difference of the sublimation points of the substances separated.

A simple laboratory device for *fractional sublimation* can be arranged in a common sublimator by gradually increasing the temperature of the substance and separating individual fractions of sublimate.

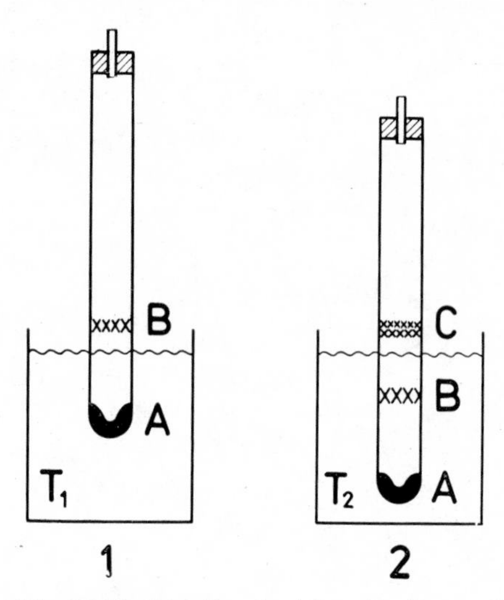

Fig. 2.1.28 Fractional sublimation in an oil bath
A — non-volatile fraction of the sublimand, B — sublimate, C — the most volatile fraction of the sublimate

The technique of fractional sublimation in a glass tube immersed in an oil bath at several different temperatures[75] is quite practical. A solution of the sample is introduced into a glass tube sealed at one end and the solvent is evaporated so that the sublimand should form a thin film on the walls of the sealed tube end. A silicone oil bath is heated to the required temperature, the tube is evacuated and immersed in a vertical position with its sealed end in the bath, 4 cm below oil level. After a short period the volatile components form a zone of sublimate in the tube, closely above the oil level, while the non-volatile components remain in the sealed end of the tube. The bath is then cooled, for example to a temperature 20 K lower and the tube is immersed into the oil so that the sublimate zone is about 3 cm below the oil level. The most volatile components form a new sublimate stripe above the bath level. The procedure of the fractionation of the sample to the unvolatile fraction *A*, volatile fraction *B* and the most volatile fraction *C* is represented schematically in Fig. 2.1.28. The sublimate fractions are isolated from the cut off parts of the tube.

An interesting solution of a vertical, air-cooled sublimator has been described by Gettler et al.[76]. The sublimation vessel is connected with a ground-glass joint with a glass tube 18.5 long which has an equal 8 mm internal diameter all along the tube. A strip of a transparent cellophane film is coiled and inserted into the tube, serving as a condensing surface (Fig. 2.1.29). During fractional sublimation the vessel with the sample

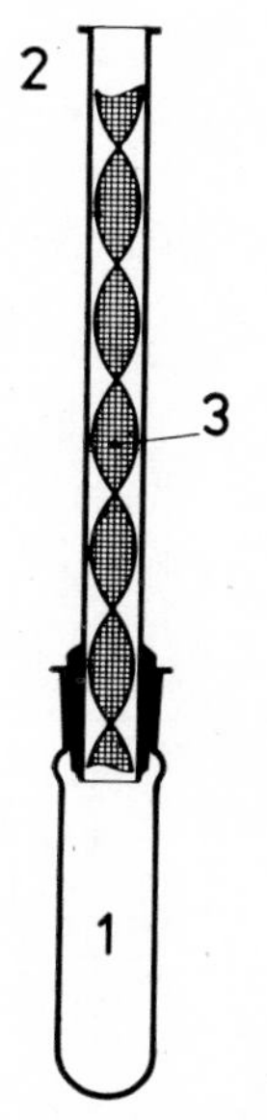

Fig. 2.1.29 Fractional sublimation with collection of the sublimate on a cellophane film
1 — sublimation vessel, 2 — glass tube, 3 — strip of cellophane film

is heated in a metallic block and the upper end of the tube is connected to a vacuum pump. The vapours of the components with different volatility condense and crystallize at various heights of the condensation tube and are deposited on the cellophane film. When the sublimation is complete the film is removed from the tube and the sublimate zones are easily inspected under a microscope, analyzing the crystals of different forms, etc. The zones of pure components are then cut off from the strip and dissolved; If the components are not separated completely, the overlapping boundaries can be cut off and resublimated.

A vertical vacuum sublimator with a controlled temperature gradient along the condensation tube is used for the separation of sublimating substances, employing the technique called *sublimatography*[77]. The components of the separated mixture condense at various heights of the tube on sites where the temperature of the wall corresponds to their so-called *vacuum condensation point*.

Horizontal sublimators provided with a temperature gradient operate by a glass tube with the sample being inserted in a horizontal position into a jacket with a non-uniformly wound heating wire. At suitable conditions (in a vacuum or in a stream of inert gas) the substance from the hot end of the tube sublimates and the sublimate of individual components condenses under the effect of the temperature gradient at more or less separated zones in the cooler part of the tube. When the sublimation is ended the components are isolated from the cut off segments of the tube.

A horizontal sublimator with a linear temperature gradient obtained by applying a special oven at one end and a water condenser at the other end of the sublimational tube was employed by Thomas et al.[78] for the separation of polynuclear aromatic hydrocarbons. They achieved, by fractional sublimation, the separation of a mixture containing anthracene, pyrene, 3,4-benzopyrene and 1,2,5,6-dibenzoanthracene. The method enables the isolation of sufficient amounts of pure hydrocarbons from the mixture, necessary for testing carcinogenic activity.

Only a relatively small number of hydrocarbons possess distinct sublimation ability. Their common features are the rigid structure and the relatively high melting points in comparison with hydrocarbons with identical vapour tension. For example, some polycyclic aromatic hydrocarbons and some highly condensed polycycloalkanes sublimate easily. The structure of some typical sublimating hydrocarbons and the conditions under which they sublimate, are given in Table 2.1.14.

TABLE 2.1.14

Conditions of sublimation of some hydrocarbons

Hydrocarbon	Formula	Melting point (°C)	Conditions of sublimation	
			Pressure (kPa)	Temperature (°C)
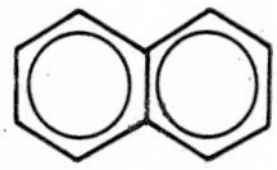 Naphthalene	$C_{10}H_8$	80	atmospheric	(36—38)[a] 50
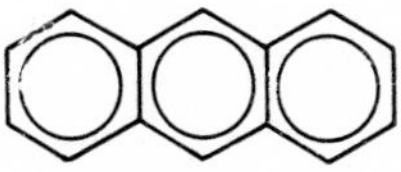 Anthracene	$C_{14}H_{10}$	218	atmospheric	(77—79)[a] 100—150
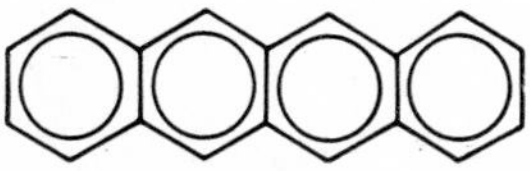 2,3-Benzanthracene	$C_{18}H_{12}$	349	0.1	175—180
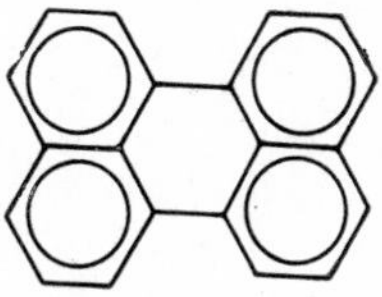 Perylene	$C_{20}H_{12}$	274	0.1 (argon)	130
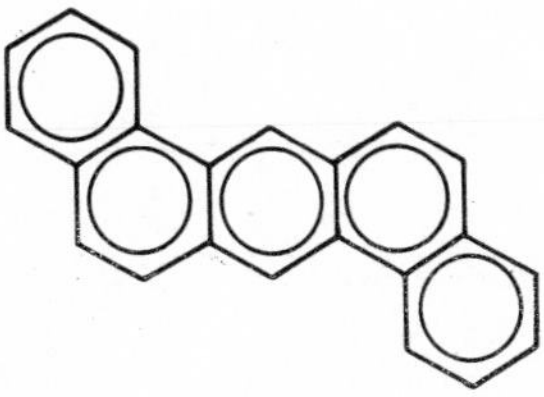 1,2,5,6-Dibenzanthracene	$C_{22}H_{14}$	260	10^{-7}	100—140

(continues)

Table 2.1.14—continued

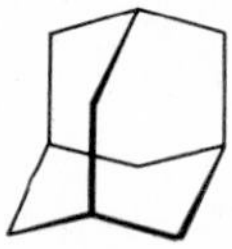 Adamantane	$C_{10}H_{16}$	270	atmospheric 1—2	(41—42)[a] 100
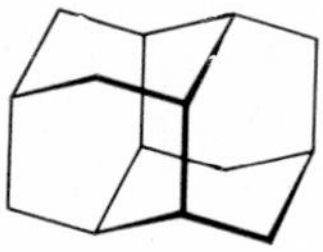 Diamantane	$C_{14}H_{20}$	245	1.3	100
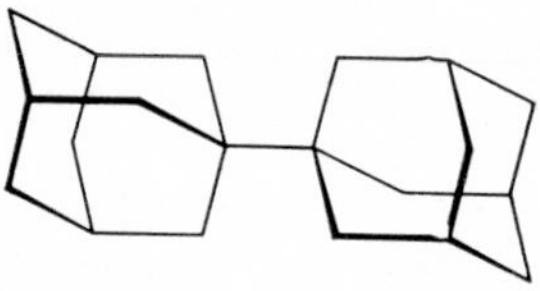 1,1′-Biadamantane	$C_{20}H_{30}$	296	2.6 1—2	(112)[a] 150—180

[a] Sublimation temperature (see text)

2.1.6 CHOICE OF DISTILLATION PROCESS

To solve some distillation problems various variants are available and often it is not easy to select an apparatus, process and operational conditions which are optimum for a given case. For easier orientation we present in this section the principles and the criteria helping a correct choice of distillation process.

When choosing a suitable distillation process the following three viewpoints should be considered: the required purpose of the separation, the amount of sample and its properties.

There may be three purposes of the separation of hydrocarbon mixtures by distillation: separation of volatile hydrocarbons from non-volatile material, separation of a hydrocarbon mixture to fractions, or the isolation of individual hydrocarbons. The following types of distillation are suitable for these purposes:

Purpose of distillation	Suitable type of distillation
Separation of volatile hydrocarbons from non-volatile material	1. Simple 2. Flash 3. Steam 4. Molecular 5. Low-temperature, vacuum 6. Sublimation
Separation of a hydrocarbon mixture to fractions	a) Non-selective 1. Simple 2. Flash 3. Rectification on column 4. Molecular b) Selective 1. Azeotropic 2. Extractive
Isolation of individual hydrocarbon	1. Rectification on column 2. Mo'ecular multistep 3. Azeotropic 4. Extractive

The size of the distillation apparatus and, to a certain extent, also the type of distillation should correspond to the amount of the sample distilled. If distilling a small sample on an excessively large apparatus the losses are usually considerable. The reverse process, i.e. working up a larger amount of sample by repeated distillation on a small apparatus takes a long time.

Current laboratory distillation apparatuses are adapted for distillations of samples of 10 – 4 000 ml volume. Certain problems arise when very small samples (below 5 ml) or when large volumes of liquid are to be distilled (above 10 l).

The smallest amount which can still be well separated by distillation is 2 – 5 ml. Even though various distillation microtechniques[27] are developed samples below 2 ml can usually be separated more conveniently by preparative gas chromatography (see Section 2.4), or by some other technique. Distillation of samples of 2 – 5 ml volume is made easy by miniature distillation apparatuses and distillation techniques with an added component.

Large volumes of liquids can be distilled in the laboratory either by repeating the process in apparatuses with a distillation pot of 4–10 l volume, or by adjusting the apparatus for a continuous process. Apparatuses and techniques suitable for distillation of small samples and for large volumes of liquids are listed in the following survey:

Distillation of small samples (2–5 ml)

Type of distillation	**Suitable apparatus and technique**
Simple	Microdistillation devices, Hickman flasks
Rectification on column	Miniature columns with a small hold-up a) spinning band b) from concentrical tubes
With an added component	a) with higher boiling inert component b) azeotrope with a low-boiling agent
Low-temperature vacuum dist. in a closed system	In a stream of carbon dioxide
Sublimation	Current sublimators

Distillation of large volumes of liquids (above 10 l)

Type of distillation	**Suitable apparatus and technique**
Simple	Rotational vacuum evaporator set for continuous operation
Rectification on column	Columns with large throughput (repeated distillation): a) with empty tube (*Vigreux*) b) spinning band c) plate
Flash	a) with stirred falling film, apparatus set for continuous work b) centrifugal, set for continuous operation

The more detailed is the knowledge of the composition and the properties of the distilled sample, the better can the conditions of distillation be selected. The most important data are the boiling points of the components, their relative volatilities and thermal stability. In analyses of hydrocarbon mixtures samples occur the composition of which is known in detail, but also samples of totally unknown composition. In the latter case it is essential to determine at least approximately the boiling point distribution of all components present before distillation, best using the method of simulated distillation, or a simple distillation test.

Hydrocarbons with boiling points of up to about 200 °C (up to $C_{10}-C_{12}$) are distilled, as a rule, under atmospheric pressure. These hydrocarbons are only exceptionally distilled under reduced pressure when especially labile hydrocarbons have to be preserved, or when the relative volatility of the separated components in a vacuum is distinctly higher than under normal pressure.

Hydrocarbons with boiling point of 200 – 250 °C and higher boiling decompose significantly on atmospheric distillation. At more accurate analyses the decomposition is disturbing, because both the distillate and the remaining heavy fraction contain artifacts that distort the original composition and the properties of the products.

The decomposition of the high-boiling hydrocarbons is prevented by distillation under reduced pressure. Each boiling point depression by 10 K restricts the rate of decomposition reactions to approximately a half. The *kinetic* aspect of the thermal decomposition is an important factor and when choosing the distillation conditions the fact should be taken into consideration that the amount of the decomposition products depends not only on the distillation temperature, but also on the time during which the fraction is exposed to high temperature. Therefore the *flash-type* stills in which the fraction remains in the hot zone for a few seconds only can operate at a relatively high temperature and still be milder than, for example, rectification columns operating at lower temperature but with up to a 1 000 times longer exposure of the fraction to heat.

The decrease of pressure in the distillation apparatus should be proportional to boiling points of the components. In distillation on columns, that pressure is considered optimum at which the temperature of the vapours of distilling hydrocarbons is within the 80 – 160 °C range. A suitable distillation pressure can be estimated from Tables 2.1.11 and

2.1.12 on the basis of normal boiling points or the number of carbon atoms in the hydrocarbons present in the fraction. If the fraction is distilled at too high a vacuum the vapours of the components have an exceedingly low temperature (below 50 °C) and cannot be condensed in the water condenser. In Table 2.1.15 the distillation pressures are given which are used for hydrocarbons with different numbers of carbon atoms, as well as the separation efficiencies and times of thermal exposure for individual types of distillations.

TABLE 2.1.15

Usual distillation conditions for hydrocarbons with various carbon number

Carbon number	Distillation pressure	Type of distillation	Separation efficiency (n)	Time of exposure to heat (s)
$C_1 — C_5$	atmospheric	low-temperature rectification on column	up to 200	unimportant
$C_5 — C_{10}$	atmospheric	1. flash 2. simple 3. rectification	1 1 10—200	10^1 10^3 $10^4 — 10^5$
$C_{10} — C_{22}$	medium vacuum 1 — 40 kPa	1. flash 2. simple 3. rectification	1 1 10—200	10^1 10^3 $10^4 — 10^5$
	high vacuum 0.1 — 10 Pa	thermal rectification	10	10^3
$C_{20} — C_{50}$	high vacuum $10^{-1} — 10^{-3}$ Pa	1. simple molecular of the film type 2. multistep molecular	1 10—20	10^1 $10^3 — 10^4$

Even at a medium vacuum the boiling point decrease of hydrocarbons is very pronounced. For example on distillation at a pressure of about 1 kPa the boiling points of hydrocarbons decrease by more than 130 K; at the same time the relative volatility of the components remains unchanged in principle, except for smaller deviations. This is the advantage of vacuum distillation over steam-distillation which also gives a similarly distinct boiling point decrease of hydrocarbons, but without the possibility of separating the components. This is clearly shown in Fig. 2.1.30, where the boiling points of *n*-alkanes $C_{13}-C_{16}$ are compared at normal pressure, at a 1.333 kPa pressure and on steam distillation.

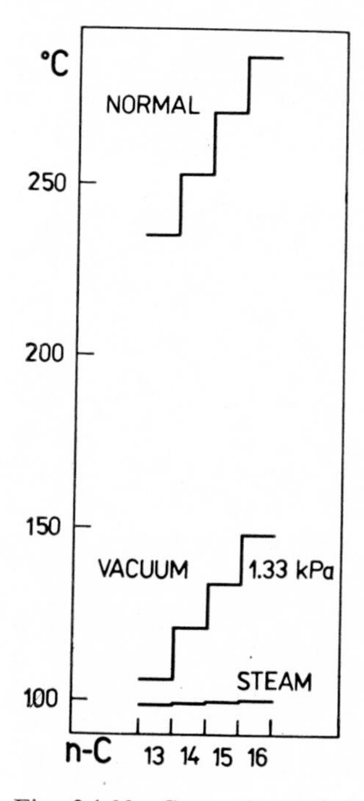

Fig. 2.1.30 Comparison of the boiling points of *n*-alkanes C_{13}—C_{16} at normal pressure, on vacuum distillation (1.33 kPa) and on steam distillation

When hydrocarbon fractions are to be separated to narrower sharp fractions with a minimum overlap, and during the isolation of pure components the main problem is a sufficient separation efficiency of the still used. The number of theoretical plates (n) necessary for the separation of a certain pair of hydrocarbons depends on their relative volatility. Rose[25] presents an equation for the computation of the optimum number of theoretical plates necessary for the separation of an ideal binary mixture under practical conditions, when the first 40 % of distillate with a content of the more volatile component exceeding 95 mol % is obtained from the starting mixture containing 50 mol % of both components:

$$n_{opt} = \frac{2.85}{\log \alpha} \qquad (2.1.22)$$

For separations of multicomponent mixtures to individual hydrocarbons the rectification section of the selected column should have as many theoretical plates as necessary for the separation of that pair of hydrocarbons which possess the lowest relative volatility. If the relative volatility is not known, it can be calculated from normal boiling points according to equation (2.1.7). In Table 2.1.16 optimum numbers of theoretical plates for pairs of hydrocarbons are given, boiling at about 100 °C at normal pressure and displaying a difference in boiling point in the range from 1 to 20 K.

TABLE 2.1.16

Optimum number of theoretical plates (n_{opt}) necessary for the separation of a pair of hydrocarbons differing in boiling points by 1 to 20 K (with relative volatility α 1.03—1.8)

Δ b.p. (in K)	α	n_{opt}	Δ b.p. (in K)	α	n_{opt}
20	1.8	10	3	1.09	75
10	1.3	23	2	1.06	113
5	1.16	45	1	1.03	225
4	1.12	56			

The optimum number of plates for the separation of a given mixture, calculated from the equation (2.1.22) or read from Table 2.1.16 is not strictly observed because the choice of the distillation columns available in the laboratory is, of course, restricted. When choosing a suitable column, the practice usually is, that optimum distillation pressure and optimum separation efficiency are determined on the basis of the sample properties and then the properties of the columns are compared that are available with the mentioned requirements and the amount of sample that is to be distilled. Simultaneously the characteristic properties of various types of rectifications sections mentioned earlier should be respected; for low pressures columns should be selected with low pressure drops, for small amounts of sample columns with a small hold-up volume are preferable. Then a column is selected which best fulfils all the requirements.

The operational conditions of the distillation are adapted in accordance with the properties of the selected column and the required optimum values. If the column possesses a higher separation efficiency than required, it is possible to operate with a higher throughput and a lower reflux ratio. In the opposite case a poor separation efficiency is compensated by slowing the withdrawal of the distillate.

If a sufficient amount of sample is available the poor separation efficiency of the column can be compensated by *repeating* the distillation. For example, on rectification of an equimolar binary mixture of hydrocarbons with a relative volatility 1.06, on a 30 theoretical plates column, a sample of the lighter component can be obtained of a purity of maximum 85 mol %. If sufficient distillate is collected, so that rectification can be repeated on the same or a smaller, but equally efficient, column, with a batch containing 80 mol % of the lighter component, then the lighter component will be concentrated in the first fractions of the distillate up to 96 mol %. A classical example of the use of this technique is the Whitmore's[79] successful isolations of pure trimethylpentenes from diisobutene by repeated rectification on 20 plate columns.

The advantages of repeated distillation are also made use of in work with highly efficient columns. From the point of view of the shortening of the distillation time, and from the point of view of a full exploitation of efficiency it is useful to rectify on a highly efficient column only a narrow distillation cut, obtained by prefractionation of the sample on another less efficient column with higher output.

Special separation problems can be solved by combining various types of distillations. Often two non-selective distillation techniques can be combined, as for example steam distillation and column rectification, or atmospheric pressure rectification followed by vacuum rectification. Non-selective and selective distillation techniques can be combined in a similar manner. For example, we have already mentioned the separations of alkanes and cycloalkanes by means of atmospheric rectification with subsequent azeotropic distillation.

The method of *monitoring* distillation also has a certain effect on the result of distillation and deserves special mention. The commonest method is the observation of the temperature of the vapours of the substances distilled. Even though the temperature in the column head can be measured with a 0.01 K accuracy, all changes in the composition of the vapours are not recorded. It is better to monitor the operation of highly efficient columns that separate components differing in boiling points by less than 1 K by analysing the distillate. Sometimes the measurement of the refractive index suffices; however, gas chromatography is more universal. Especially when multicomponent mixtures are distilled only gas chromatography gives an accurate picture of the representation of all components of the distillate. However, the fact should not be forgotten that some hydrocarbons with close boiling points, but differing in polarity, have a reversed order of elution on polar stationary phases (but also on Apiezon L) than is the order of their boiling points. For example diphenyl, although it has a lower boiling point than 2,2′-dimethyldiphenyl, is eluted on a column with Apiezon L (at 200 °C) after this homologue. Overlooking such inversions can lead to a false interpretation of the results of distillation.

In vacuum distillation frequent withdrawals of the samples for analysis can impair the process of distillation; in such a case it is better to collect the distillate in a large number of small fractions and analyse them afterwards. Suitable fractions are then selected and combined on the basis of the results of the analysis and according to the required purity of the product.

REFERENCES (2.1)

1. B. L. Karger, L. R. Snyder and C. Horváth, An Introduction to Separation Science, Wiley-Interscience, New York, 1973.
2. J. M. Miller, Separation Methods in Chemical Analysis, Wiley-Interscience, New York, 1975.
3. E. S. Perry, Separation and Purification Methods, Vol. 1, Marcel Dekker, New York, 1973.
4. E. W. Berg, Physical and Chemical Methods of Separation, McGraw—Hill, New York, 1963.

5. C. J. King, Separation Processes, McGraw—Hill, New York, 1971.
6. R. Bock, Methoden der Analytischen Chemie, Band 1, Trennungsmethoden, Verlag Chemie GmbH., Weinheim 1974.
7. F. D. Rossini, B. J. Mair and A. J. Streiff, Hydrocarbons from Petroleum, Reinhold Publishing Corp., New York, 1953.
8. J. C. Giddings, *Anal. Chem.* **39** (1967) 1027.
9. Selected Values of Hydrocarbons and Related Compounds, API Project 44, 1956.
10. S. W. Ferris, Handbook of Hydrocarbons, Academic Press, Inc., New York, 1955.
11. Vapor Pressures and Boiling Points of High Molecular Weight Hydrocarbons C_{11} to C_{100}, Report of investigation of API Project 44, 1965.
12. R. D. Obolencev, Fizicheskie konstanty uglevodorodov zhidkikh topliv i masel, Gostoptekhizdat, Moscow, 1953.
13. ASTM E 133—71, 1975 Annual Book of ASTM Standards, Part 25, Amer. Soc. for Testing and Materials, 1916 Race St., Philadelphia, Pa. 19103.
14. ASTM D 86—67 (1972) 1975 Annual Book of ASTM Standards, Part 23, Amer. Soc. for Testing and Materials, 1916 Race St., Philadelphia, Pa. 19103.
15. L. E. Green, L. J. Schmauch and J. C. Worman, *Anal. Chem.* **36** (1964) 1512.
16. E. J. Hoffman, *Chem. Eng. Sci.* **24** (1969) 113.
17. W. L. Nelson, Petroleum Refinery Engineering, McGraw—Hill, 1958.
18. ASTM D 2887—73, 1978 Annual Book of ASTM Standards, Part 24, Amer. Soc. for Testing and Materials, 1916 Race St., Philadelphia, Pa. 19103.
18a. B. W. Jackson, R. W. Judges and J. L. Powell, *J. Chromatogr. Sci.* **14,** (1976) 49.
19. J. C. Worman and L. E. Green, *Anal. Chem.* **37** (1965) 1620.
20. T. H. Gouw, *Anal. Chem.* **45** (1973) 987.
21. T. H. Gouw, R. L. Hinkins and R. E. Jentoft, *J. Chromatogr.* **28** (1967) 219.
21a. C. L. Stuckey, *J. Chromatogr. Sci.* **16** (1978) 482.
22. R. J. O'Donnell, *Ind. Eng. Chem. Process Des. Develop.* **12** (1973) 208.
22a. T. Aczel and H. E. Lumpkin, paper presented at the 18th Annual Conference on Mass Spectrometry, San Francisco, California, 1970.
23. F. W. Melpolder and C. E. Headington, *Ind. Eng. Chem.* **39** (1947) 763.
24. H. Kolling, *Chemie—Ing.—Techn.* **22** (1952) 405.
25. A. Rose, *Ind. Eng. Chem.* **33** (1941) 594.
26. Technique of Organic Chemistry, Vol. IV, Distillation, A. Weissberger, editor, Interscience Publishers, 1965.
27. E. Krell, Handbook of Laboratory Distillation, Elsevier Publ. Comp., 1963.
28. W. G. Fischer, *Fette—Seifen—Anstrichmittel* **72** (1970) 444.
29. J. C. Winters and R. A. Dinerstein, *Anal. Chem.* **27** (1955) 546.
30. A. G. Norheim and R. A. Dinerstein, *Anal. Chem.* **28** (1956) 1029; 29 (1957) 1546.
31. ASTM D 2892/73, 1975 Annual Book of ASTM Standards, Part 24, Amer. Soc. for Testing and Materials, 1916 Race St., Philadelphia, Pa. 19103.
32. T. J. Walsh, G. H. Sugimura and T. W. Reynolds, *Ind. Eng. Chem.* **45** (1953) 2629.
33. A. C. Bratton, W. A. Felsing and J. R. Bailey, *Ind. Eng. Chem.* **28** (1936) 424.
34. R. Bock and M. Fariwar-Mohseni, *Z. Anal. Chem.* **215** (1966) 324.
35. G. J. Pierotti, C. H. Deal and E. L. Derr, *Ind. Eng. Chem.* **51** (1959) 95.

36. C. H. Deal, E. L. Derr and M. N. Papadopoulos, *Ind. Eng. Chem. Fundamentals* **1** (1962) 17.
37. G. M. Wilson and C. H. Deal, *Ind. Eng. Chem. Fundamentals* **1** (1962) 20.
38. W. A. Scheller, *Ind. Eng. Chem. Fundamentals* **4** (1965) 459.
39. R. F. Weimer and J. M. Prausnitz, *Hydrocarbon Process.* **44** (1965) 237.
40. E. Erdös, *Coll. Czech. Chem. Commun.* **21** (1956) 1528.
41. R. Gilmont, D. Zudkevitch and D. F. Othmer, *Ind. Eng. Chem.* **53** (1961) 223.
42. J. H. Purnell, Gas Chromatography, John Wiley, New York, 1962.
43. H. Röck, *Chem.—Ing.—Tech.* **28** (1965) 489.
44. J. Griswold, D. Andres, C. F. Van Berg and J. E. Kash, *Ind. Eng. Chem.* **38** (1946) 65.
45. C. S. Carlson, The Chemistry of Petroleum Hydrocarbons, Vol. I, edit. B. T. Brooks et al., New York, 1954.
46. R. S. Porter and J. F. Johnson, *Ind. Eng. Chem.* **52** (1960) 691.
47. G. W. Warren, R. R. Warren and V. A. Yarborough, *Ind. Eng. Chem.* **51** (1959) 1475.
48. D. Tassios, *Hydrocarbon Process.* **49** (1970) 114.
49. W. A. Spelyng and D. P. Tassios, *Ind. Eng. Chem., Process Des. Develop.,* **13** (1974) 328.
50. H. Stage, *Erdöl u. Kohle* **3** (1950) 377, 478.
51. L. H. Horsley, *Anal. Chem.* **19** (1947) 508; **21** (1949) 831.
52. L. H. Horsley, Azeotropic Data-III, Advances in Chemistry Series 116, American Chemical Society, Washington, D.C., 1973.
53. B. J. Mair, *Anal. Chem.* **28** (1956) 52.
54. H. S. Nutting and L. H. Horsley, *Anal. Chem.* **19** (1947) 602.
55. ASTM D 1160—61 (1973), 1975 Annual Book of ASTM Standards, Part 23, Amer. Soc. for Testing and Materials, 1916 Race St., Philadelphia, Pa. 19103.
56. J. Feldman, M. Myles, I. Wender and M. Orchin, *Ind. Eng. Chem.* **41** (1949) 1032.
57. J. E. Hawkins and J. A. Brent Jr., *Ind. Eng. Chem.* **43** (1951) 2611.
58. M. Myles, J. Feldman, I. Wender and M. Orchin, *Ind. Eng. Chem.* **43** (1951) 1542.
59. M. S. Peters and M. R. Cannon, *Ind. Eng. Chem.* **44** (1952) 1452.
60. H. J. Coleman, J. E. Dooley, D. E. Hirsh and C. J. Thompson, *Anal. Chem.* **45** (1973) 1724.
61. W. E. Haines and C. J. Thompson, API RP 60 Published Report No. 37, July 1975.
62. H. T. Rall, R. L. Hopkins, C. J. Thompson and H. J. Coleman, *Proc. Amer. Petrol. Inst., sect.* **8,** 42 (1962) 46.
63. F. W. Melpolder, T. A. Washall, and J. A. Alexander, *Anal. Chem.* **27** (1955) 974.
64. B. J. Mair, A. J. Pignocco and F. D. Rossini, *Anal. Chem.* **27** (1955) 190.
65. V. A. Malyusov, N. N. Umnik, N. A. Malafeev and N. M. Zhavoronkov, *Dokl. akad. nauk SSSR* **109** (1965) 828.
66. V. A. Malyusov and N. M. Zhavoronkov, *Coll. Czech. Chem. Commun.* **23** (1958) 1720.
67. P. R. Watt, *Chem. a. Ind.* (1960) 1207.
68. W. J. Podbielniak and S. T. Preston in Physical Methods in Chemical Analysis, Berl W. G. (Ed.), Vol. 3, Academic Press, New York, 1956.

69. E. S. Byron, J. R. Bowman and J. Coull, *Ind. Eng. Chem.* **43** (1951) 1002.
70. E. S. Perry and D. S. Cox, *Ind. Eng. Chem.* **48** (1956) 1473, 1479.
71. C. R. Enzell, B. Kimland and A. Rosengren, *Acta Chem. Scand.* **24** (1970) 1462.
72. B. Riegel, J. Beiswanger and G. Lanzl, *Ind. Eng. Chem. Anal. Ed.* **15** (1943) 417.
73. M. H. Hubacher, *Ind. Eng. Chem. Anal. Ed.* **15** (1943) 448.
74. Encyclopedia of Industrial Chemical Analysis, Vol. 3, p. 572, F. D. Snel and G. L. Hilton, eds., Interscience Publishers, New York, 1966.
75. S. Kaufmann, J. C. Medina and C. Zapata, *Anal. Chem.* **32** (1960) 192.
76. A. D. Gettler, C. J. Umberger and L. Goldbaum, *Anal. Chem.* **22** (1950) 600.
77. H. Sugisawa, K. ASO, *Chem. Ind. (London)* (1961) 781.
78. J. F. Thomas, E. N. Sanborn, M. Mukai and B. D. Tebbens, *Anal. Chem.* **30** (1958) 1954.
79. F. C. Whitmore and S. N. Wrenn, *J. Amer. Chem. Soc.* **53** (1931) 3136; **54** (1932) 3706, 3710.

2.2 Liquid chromatography

Liquid chromatography (LC) is a process in which the liquid phase moves over the stationary phase in close contact, during which the sample (*solute*) is distributed between the moving (mobile) and the stationary phase.

Liquid chromatography plays a prominent role among separation methods because it is capable of rapidly separating mixtures of substances, both in gram quantities and in quantities as low as fractions of milligrams. As a rule gas chromatography is more rapid and capable of separating more complex mixtures of substances. On the other hand, its use is limited by the non-volatility of some substances, and this limiting factor is evident mainly in preparative gas chromatography. In liquid chromatography no such obstacle exists and both macromolecular and ionic substances also can be separated by this method. Hence, it may be stated that both types of chromatography are irreplaceable in their specific areas, and in cases when both can be employed they can be complementary.

Liquid chromatography cannot surpass gas chromatography in speed, because the diffusion coefficients of the substances separated in the liquid phase are smaller by several orders of magnitude than in the gas phase. On the other hand, in some instances liquid chromatography can solve many a separation problem by making use of specific interactions, changing of eluents, etc., which gas chromatography could solve only with difficulty.

An important advantage of the use of liquid chromatography consists in the ability to recover the separated substance in an almost 100 % yield.

Liquid chromatography procedures can be classified into large groups according to various viewpoints. Most common is the classification according to the shape or type of bed:

1. flat, open bed – *thin-layer chromatography* (TLC), or *paper chromatography* (PC)
2. packed column – *liquid column chromatography* (LCC).

Further classification can be made according to the principles of interactions between the sample and the stationary phase:

1. liquid-liquid or *partition chromatography* (LLC) – sample distribution between two liquid phases
2. liquid-solid or *adsorption chromatography* (LSC) – distribution of the sample between the liquid and the solid phase
3. *gel permeation chromatography* (GPC) or gel chromatography—separation of the sample according to the accessibility of the pores for molecules of different size
4. *ion exchange chromatography* – separation on the basis of ion exchange between the sample and the stationary phase.

Classification of liquid chromatographic techniques on the basis of the elution technique is the following:

1. *linear* elution technique – mobile phase is a single solvent
2. *stepwise* elution with several eluents of increasing solvent strength
3. *gradient* elution with a mobile phase the solvent strength of which continuously increases
4. *displacement* – the eluent is considerably stronger than any component of the sample – the sample moves in front of the eluent
5. *frontal* elution and *percolation* – the sample (also diluted with a solvent) serves as the mobile phase

The complexity and the extent of the problems of liquid chromatography are evident from this review: In this section only the problems of *preparative technique* are discussed, which enables the isolation of at least a few milligrams of hydrocarbon fractions. Thin-layer chromatography (TLC) is not discussed here even though it is widespread and popular. The flexibility and the speed of thin-layer chromatography is considerable, but the majority of the problems mentioned can be solved successfully by column chromatography as well. The preparation of groups of hydrocarbons can be carried out more easily on a column than on a thin layer, the yields are higher and the purity of the preparations is usually better. For orienting purposes and for the selection of the most suitable separation system TLC is undoubtedly utilizable. Thin-layer chromatography is most often used in the analysis of hydrocarbon mixtures for individual identification and determination of polycyclic aromatic hydrocarbons (see Section 4.2.3).

The importance of column chromatography (LCC) is stressed by the present rapid development of this technique. The common feature of all types of column chromatography is the introduction of the sample into a column packed with fine particles of a solid material and its transport by a stream of liquid – the mobile phase. The separation of the substances takes place on the basis of their different migration rate in the column, i.e. on the differences in specific interactions between the separated substances and the stationary phase. Gel chromatography and ion exchange chromatography will be discussed in greater detail in Chapter 2.3, and therefore only the principles of adsorption and partition chromatography will be presented here.

Adsorption chromatography (LSC) makes use of the interactions between the sample and the solid phase – the adsorbent – in the surrounding mobile liquid medium – the eluent. The sample moves in the direction of eluent flow, the more rapidly the less it is adsorbed. The forces responsible for the interaction between the sample and the adsorbent are the following[1]:

Dispersion forces – that occur practically in all adsorption systems and which contribute substantially to the interaction of the non-polar adsorbent and the non-polar sample. They are relatively weak and their part in the interactions of hydrocarbons and polar sorbents is small.

Inductive forces – occur mainly on polar adsorbents. It is assumed that the adsorbent surface, for example alumina, contains centres with a strong electrostatic field; these fields induce a dipole moment in the molecule of an aromatic hydrocarbon in their close proximity.

Acid-base interactions – have been observed in separations of basic substances on silica gel, or acid substances on alumina. It was also observed on alumina[2] that the adsorption energy of some polycyclic aromatic hydrocarbons depends on the acidity of the alumina surface.

The formation of hydrogen bonds – is the most substantial contribution in the adsorption of solutes on silica gel. The adsorbent surface is covered with hydroxyl groups that react as proton donors with weakly basic substances, such as aromatic hydrocarbons. In fact it is a type of acid-base interaction.

The formation of molecular complexes between the adsorbent and the sample has predominantly a covalent character. The separation of olefins on silica gel impregnated with silver nitrate is enabled by the *charge transfer* between the silver ion and the olefins. Similar complexes are formed during the separation of aromatic hydrocarbons on adsorbents impregnated with 2,4,7-trinitrofluorene, picric acid, etc.

Partition chromatography (LLC) is based on the partition of substances between two immiscible or restrictedly miscible liquid phases. The column is packed with solid particles the surface of which is coated with a *liquid stationary phase*. This phase can also be bound chemically on the particle surface. When the sample is introduced into the column and the column is washed with the eluent, equilibrium is attained between the stationary and the mobile phase. The separation depends on the differences in distribution constants of individual sample components for these two liquids. The whole process of separation between the stationary and the mobile phase can be represented as a multistep separation of a sample in a separation funnel between two immiscible liquids. Further details on partition chromatography are given in Section 3.5.3.

Nowadays adsorption chromatography leads in the separation of hydrocarbon mixtures, because it is suitable for preparative separations and enables the separation of milligram to gram quantities of sample. These quantities are usually sufficient for further analyses, carried out by mass spectrometry, nuclear magnetic resonance, infrared spectrophotometry, etc.

2.2.1 TECHNIQUES OF COLUMN CHROMATOGRAPHY

The column is the basic element of the chromatographic system that may be either open – as in classical liquid chromatography, or closed – as in *high performance* liquid chromatography (HPLC).

The open chromatographic system is composed of a column packed with an adsorbent, provided at its upper part with a reservoir with the eluent, and at its bottom part with a receiver for eluent or also a fraction collector. No detector of the eluted compounds is inserted between the column and the receiver. The sample is introduced directly onto the column bed and the flow of the eluent through the column takes place through force of gravity.

In a closed chromatographic system the eluent flow is provided by a pump connected in front of the column. A device for sample introduction, called an injection port, is connected with the column inlet. A detector is connected with the column outlet behind which a fraction collector can be inserted.

The column is packed with an adsorbent; the quality of the adsorbent and the excellence of the column packing affect the separation efficiency decisively.

The basic equipment is composed of a pump, sample injection device, column and detector. The quality of single elements can affect the efficiency of the whole chromatographic system distinctly. The technique of column chromatography is described in detail in monographs[3, 4, 5].

Pumps. The closed systems of liquid chromatography require the following of pumps:

They should provide a constant flow of the mobile phase with an accuracy better than 2 %.

This through-flow should be continuously adjustable within the 0.1 to 5 ml/min range.

All parts of the pump that are in contact with the mobile phase should be made of such materials as would not contaminate the mobile phase or corrode.

The mobile phase flow should be pulse-free.

The condition of a constant mobile phase flow and its continuous regulatability should be fulfilled even when the operating pressure is minimum 10 MPa.

The pumps can be classified according to whether they deliver the mobile phase without pulsation.

Both piston and membrane reciprocating pumps are pulsating. At each stroke of the piston or of the membrane a small volume of the mobile phase is expelled into the chromatographic system, and during the back-movement the chamber is again filled. The delivery of mobile phase depends on the height of the stroke of the piston or membrane. The advantage of these pumps is the uninterrupted feed of the mobile phase, and their disadvantages are the pulsating flow that disturbs sensitive detectors and small changes of through-flow at pressure change. The pulsation can be restricted considerably by using doubled pumps and insertion of pulse dampers.

Pulseless pumps are constructed either as constant displacement pump or as pneumatic amplifiers.

Constant displacement pumps are a very widely used type of pump in liquid chromatography. In principle they consist of a cylindrical chamber of 200 to 500 ml volume in which a perfectly sealed piston moves. The chamber is filled with the mobile phase and the rate of delivery is regulated by the rate of movement of the piston. The disadvantage of these pumps consists in their limited capacity, because at a complete expulsion of the mobile phase the injection must be interrupted and the chamber refilled.

A pneumatic amplifier is based on a similar principle, but the piston movement is provided by a large area membrane, moved by air pressure. The air pressure then determines the rate at which the mobile phase is expelled. The constant flow of the mobile phase is guaranteed only if the resistance of the whole chromatographic system (i.e. the column, connections, etc.) does not change.

Peristaltic pumps that are frequently used in laboratories for other purposes are not suitable for adsorption chromatography of hydrocarbons, because flexible tubes are not available which would stand contact with the eluents employed and retain the required mechanical properties.

Equipment for gradient elution. This equipment enables a continuous change of the mobile phase composition during the analysis, at constant flow. The simplest device, enabling an exponential course of the gradient, consists of two vessels connected with a small diameter tube. In the first vessel the less polar and in the second one the more polar eluent is placed. A piston or membrane pump sucks the eluent from the first vessel and delivers it into the column. The more polar eluent flows into the first vessel where the stirrer mixes it.

Gradient forming devices consist of two constant displacement pumps that are controlled electrically so that the individual pumps should take the necessary part during the process in the total output.

Other devices employ an electrically controlled valve, on the pump suction. Then the valve switches the suction over either to one or the other eluent and the length of the switching-over period regulates the volume ratio of both eluents. Before entrance into the pump the mixture is perfectly mixed in the mixing chamber. In these gradient forming devices any gradient course can be set for two or three eluents in dependence on time.

Sample injection. The sample injection is simplest in open chromatographic systems where the sample is introduced with a pipette or a syringe on the bed top. Care should be taken that the sample should not swirl the adsorbent particles and that its penetration into the bed should be uniform throughout the whole column cross-section.

In closed chromatographic systems the sample injection is the more complicated the higher the operating pressure. The sample is introduced into the column either by a direct injection with a syringe or through an injection loop.

For direct injection the column is provided with an injection port with a septum. The sample is introduced by perforating the septum with the syringe, exactly as in gas chromatography. Syringes for liquid chromatography are of special construction, gas tight, and allowing the injection of samples against up to 10 MPa pressure. The septum is usually made of silicone rubber, but in the case of some aggressive solvents a septum is used the one side of which, that is in contact with the mobile phase, is coated with a layer of PTFE. The syringe is provided with a long needle so that the sample can be introduced at the beginning of the bed, in the centre of the column section. The injection port should have a small volume and the whole space should be perfectly rinsed with the mobile phase. When the injection is to be made against a pressure higher than 10 MPa, the pumping of the eluent into the column is stopped and the sample is injected almost at atmospheric pressure.

The second method of sample introduction makes use of a sixway valve provided with a sample loop. For injection the loop is first filled with the sample, the valve is switched over to the second position when the eluent flows through the loop and carries the sample into the column. Injection loops enable the injection of larger amounts of sample against up to a 40 MPa pressure. The disadvantage of the sample loop consists in its constant volume and consequently the necessity of adjusting the sample concentration. Sample loops can be interchanged, which requires a small adjustment. High-performance systems require that the connections between the loop and the column should be of minimum volume.

Columns. As regards column construction preference is given mainly to straight tubes the length of which varies between 10 and 200 cm. The ratio of the column diameter and its length is usually kept between 1 : 20 and 1 : 100, and the commonest diameters are 0.2 to 4.0 cm. In some instances, when poorly separable substances are chromatographed the columns are connected in series so that their total length can be, for example, up to 10 m. Preferably identical columns are connected, i.e. those of equal diameter and length, and it is important that the spaces in the connections between the columns be minimum.

Columns for open systems are made of ordinary glass tubes, for pressures up to 5 MPa special strong-glass columns may be used. The commonest material for columns operating under pressure is stainless steel, less often copper or aluminium is used. In high-performance columns it is important that their inner diameter should be accurately identical along their whole length, and that the inner surface be perfectly smooth, or even polished.

The adsorbent in the column is placed on a small layer of glass wool (for low-pressure columns) or on a porous disc, made either of sintered glass, stainless steel sponge, or porous PTFE.

The technique of column packing depends on the particle size of the adsorbent. If the particles are larger than 30 μm, which is the commonest case in preparative chromatography, the columns are packed in a dry state. The adsorbent is put into the column in small doses under constant tapping on the column walls or by tapping or vibrating the column along its axis. Smaller particles used for analytical columns are packed in suspension. A suspension of the adsorbent particles in an organic liquid of high density (1 g of adsorbent in 10 ml of liquid) is prepared using ultrasonic stirring, and the obtained suspension, for example in a mixture of tetrachloroethane and tetrachloromethane or tetrabromoethane etc., is then packed, using a pressure pump, into the column at maximum possible pressure and speed. It is important that the liquid used is of such density that the particles float in it. The uniformity of the packing and the technique of packing affect the resulting column efficiency.

Detectors. The detector is a very important and also limiting factor in modern liquid chromatography. Preparative separations can be carried out sometimes without a detector, but analytical separations practically always require a sensitive detector. The technique of liquid chromatography has no such universal detector as the flame ionization detector in gas chromatography. Modern detectors for liquid chromatography are unable in a number of cases to record all the eluted substances; no detector exists for which the response could be predicted generally in dependence on the concentration of the components eluted. Individual detectors have further limitations. Thus a relatively poor possibility is given of quantitative analysis from the detector response. In the majority of cases the quantitative analysis is possible only for a certain component known in advance, and the apparatus must be calibrated for this purpose. Seldom can the data be quantified for groups of structurally related substances, but in such instances the accuracy of the analysis is usually lower.

For the analysis of hydrocarbons the following detectors are used predominantly:

1. *optical* – ultraviolet, fluorescence and differential refractometer,
2. *solute transport detectors.*

The *ultraviolet detector* is most widely used in column chromatography. It is based on the same principle as the UV spectrophotometer (see Chapter 3.4). The eluate flows through a measuring cell that should have a small volume and a large optical length for analytical purposes. These cells are of 5 – 10 μl volume and 10 mm optical length. For preparative purposes cells of higher volume are employed, i.e. 30 – 500 μl. Cheaper devices have a low-pressure mercury lamp as UV light source and they operate at a single wave-length of 254 nm or, when filters are used, at 280 nm. More recent apparatuses are also provided with a deuterium lamp and a grating monochromator and they can operate at any selectable wave-length within the 220 – 600 nm range. Modern apparatuses have full-scale sensitivity at the absorbance of 0.005 absorbance units, with the linearity in the 0.005 – 1.0 absorbance units range. The main disadvantage of UV detectors is their inability to record saturated hydrocarbons, but they are excellent for the analysis of aromatic hydrocarbons. UV detectors can also be used for gradient elution if the eluents do not have an excessively high absorption in the UV light of selected wave-length. The response depends both on the substance detected and on the operating wave-length. In the case of aromatic hydrocarbons the response may vary by three orders of magnitude.

Fluorescence detectors are constructionally very similar to the UV detectors, with the difference that instead of the radiation passed through, fluorescence of the excited substance is recorded (see Chapter 4.3). They can be constructed as fluorimeters for one excitation wave-length, or as spectrofluorimeters with a flow-through cell, permitting the change both of the excitation and of the emission wave-length. The advantage of this type of detector consists in its sensitivity which is up to 100 times higher for some aromatic hydrocarbons than that of the UV detector. Its use is limited to substances that fluoresce and the detector response depends both on the substance detected and on the excitation and emission wave-length. In this type of detector it is important that substances should not be present which *quench* the fluorescence. As fluorescence quenching is also observed at higher concentrations of the substance in the eluate, fluorescence detectors are not applicable for preparative separations. In the analysis of hydrocarbons their use is limited to the determination of trace amounts of polycyclic aromatic hydrocarbons, and they can also be used in gradient elution.

Differential refractometers record the difference in the refractive index between the pure mobile phase, located in the reference cell, and the eluate from the column, flowing continually through the measuring cell. Two different systems are used for the construction of these detectors. The Fresnel type is constructed so that the light enters onto the liquid-glass boundary at an angle which is slightly smaller than the critical angle, and the part of the reflected light depends on the refractive index of the liquid. The other type makes use of the fact that a change in the deviation of the light beam passing through the cell takes place when the mobile phase composition in the measuring cell is changed (and thus also the refractive index). The light beam hits the photomultiplier and the electric signal changes with its deviation. The main advantage of differential refractometers consists in their ability to record any type of compound. They require good thermostating and cannot be used for gradient elution. They are also less sensitive than UV detectors and can be used for preparative separations. In dependence on the eluent used and the substance measured the record of the eluted component is obtained in the positive or negative direction from the zero line. In analyses of complex mixtures containing components that considerably differ in their refractive index (for example oils), some component may have the same refractive index as the eluent and it is then not recorded by the detector. A quantitative evaluation of the records is mostly very difficult, and in more complex mixtures almost impossible.

All optical detectors are non-destructive and permit a quantitative recovery of the substance from the eluate or its determination by some analytical procedure. In these detectors the volatility of the substance determined is also irrelevant.

Solute transport detectors withdraw a negligible part of the eluate and put it on a transporter system consisting of a wire or chain. The part of the eluate which adheres to the wire is led into the evaporation part where the solvent is evaporated at elevated temperature in a stream of gas. The wire coated with the sample then proceeds into the pyrolysis oven where the sample is thermally destructed either in a stream of oxygen or argon. In oxygen organic compounds are burnt to carbon dioxide which is converted to methane in the next part of the detector, and the methane is then led into the flame ionization detector. In argon pyrolysis takes place and the pyrolytic products enter the argon ionization detector. The main disadvantage of transport detectors is their low sensitivity, because they utilize only a negligible part of the sample. They cannot be used for the determination of low-boiling substances and the difference in the boiling points of the solvent and the sample must be at least 200 K. The response of the argon ionization detector depends on how easily the substance undergoes pyrolysis. In the analysis of hydrocarbons the aromates gave a substantially weaker response on this detector than paraffins or cyclanes. Transport detectors have a broad range of linearity and they can be used even for preparative separations. They are also useful in gradient elution, because they are insensitive to changes of solvents.

2.2.2 ELUTION DATA AND COLUMN EFFICIENCY

In adsorption-elution chromatography the sample is introduced onto the beginning of the column packing that is washed with the eluent until the sample is gradually eluted. The compounds present in the sample move through the column at various rates, depending on their adsorption *distribution constant* K, defined by the equation

$$K = \frac{(X)_s}{(X)_m} \tag{2.2.1}$$

where $(X)_s$ is the concentration of the solute on the adsorbent phase, expressed in moles per gram of adsorbent, and $(X)_m$ is the concentration of the solute in the mobile phase, expressed in moles per millilitre of mobile phase. For a given system K is a function of temperature only in dilute solutions, and its dimension is ml/g. It is evident that a higher value of K means a stronger interaction between the solute and the adsorbent, so that the substance will be eluted from the column later. In a concrete chromatographic system usually the weight of the adsorbent in the column is known; this is represented by W (in grams), while the volume of the mobile phase in the column is indicated by V_M (in millilitres). The V_M value is also called dead volume of the column, and it is measured as the elution volume of a component which is not adsorbed at all. The total number of moles of the solute on the adsorbent, m_s, and in the mobile phase, m_m, can be calculated from the equation

$$K\frac{W}{V_M} = \frac{(X)_s \cdot W}{(X)_m \cdot V_M} = \frac{m_s}{m_m} = k'. \tag{2.2.2}$$

From equation (2.2.2) the value k' can be calculated for the measured compound, called the capacity factor. For a given substance and a given chromatographic system the capacity factor is a constant. The k' value is computed from the chromatogram where the dependence of the concentration of the eluted component on the elution volume is recorded, according to equation:

$$k' = \frac{V_R - V_M}{V_M} \tag{2.2.3}$$

V_R is such a volume of the eluent as is just necessary for the elution of the centre of the chromatographic peak from the column. The difference $V_R - V_M$ is also called corrected elution volume of the substance, and it is represented by V_R' (ml). If instead of the elution volumes elution times are recorded (see Fig. 2.2.1), i.e. t_0 as the time at which the centre of the peak of an unadsorbed substance and t_R as the time when the centre of the peak of the measured substance just leave the columm k' can be calculated according to equation

$$k' = \frac{t_R - t_0}{t_0} \tag{2.2.4}$$

It is evident that from the known value of the capacity factor of the given component, k', and the dead volume of the column, V_M, the elution volume V_R can be calculated according to equation

$$V_R = V_M(1 + k') \tag{2.2.5}$$

A similar procedure can be assumed if the adsorption distribution constant K and the mass of the adsorbent, W, is known:

$$V_R = V_M + K \,.\, W \tag{2.2.6}$$

$$V'_R = K \,.\, W \tag{2.2.7}$$

When operating in the range of linear capacity of the column (see Section 2.2.3) the adsorption distribution constant value K is identical with the so-called specific elution volume R^0 (ml/g) that is very often referred to in literature (cf. ref.[6]).

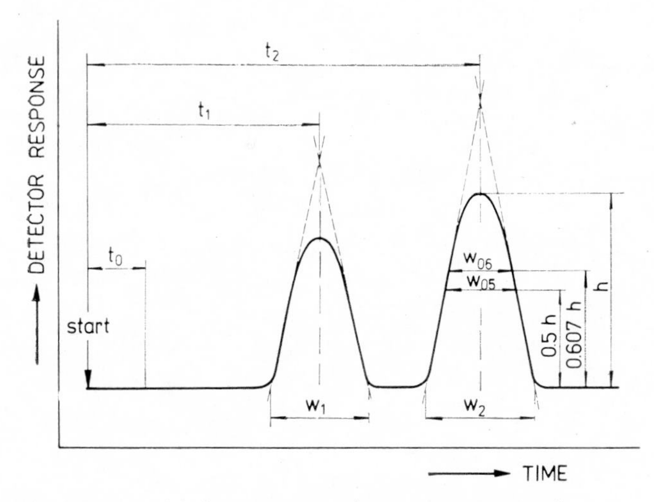

Fig. 2.2.1 Model chromatogram of a two-component mixture
On the x-axis either elution time t_R or elution volume V_R can be recorded

Elution data of hydrocarbons can be presented either as capacity factors k', or adsorption energy S°[7], or as retention indices I_x[2,8].

Capacity factor k' is a value commonly used for any substance. For its determination only the elution time of the measured component, t_R, and the elution time of the unadsorbed component, t_0, need be known.

It is a disadvantage that this data is valid for a single chromatographic system only, and that it changes when any parameter of the system is changed, for example the activity of the adsorbent or the humidity of the solvent.

Dimensionless adsorption energy S^0, introduced by Snyder[7], can be calculated for any mobile phase from the corrected elution volume V_R':

$$\log \frac{V_R'}{W} = \log V_a + \alpha' S^0 - \alpha' \varepsilon^0 A_s \tag{2.2.8}$$

or, when pentane is used as eluent, for which it is taken that $\varepsilon^0 = 0$:

$$\log \frac{V_R'}{W} = \log V_a + \alpha' S^0 \tag{2.2.9}$$

where α' and V_a are parameters characterizing the adsorbent, ε^0 is the parameter of the elution strength of the mobile phase and A_s is the area occupied by the molecule of the substance measured on the adsorbent surface (unit 0.085 nm^2) and W is the weight of the adsorbent. For a number of substances the S^0 values have been measured and published[9, 10, 11] and the values for further substances can be calculated from the group contributions. When S^0 value is determined, first the corrected elution volumes of two compounds are measured the adsorption energies of which are known, and parameters of the adsorbent, V_a and α' are then calculated from the equation (2.2.8) or (2.2.9). Then adsorption energy of any substance can be calculated from the known elution volume, or, on the contrary, elution volume may be calculated from the known value of adsorption energy. The adsorption energy value does not change with the change of adsorbent activity and the term $\alpha' \varepsilon^0 A_s$ in equation (2.2.8) also eliminates the changes caused by the eluent. The ε^0 values are then sought in tables (see Section 2.2.4); A_S values have also been published[6]. The adsorption energy is constant only for a certain adsorbent. The choice of two substances from the elution data of which the parameters of the adsorbent α' and V_a are calculated is of great importance.

Retention indices used for elution data of aromatic hydrocarbons follow from the selected series of standards: benzene, naphthalene, phenanthrene, benz(a)anthracene and benzo(b)chrysene, where a retention index equal to 10 has been assigned to the first benzene, 100 to naphthalene, 1 000 to phenathrene, 10 000 to benz(a)anthracene and

100 000 to benzo(b)chrysene. The calculation of the retention index of the unknown aromatic hydrocarbon I_x is then carried out similarly as in the case of Kováts indices in gas chromatography, according to equation:

$$\log I_x = \log I_n + \frac{\log V'_x - \log V'_n}{\log V'_{n+1} - \log V'_n} \tag{2.2.10}$$

where the symbols x, n, $n + 1$ correspond to the measured substance, to the closest lower or higher standard, and V' are the corresponding corrected elution volumes. Retention indices of substances the elution volumes of which are lower than the elution volume of benzene or higher than the elution volume of benzo(b)chrysene are determined by extrapolation. Retention indices of aromatic hydrocarbons on alumina can also be calculated easily from the published values of adsorption energies S^0, according to equation

$$\log I_x = \log I_n + \frac{S_x^0 - S_n^0}{S_{n+1}^0 - S_n^0} \tag{2.2.11}$$

under the assumption that the following values have been used for the adsorption energy of standards: benzene = 1.86, naphthalene = 3.10, phenanthrene = 4.50, benz(a)anthracene = 6.1 and benzo(b)chrysene = 7.2.

Efficiency of chromatographic columns. When a substance passes through a column a phenomenon is observed called *spreading* of the chromatographic peak. The substance is introduced onto the beginning of the column bed in a small volume of solvent, and the same substance is eluted from the column in a much larger volume of eluate. This spreading of the chromatographic peak is the higher the less efficient the column and the later the substance is eluted. The main reasons for the peak spreading during the adsorption process are the following[3, 12, 13]:

1. *irregularities in the flow* of the mobile phase between the adsorbent particles, called *eddy diffusion*. In broader spaces between the particles the mobile phase has a higher velocity and the sample molecules move more rapidly. In the smaller spaces between the particles the reverse is true. Eddy diffusion depends on the size of the adsorbent particles only, and it can be substantially restricted by using adsorbents with fine particles.

2. insufficient *mass transfer* inside each stream of the *mobile phase.* The eluent molecules that are very close to the adsorbent surface move only negligibly, while molecules in the centre of the stream move substantially faster. In the same manner the sample molecules near the adsorbent surface move slower than the molecules in the centre of the stream. The higher diffusion coefficient of the measured substance in the mobile phase and the lower linear velocity of the eluent decrease this negative effect.
3. insufficient *mass transfer* in the *adsorbent pores.* The pores of the adsorbent are filled with the eluent that practically does not move, but the sample molecules enter these pores and diffuse back into the mobile phase stream. The molecules that diffuse more deeply into the pores move more slowly in comparison with those that remain on the adsorbent surface. This effect is smaller in fine-particle and pellicular adsorbent because the distance is shortened which has to be overcome by the substance molecule in the adsorbent pores. The low linear velocity of the eluent decreases this negative effect substantially.
4. *molecular diffusion* along the column (in the direction of the eluent flow). This occurs only at linear velocities of the eluent which are below 0.2 mm/s[14]. The sample molecules diffuse according to Fick's law from the place of the highest concentration to the places where the concentration is lower. Under normal conditions it does not apply.

The spreading of chromatographic peaks always takes place during the adsorption process, but it can be restricted substantially by using adsorbents with fine particles, and it is also advantageous if these particles are spherical. Linear flow-rate of the eluent plays an important role, especially for packings with larger particles. Larger particles of the adsorbent mean a longer distance that the solute molecules must migrate in the adsorbent pores. In these cases the lower linear velocity of the eluent enables a mass transfer limited by the molar diffusion coefficient value and restricts the spreading of the chromatographic peak substantially. Qualitatively the part of individual processes causing the spreading of peaks is represented in Fig. 2.2.2, as a plot of H (plate height) versus linear velocity of the eluent. Even when the absolute values change with the particle size, the character of the dependence remains unchanged.

The low viscosity of the eluent increases the value of the molar diffusion coefficient and thus the rate of mass transfer in the mobile phase and in the pores of the adsorbent. Mathematically the *column efficiency* is expressed

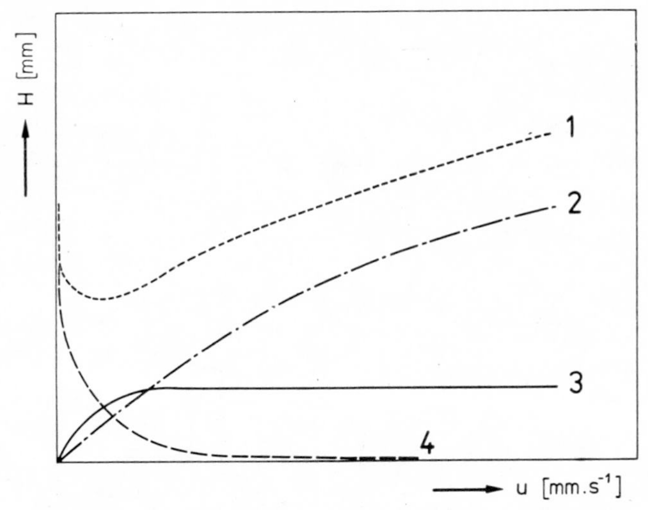

Fig. 2.2.2 Dependence of individual contributions to the plate height H on the linear velocity of the mobile phase u
1 — total plate height, 2 — contribution of mass transfer in the mobile phase, 3 — contribution of eddy diffusion, 4 — contribution of molecular diffusion

as the *number of theoretical plates n* that can be calculated for a given component from the chromatogram in Fig. 2.2.1 according to the equation

$$n = 16\left(\frac{t_R}{w}\right)^2 \tag{2.2.12}$$

where t_R is measured as the distance from the start in length units (cm, mm), and w in the same units, which means the peak width at the base. The width w is obtained when tangents are drawn on each side of the chromatographic peak and extrapolated to the baseline. The number n can be calculated in a different manner by measuring the peak width $w_{0.6}$ at the 0.607 part of the peak height, or $w_{0'5}$ at half the height of the peak:

$$n = 4\left(\frac{t_R}{w_{0.6}}\right)^2 = 5.54\left(\frac{t_R}{w_{0.5}}\right)^2. \tag{2.2.13}$$

From the number of theoretical plates n the height equivalent to a theoretical plate H can be calculated:

$$H = \frac{L}{n} \tag{2.2.14}$$

where L is the column length in centimeters or millimeters, and H has then the same dimension.

The column efficiency increases with decreasing diameter of the adsorbent particles. Lately attention has been focussed on adsorbents with a 10 – 30 μm diameter. However, for preparative purposes adsorbents still prevail with particle size about 100 μm. Jardy and Rosset[15] proposed an empirical relationship for the dependence of the column efficiency on the particle size of the adsorbent at a constant eluate velocity:

$$H = k \, . \, d_p^{\beta} \tag{2.2.15}$$

where d_p is the diameter of the adsorbent particle and k is a constant including all factors except the particle size, physically corresponding to the height H (in mm) for $d_p = 1$ mm. The empirical exponent β is a function of the linear eluent velocity and it decreases slowly with increasing velocity. For very high velocities its value is 1.4. A further parameter that affects the column efficiency decisively is the linear eluent velocity u (cm/s or mm/s). The dependence of H on u was expressed by Snyder[16] with the empirical equation

$$H = D \, . \, u^a \tag{2.2.16}$$

where D is a constant for a given system. Physically it corresponds to the height H in cm for $u = 1$ cm/s. The value of the second constant a varies within the 0.3 to 0.7 interval. For porous adsorbents the value of $a = 0.6$ is usually convenient. An illustration of both these effects is shown in Fig. 2.2.3, where the dependence of H on the linear velocity of the eluent u is shown for various particle diameters of the adsorbent. In Fig. 2.2.3 the differences in the slopes of the curves for various d_p values deserve mention in the first place. Adsorbents with fine particles are characterized by a flat curve, where even a high linear velocity of the eluent does not substantially worsen the column efficiency. However, it should be noted that these velocities can be achieved with fine-particulate adsorbents only when high-pressure pumps are employed, because the column permeability is directly proportional to the square of the particle diameter[17].

Among other variables that can affect the column efficiency the size of the sample deserves mention. When analytical separation is carried out the total charge of a mixture of substances, except the solvent, should not exceed the linear column capacity (see Section 2.2.3). Further the total volume of the charge, including the solvent, should be taken into consideration. Best results can be achieved when the total volume of the injection does not exceed 5 % of the dead volume of the column, V_M; in no case should the injection exceed 15 % of this volume. In preparative separations a reasonable compromise should be accepted, so that the viscosity of the charge should not be either too high in comparison with the eluent used, or the injection volume excessively large.

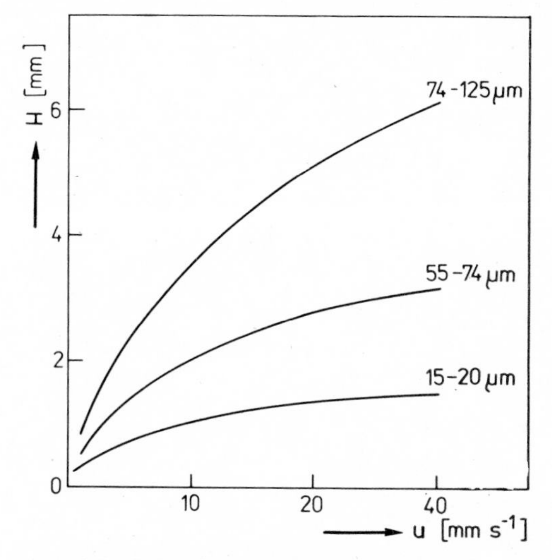

Fig. 2.2.3 Dependence of the plate height H on the linear velocity of the eluent u for various diameters of adsorbent particles
Irregular, fully porous alumina

The effect of temperature at which the chromatographic separation is carried out is not yet quite clear, but it does not seem very important. A higher efficiency has been observed at elevated temperatures for substances with a capacity factor $k' < 1.0$, and a worsened efficiency for substances with $k' > 5.0$.

The determination of column efficiency is not totally independent of the solute that was used for measurement. Differences in the chemical character of solutes are the primary factor which can produce another type of interaction between the solute and the adsorbent. Further it was observed even within the same group of compounds that substances with a lower k' value afford lower H values and thus a higher column efficiency. For these reasons the concept of effective plate n_{eff}[1] has been introduced in which the column efficiency n and the capacity factor k' of the measured substance are included:

$$n_{eff} = n\left(\frac{k'}{1 + k'}\right)^2 \qquad (2.2.17)$$

In chromatography it is also important to express the degree of the separation of two close peaks quantitatively. The degree of separation is indicated as *resolution, R,* and it can be calculated from the chromatogram according to equation:

$$R = \frac{2(t_2 - t_1)}{w_1 + w_2} \qquad (2.2.18)$$

where t_1 and t_2 are the retention distances from the start of the first or the second eluted peak, respectively, expressed in length units, w_1 and w_2 mean the corresponding peak widths at the peak base in the same units (Fig. 2.2.1). Equation (2.2.18) gives instruction on how to calculate resolution, but it does not say anything on the experimental conditions that affect the resolution. These conditions follow from the next equation which shows that resolution can also be expressed by means of the capacity factor and the column efficiency[12]:

$$R = \frac{1}{4}\sqrt{n_2}\left(\frac{k'_2}{1 + k'_2}\right)\left(\frac{\alpha - 1}{\alpha}\right) \qquad (2.2.19)$$

where n_2 is the number of plates of the column, calculated for the later eluted component, k'_2 is the capacity factor of this component, and α is the separation factor k'_2/k'_1. From equation (2.2.19) it is evident that it involves three elements that can be varied to a certain extent independently. Column efficiency has been discussed earlier. The capacity factor k'_2 and the separation factor α can be influenced by changing the eluent, or the adsorbent, or its activity. On comparison of equations (2.2.17) and (2.2.19) it becomes evident that at constant α, R is proportional to the square root of n_{eff}.

2.2.3 ADSORBENTS AND SUPPORTS

Commercial adsorbents are supplied in various modifications that have certain limitations. One of the important parameters, especially for preparative chromatography, is the *linear capacity* of the adsorbent. When a sample is eluted on a column it may be observed that the elution volume or the elution time are not independent of the sample size, and that—from a certain amount—they begin to decrease. In analytical separations it is imperative that the elution volumes be reproducible, and therefore very small samples are applied. The amount of sample that causes a decrease in the capacity factor k' of 10 % with respect to the constant k' value, determined for smaller samples, is called linear capacity of adsorbent[6] and it is expressed in grams of sample per 1 g of adsorbent. It was observed that common adsorbents (silica gel, alumina) containing several percent of water have a higher linear capacity than fully activated adsorbents. In preparative separations the linear capacity of the adsorbent is practically always exceeded and a worsened separation should be reckoned with.

The most current and cheapest adsorbents are those supplied as irregular porous particles. These adsorbents are also most used in preparative separations. Their surface area ranges from 100–500 m^2/g, and they have a high linear capacity. Columns packed with these adsorbents have a lower efficiency in comparison with other types, and also a lower permeability. An increase in efficiency may be achieved by using adsorbents with small diameter particles, and a narrow range of granulation.

Another type is represented by spherical adsorbents with solid cores, that are coated with a *porous layer*[18]. These adsorbents have a specific surface area of 5–15 m^2/g, and they are used for analytical separations exclusively because their linear capacity is very low. Columns packed with these adsorbents have a high efficiency and a low resistance to the eluent flow. They are used with advantage for high-performance liquid chromatography. These adsorbents are sold under the commercial names Corasil, Pellosil, Vydac, Pellumina, etc.

The third type is represented by adsorbents with *spherical particles* which are *fully porous*[18]. These adsorbents combine the advantages of both preceding types, since they have a high linear capacity, high permeability, and they are supplied in relatively fine granulation, permitting the achievement of high efficiency. They are produced under the commercial names Spherosil, Porasil, etc., and they are used for analytical separations exclusively.

For adsorption chromatography a number of adsorbents have been used, but only few of them are used generally. *Silica gel* and *alumina* are the commonest.

Silica gel is a polar adsorbent with a semicrystalline or amorphous structure, of the general formula $SiO_2 . H_2O$. According to the value of the specific surface area and the pore diameter two forms of silica gel can be distinguished. Silica gel with large pores above 10 nm has a specific surface area of between 100 – 500 m^2/g. Silica gel with small pores below 10 nm has a specific surface area above 500 m^2/g, and it is less used than the former. The silica gel surface is covered with hydroxyl groups that react in adsorption chromatography with the adsorbate under formation of hydrogen bonds. This base of adsorption of substances on silica gel has important consequences in practical separations of hydrocarbons on this adsorbent. The planarity of the molecule plays an unimportant role on silica gel, because planar adsorption does not take place. Hydrocarbons that differ only in the spacial arrangement of their molecules can as a rule be separated better on alumina than on silica gel. A further characteristic feature of silica gel is the acidity of its surface, ranging between pH = 3 – 5. This acidity causes a very strong retardation of substances of basic character, for example nitrogen bases in the separation of tars. In air silica gel takes in water easily, and therefore commercially supplied silica gel should be *activated*. Activation is carried out at temperatures of 150 – 200 °C. Most commonly heating at 180 °C for about three hours is recommended. Active silica gel is treated by addition of water, adding the required amount of water, to a weighed amount of silica gel and shaking thoroughly in order to eliminate all lumps. Then it is allowed to stand for 24 h which provides for an uniform distribution of water throughout the whole material. Silica gel is characterized by a relatively high linear capacity of up to 10^{-3} g/g.

A further adsorbent which is currently used is alumina, representing the crystalline γ-form of aluminum oxide. Depending on the degree of deactivation its surface is covered with hydroxyl groups or also adsorbed water. The specific surface area of alumina used for chromatography ranges between 100 and 250 m^2/g. The surface of alumina is covered with active centres that form a strong electrostatic field (positive and negative). When the adsorbate approaches the surface of alumina a dipole moment is induced in the molecule. Interaction of the polarized molecule and the surface field of the adsorbent brings about adsorption of the molecule on the alumina surface. It is assumed that inductive interaction represents a substantial contribution in the adsorption of aromatic hydrocarbons on alumina. In this interaction planar adsorption takes place, i.e. the molecule of the aromatic compound assumes a plane-parallel position with respect to the alumina surface. Non-planarity of the adsorbate molecule decreases adsorption forces, since the energy of inductive interaction depends on the square of the electrostatic field intensity. A practical consequence is a good separation of isomers that differ in the degree of non-planarity of the molecule. It is obvious that other types of substances can be adsorbed on other centres, or they can react with hydroxyl groups. An important feature of alumina is the basicity of its surface, usually pH $= 8-11$. This can be made use of in the separation of weakly acid components from neutral ones. In the case of stronger organic acids their chemisorption takes place so that these substances cannot be separated on alumina. So-called acid aluminas appear on the market the surface of which has pH ≤ 5. These adsorbents are specially treated and in some instances they possess different adsorption properties even for hydrocarbons[2]. Activation of alumina is commonly carried out by heating at 400 °C for $6-16$ h. The activity of alumina is then adjusted by addition of water, as in the case of silica gel. Linear capacity of alumina is much lower than that of silica gel, it usually does not exceed $5 \,.\, 10^{-5}$ g/g.

Among other adsorbents *Florisil* and *active carbon* are also rather widespread.

Florisil is a polar adsorbent, basically magnesium silicate, the properties of which are very similar to those of alumina. For the separation of hydrocarbons it usually gives the same sequence of the eluted components as alumina.

Active charcoal is a non-polar adsorbent that is not much used nowadays. It has found a certain application in the purification of natural mixtures. Among its adsorption properties the most important is the retention of substances with large molecular weight. Substances with a low molecular weight are less adsorbed and they can be eluted from the column easily. Non-polar substances are adsorbed preferentially, polar non-ionogenic substances follow, while salts are not adsorbed.

Silver nitrate impregnated silica gel is a special adsorbent used for the separation of olefins or for the elimination of acetylenes. On this adsorbent olefins react with silver ions under formation of a complex, which enables a good separation of olefins from other saturated hydrocarbons. The reaction is reversible and olefins can be eluted from the adsorbent. Acetylenes react with silver ions under formation of silver acetylides which are retained on the column.

Preparation of silver nitrate impregnated silica gel

Silver nitrate (75 g) is dissolved in 525 ml of distilled water and the solution is mixed with 400 g of silica gel. The paste formed is freed from water in a rotatory evaporator under reduced pressure at 50—60 °C. The free-flowing material is then activated in a vacuum at 120—130 °C for 8 h. This adsorbent should be protected from light both during storage and operation, because it blackens rapidly.

Similarly, alumina or Florisil can also be impregnated with silver nitrate. On silver impregnated adsorbents *cis*-olefins react preferentially, as well as those the double bond of which is not sterically hindered by other alkyl groups.

2.2.4 ELUENTS

Basic principles of the interaction of the eluents in the adsorption process should be elucidated qualitatively in this introduction. First of all interactions between the eluent molecules and the adsorbent surface should be mentioned, further the interactions between the solute and the eluent in the adsorbed phase.

At the beginning of the process the adsorbent surface is occupied by the molecules of the eluent, and after the introduction of the sample into the column the sample molecule must displace a corresponding number of eluent molecules from the adsorbent surface. Adsorption energy of the eluent, i.e. the value of the interaction between the eluent and the adsorbent surface, has a decisive role. The interactions between the sample and the eluent both in the liquid and in the adsorbed phase are less important. Only in cases when the eluent and the sample form a complex in the liquid or in the adsorbed phase, is the adsorption process affected by the properties of the complex formed, mainly by its size and stability.

The strength of the interaction between the eluent and the adsorbent surface is expressed as *solvent strength* ε°. This is an empirical parameter[19] the values of which are determined experimentally from the chromatographic behaviour. The scale of the values is relative, referred to the solvent strength of *n*-pentane for which it has been defined that $\varepsilon^\circ = 0$.

This definition followed from earlier experiences when n-pentane was considered to be the weakest eluent. Later investigations have shown that *perfluorinated* alkanes possess a weaker solvent strength than n-pentane and their ε^0 value is negative with respect to that of n-pentane. The higher ε^0 the stronger is the eluent adsorbed, while the given sample is adsorbed less. The ε^0 values change with the type of adsorbent, in some instances even according to the type of the substance eluted. This occurs especially in cases when the solvent occupies preferentially some type of adsorbent centres and increases or suppresses the acid-base interactions. The size of the adsorbent surface occupied by the sample molecule affects the adsorption energy of the sample. If two substances differing in their molecular size are eluted equally with an eluent with a weak solvent strength, then the use of an eluent with a higher ε^0 value will cause those substances to be eluted earlier the molecule of which occupies a larger surface area in the adsorbed state. However, a practical exploitation of this phenomenon is limited because if the substances are of the same character and are eluted equally, they usually do not differ in their

TABLE 2.2.1

Properties of some eluents[6,17,19]

Eluent	Boiling point °C	Viscosity 10^{-3} Pa . s at 20 °C	Area of the adsorbed molecule n_b	Parameter of solvent strength ε^0	
				silica gel	alumina
n-Pentane	36.1	0.24	5.9	0.00	0.00
2,2,4-Trimethyl-pentane	99.2	0.50	7.6	0.01	0.01
Cyclopentane	49.3	0.47	5.2	0.04	0.05
Benzene	80.1	0.65	6.0	0.25	0.32
Dichloromethane	40.2	0.44	4.1	0.32	0.42
Diethyl ether	34.6	0.24	4.5*	0.38	0.38
Methyl ethyl ketone	79.6	0.43	4.6*	0.39	0.51
Acetone	56.2	0.33	4.2*	0.47	0.56
Methanol	64.7	0.59	8.0*	0.73	0.95

* For these eluents and silica gel the value $n_b = 10$ is used

molecular size. In Table 2.2.1 some current eluents and their physical properties and ε^0 values for silica gel and alumina are listed.

A much more complex situation arises when a solvent is used that has two components. Several undesirable effects can take place of which the most important is the so-called *demixing effect* which is observed in binary mixtures the components of which differ substantially in their ε^0 values. When such an eluent is injected into a dry column, first the pure component of the lower solvent strength leaves the column. After a certain time the adsorbent is saturated with the more polar component and only then the eluent of the same composition as that injected begins to flow out of the column. When binary eluents are employed the column should always be washed for a longer period before use, until equilibrium is attained. The calculation of the solvent strength of the binary eluent is possible only when some simplifications are assumed:

1. demixing does not take place,
2. the interactions between the eluent and the solute are negligible both in the liquid and the solid phase,
3. the molecules of both eluents are approximately of equal size.

Under these assumptions the following equation applies for the elution strength of the binary mixture ε^0_{ab}:

$$\varepsilon^0_{ab} = \varepsilon^0_a + \frac{\log (x_b \,.\, 10^{\Delta\varepsilon^0 n_b \alpha'} + x_a)}{n_b} \qquad (2.2.20)$$

where ε^0_a is the parameter of the elution power of the weaker eluent, x_b and x_a are molar fractions of eluent A or eluent B, respectively, in the mixture, $\alpha'$is the parameter of the adsorbent, n_b is the effective area of the adsorbed molecule of eluent B (Table 2.2.1), $\Delta\varepsilon^0$ is the difference of the solvent strengths $\varepsilon^0_b - \varepsilon^0_a$ of both eluents. The calculation of the solvent strength of binary mixture is rather complex. For some binary mixtures of solvents the calculated ε^0_{ab} values are listed in Table 2.2.2. Equation (2.2.20) fails when such binary mixtures are calculated where the difference $\varepsilon^0_b - \varepsilon^0_a$ is larger, for example in the case of the binary mixture pentane-methanol. In such instances it suffices when the binary mixture contains 2 vol. % of the more polar eluent and a complete covering of the adsorbent surface with the more polar eluent takes place.

The parameter of the solvent strength is most important for the calculation of the change of the elution volume when the eluent is changed. When working with the same adsorbent of the same activity, the elution volume of a given substance can be calculated according to equation

$$\log V_2' = \log V_1' + \alpha' A_s(\varepsilon_1^0 - \varepsilon_2^0) \qquad (2.2.21)$$

where V_1 and V_2 are corrected elution volumes of a given substance, obtained with the first or the second eluent, α' is the parameter of adsorbent activity, A_s is the area occupied by the molecule of the given substance on the adsorbent surface, ε_1^0 and ε_2^0 are the parameters of the solvent strength of the first (less polar) or the second, more polar eluent. Analogously the following equation also applies if the elution data are given as capacity factors k_1' and k_2':

$$\log k_2' = \log k_1' + \alpha' A_s(\varepsilon_1^0 - \varepsilon_2^0) \qquad (2.2.22)$$

Finally, let us mention several general requirements of the eluent:

TABLE 2.2.2

Values of the parameter of solvent strength ε_{ab} for some binary mixtures of eluents

	Alumina ($\alpha' = 0.6$)				Silica gel ($\alpha' = 0.7$)			
Vol. % of the more polar eluent	pentane-diethyl ether	pentane-methyl ethyl ketone	diethyl ether-acetone	diethyl ether-methanol	pentane-benzene	pentane-diethyl ether	benzene-acetone	benzene-methanol
0	0	0	0.38	0.38	0	0	0.25	0.25
2	0.031	0.077	0.389	0.683	0.024	0.149	0.287	0.535
5	0.068	0.148	0.401	0.759	0.052	0.202	0.318	0.590
10	0.115	0.221	0.419	0.815	0.086	0.244	0.350	0.629
15	0.152	0.270	0.434	0.846	0.110	0.268	0.370	0.651
25	0.205	0.334	0.460	0.881	0.146	0.299	0.397	0.677
35	0.245	0.378	0.481	0.901	0.171	0.319	0.415	0.692
100*)	0.380	0.510	0.560	0.950	0.250	0.380	0.470	0.73

* The interval of 35—100 vol. % of the more polar eluent can be approximated with good accuracy by a straight line connecting these two points.

The solvent strength of the mobile phase should give the values of the capacity factors of the analyzed components on a given adsorbent that are within the interval $k' = 1-10$. Preference should be given, if possible, to eluents with low viscosity and boiling point, which guarantees a high column efficiency and facilitates the isolation of the substance from the eluate. The eluent should not corrode or dissolve the materials with which it is in contact, and it should fulfil safety requirements. The eluent should suit the detector used.

Gradient elution chromatography. In the analysis of complex mixtures, when individual substances differ substantially in the values of the capacity factors, a feasible separation is impossible if a single eluent is used. In such instances either unresolved components would be obtained, or the elution volumes of the components with the highest k' values would be eluted only with a large volume of the eluent. This would excessively increase the time of elution on the one hand, and the detection of later eluted substances would thus be deteriorated. In such instances several solvents of increasing strength can be used gradually. A stepwise change in the solvent strength can be substituted by a continuous change, i.e. the composition of the mobile phase is changed during the elution by increasing its solvent strength, while keeping the flow-rate constant. This technique is called *gradient elution chromatography*. Its use means a similar improvement in liquid chromatography as programmed temperature in gas chromatography. The only disadvantage consists in the difficulty of prediction of elution volumes for any course of gradient in dependence on time or the eluate volume. The calculation of elution volumes of the components of the analyzed mixture at a selected gradient enables a preliminary evaluation of the separation possibilities and optimalize thus the course of the gradient. In many instances preference is given to a *linear gradient,* i.e. a linear increase in the solvent strength in dependence on the eluate volume:

$$\alpha'\varepsilon^0 = a + bV \qquad (2.2.23)$$

where a and b are constants for a given separation, of which b characterizes the gradient slope and V is the eluate volume and ε^0 is a variable. For this gradient course the elution volume of a certain component[6] can be calculated beforehand under the assumption that the elution volume of this component with a weaker eluent is known, and that certain secondary effects mentioned for binary mixtures do not take place. For aromatic hydrocarbons it was shown that the calculation of elution volumes for any gradient course is possible[20], but the calculation is rather time-consuming. Snyder and Saunders[21] recommend an increase of ε^0 by 0.04 units as optimum per an increase in eluate volume equal to the column dead volume.

2.2.5 SEPARATION OF HYDROCARBONS BY ELUTION TECHNIQUE

Adsorption chromatography enables the separation of petroleum distillates, distillation residues and some other products into the following groups:

saturated hydrocarbons
olefins
monoaromates
diaromates
triaromates and higher
polar non-hydrocarbon substances

This separation facilitates further analyses substantially, for example in combination with mass spectrometry or nuclear magnetic resonance. In cases when heavy residues are analyzed, this separation is indispensable.

A group separation of petroleum products can be carried out by very different techniques, according to the requirements of the results of separation, for example according to the number of the groups determined, accuracy of the determination, need of the sample for subsequent analyses, etc.

For the separation of *saturated hydrocarbons* from other groups silica gel or combined packings, containing silica gel, are used as a rule. In one fraction three types of saturated hydrocarbons are obtained, i.e. *n*-alkanes, isoalkanes and cycloalkanes. Sometimes *n*-alkanes are already eliminated, for example with urea (see Section 2.5.1), before separation by adsorption chromatography. This is done in cases when the content of *n*-alkanes is high, the sample is semisolid or solid, dissolves with difficulty, and the content of other groups is low. The separation of saturated hydrocarbons to other subgroups (for example alkanes, cyclanes) by adsorption chromatography is rather exceptional; in most instances these substances move with the solvent front. Using alumina with pentane as eluent cycloalkanes were successfully separated according to the number of saturated rings[22, 23].

Olefins are not common in petroleum products, but they occur in the products of thermal and catalytic cracking. They are separated rather exceptionally and can be isolated mostly together with the group of saturated hydrocarbons and then determined in that fraction by some other method. If it is necessary to isolate olefins, they can be separated from saturated hydrocarbons by adsorption chromatography on adsorbents impregnated with silver nitrate. On alumina or silica gel containing a few percentage of water olefins are eluted simultaneously with saturated hydrocarbons.

Aromates can be separated together, i.e. mono-, di- and triaromates all in one fraction, or they can be separated into individual groups. The separation into single groups is not quite perfect, especially in the heaviest hydrocarbon fraction where a certain overlapping with neighbouring groups is observed. Simultaneously with the aromates aromatic substances containing sulphur or oxygen in the molecule are also eluted. For example, benzothiophenes and benzofurans are eluted together with diaromates, and dibenzothiophenes and dibenzofurans together with triaromates. Oxygen-containing compounds are present in petroleum in small amounts, but sulphur-containing compounds represent in some instances an important part, especially in the groups of di- and triaromates. Sulphur compounds in the form of sulphides, either aliphatic, cyclic or aromatic are eluted on alumina predominantly with the group of diaromates, and to a lesser extent with the group of triaromates and higher. In the group of triaromates some polar substances are also eluted, for example some derivatives of carbazole, sterically shielded nitrogen bases, acids, etc., but this contamination depends on the overall separation scheme.

The simplest separation scheme consists in the separation of the sample to saturated hydrocarbons (+ olefins), aromates and polar substances. For this separation silica gel is used almost exclusively because it has a higher linear capacity and enables a substantially better separation of saturated hydrocarbons and aromates. Usually activated silica gel is used, or, when the sample contains olefins, it is deactivated by addition of up to 5% of water. The loading of the column is usually 1 : 50 – 1 : 200. The separation is started with the elution of saturated hydrocarbons, or olefins, with *n*-pentane. For the elution of aromates dichloromethane, chloroform or diethyl ether are used. Final elution of the group of polar substances, sometimes known as asphaltic substances, can be achieved with methanol, 50 vol. % of methanol + 50 vol. % of benzene, or 75 vol. %

of methanol + 25 vol. % of benzene. Currently glass columns and an open chromatographic system are used for this separation, without a detector. The subsequent example illustrates the procedure:

Group separation of higher petroleum fractions on silica gel

A glass column of 100 × 1 cm I.D. dimensions, fitted above its upper part with a reservoir of 200 ml volume and provided at the bottom with a plug of glass wool, is packed with 50 g of silica gel of 100—200 mesh with 2 % (by weight) of water. The reservoir is filled with 100 ml of *n*-pentane, its level is then allowed to drop up to the top of the bed (about 5 mm above it), but the adsorbent should not be uncovered. Sample (0.5—0.6 g) dissolved in 2 ml of *n*-pentane or cyclohexane is then introduced onto the adsorbent with a pipette. The sample is allowed to soak into the packing and two 2 ml portions of *n*-pentane are also allowed to enter the bed. Then 150 ml of pentane are introduced into the reservoir and elution is started. The first 50 ml of eluate are discarded and the *n*-pentane fraction is then collected. When the *n*-pentane level has dropped to 1 cm above the column packing level, 150 ml of ethyl ether are introduced, the container for the eluate fractions is exchanged, and the ether fraction is collected until the level of ether has dropped 1 cm above the packing level. Then the reservoir is filled with 150 ml of the mixed eluent (50 vol.% of methanol + 50 vol.% of benzene), the eluate container is exchanged and the last fraction is collected. Individual fractions are then evaporated and weighed. The calculation of the content of individual groups in weight % is carried out using the following equations:

$$\text{saturated hydrocarbons} = \frac{S \,.\, 100}{S + A + P}$$

$$\text{aromatic hydrocarbons} = \frac{A \,.\, 100}{S + A + P}$$

$$\text{polar substances} = \frac{P \,.\, 100}{S + A + P}$$

$$\text{losses} = \frac{(W - S - A - P) \,.\, 100}{W}$$

where S, A and P are the weights of the residues after the evaporation of the pentane, ether and methanol-benzene fractions, and W is the weight of the sample in grams.

A certain modification is the normalized method[24] that is used for the separation of petroleum distillates boiling between 232 and 340 °C to non-aromatic and aromatic fractions. A column of 50 × 1 cm I.D. dimensions is packed with active silica gel up to 35 cm height which is then overlayered with 15 cm of activated bauxite. The sample of 2 ml volume is first eluted with *n*-pentane. Thus a non-aromatic fraction is obtained, containing alkanic and naphthenic hydrocarbons. In cases when the sample is a product of cracking the non-aromatic fraction also contains aliphatic and cyclic olefins. The aromatic fraction is gradually eluted with ether, chloroform and methanol. In addition to aromates this fraction also contains polar substances that do not separate.

Another method,[25] used for high-boiling oil fractions, oil extracts etc., separates these substances to a saturated and a polar fraction. The oils are separated on two combined columns where the upper column is packed with diatomaceous earth and the lower with diatomaceous earth and silica gel. In this separation the aromates remain adsorbed in the lower column and must be calculated from the difference.

Further separations of aromatic hydrocarbons to the groups of mono-, di- and triaromates are carried out on alumina. In view of the fact that alumina has a lower linear capacity, the operation should be performed with a substantially lower amount of sample and it must be reckoned with the fact that the separation is not perfect. In this separation it is advantageous to operate with a closed chromatographic system and connect a detector to the column outlet so that optimum conditions for the separation of individual groups can be selected. For this purpose a transport or a UV detector with a cell of 1 mm optical length were found suitable. As example the separation of the aromatic fraction from a low-sulphur oil, with the distillation range 280–420 °C, may serve.

Separation of the aromatic fraction from higher petroleum distillates on alumina

A stainless-steel column of 100 cm × 1 cm I.D. dimension is packed with alumina of 100—200 mesh particle size. Before use alumina should be calcinated at 400 °C for 8 h and then deactivated with 4.5 % of water. The flow of the eluent at a 160 ml/h rate is carried out by a pump. A transport detector is connected to the column outlet. After prewetting the column with *n*-pentane 100 mg of aromates is charged dissolved in 0.3 ml of iso-octane. Elution is carried out with *n*-pentane which elutes mono- and diaromates. A mixture of 90 vol.% of *n*-pentane and 10 vol.% of diethyl ether elutes simultaneously tri- and tetraaromates. Pure ether elutes the remains of the polar substances which passed into the aromatic fraction during the group separation of the original oil. The chromatographic record is shown in Fig. 2.2.4 where the changes in the eluent composition are also indicated. The position of the fraction collector is changed at the minimum between two peaks. After evaporation the fractions are weighed.

Jewell et al.[26] used freshly calcined alumina for the separation of the aromatic fraction freed from polar substances. Using gradient elution with a series of solvents, *n*-hexane-cyclohexane-chloroform-methanol, with an exponential increasing of the elution power parameter they achieved a separation to monoaromates, diaromates + triaromates, and polyaromates. Monoaromates were eluted by addition of cyclohexane, diaromates + triaromates were eluted together during the addition of chloroform, and the last group, polyaromates, was eluted with methanol. As chloroform and methanol have to be evaporated in order to concentrate the samples, the method is limited to poorly volatile samples only.

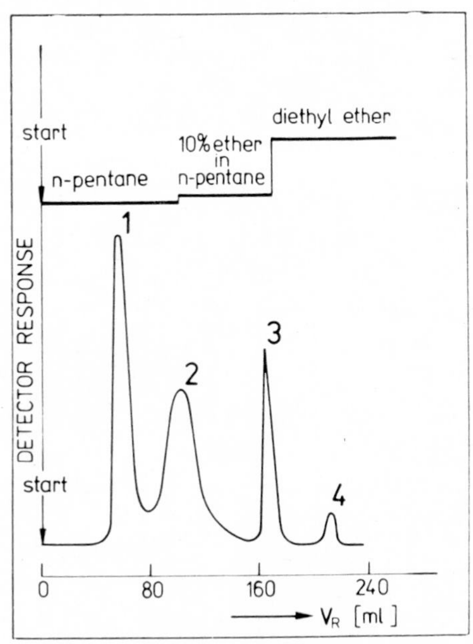

Fig. 2.2.4 Separation of aromates from a low-sulphur petroleum distillate (280—420 °C) Column 100 × 1 cm, alumina with 4.5 % of water, 1 — monoaromates, 2 — diaromates, 3 — triaromates, 4 — polar substances

The most complete separation scheme for high-boiling petroleum distillates is a part of the procedure elaborated in collaboration between the *US Bureau of Mines* and the *American Petroleum Institute*[27]. This procedure is very perfect, but requires time and large samples. The fundamental procedure begins with 250 – 500 g of oil distillate, but a procedure has also been elaborated for 10 – 20 g of sample only. In the first step the acids, phenols and amines present are eliminated using an anion-exchange resin which enables operation in organic solvents (in pentane or cyclohexane). Then the elimination of basic components follows on a cation-exchange resin in pentane. Neutral nitrogen-containing substances are retained on a column containing ferric chloride fixed on diatomaceous earth or powdered cellulose, while the oil, free of acids, bases and neutral nitrogen-containing substances, is eluted with pentane. The hydrocarbon fraction is then separated on a 244×2.54 cm I.D. column packed in its lower part with active alumina 80–200 mesh and in its upper part with active silica gel of 28–200 mesh. The column is loaded with 1 g of aromates and polar substances per 100 g of adsorbent, which means that this column can be currently loaded with 20–30 g of sample. Before the beginning of the operation the column should be prewetted with pentane and the rate of eluent flow should be maintained at about 200 ml/h. The column is then eluted with the following eluents:

2.5 l of *n*-pentane	-fraction of saturated hydrocarbons
3.0 l of benzene–*n*-pentane (5 : 95 v/v)	-fraction of monoaromates
3.0 l of benzene–*n*-pentane (15 : 85 v/v)	-fraction of diaromates
0.5 l of methanol–diethyl ether–benzene (60 : 20 : 20 v/v)	-fraction of polyaromates and polar substances
1.0 l of methanol	

This scheme was elaborated for fractions distilling within the 330 to 535 °C range, and for the fractions boiling above this range a modified procedure was worked out. The finding is important that on an anion-exchange resin together with acid substances some polycyclic aromates are also retained. Sawatzky et al.[33] modified the described method and reduced the time of analysis by applying pressure on the column and eliminating the ion-exchange pretreatment.

It should be noted in connection with the separation of petroleum distillates and the residues that the difficulty of separation increases with increasing distillation range of the sample. Retention volumes of alkyl-aromates on alumina increase with the number of substituents and with the length of the alkyl chain. In polycyclic aromates the inequality of the positions and the increasing number of isomers complicates the situation further. A result of these effects is that in higher boiling fractions a certain overlapping with neighbouring groups may take place. In the case of distillates containing up to several percent of sulphur, thiophenes, sulphides and other sulphur compounds are separated together with aromates which they can exceed in percentual content in some fractions.

Different problems are met during the separation of gasolines to saturated hydrocarbons, olefins and aromates. In these substances the low boiling point of the sample does not permit the use of an eluent, because it would be impossible to eliminate it without a simultaneous loss of the sample. In such cases *displacement chromatography* may be useful (see Section 3.5.1). In the separation of saturated hydrocarbons either from olefinic[28] or non-olefinic gasolines[29] displacement chromatography on active silica gel is used, where the sample is displaced with isopropyl alcohol or ethanol. For these separations columns composed of three to four tubes are produced, the diameters of which gradually decrease. At the top the column is provided with an eluent container. Several centimeters before the end of the column a small layer of silica gel with an indicator dye is introduced into the activated silica gel column. The sample is allowed to soak into the silica gel, the container is filled with isopropyl alcohol, and the displacement of the sample is accelerated by a mild overpressure of nitrogen gas. During the passage through the column the sample is separated to saturated hydrocarbons, moving with the solvent front, and olefins and aromates. Saturated hydrocarbons are collected into the receiver until a zone of light colour under the UV light indicates the moving front of olefins, or – in the case of non-olefinic gasolines – the front of aromates. The columns have a large capacity, for example up to 30 ml can be separated, but for olefinic gasolines the total column length goes up to 250 cm. The collection of other fractions, i.e. the olefinic and the aromatic, is possible in principle, but as the process is based on the displacement of the sample, an increased overlapping with neighbouring groups may be expected and should be taken into account.

The method of determination of aromates in non-olefinic gasolines[30], permitting the determination of the content of aromates from the refractive index of the collected fractions, is similar, but during the isolation of aromates a contamination both with saturated hydrocarbons and with ethanol or isopropyl alcohol should be reckoned with.

Separation of olefins and terpenes. In the products of cracking of petroleum fractions and residues olefins are present in higher concentrations. In natural essential oils saturated hydrocarbons and terpenes are present in addition to other substances. In both cases preference is given to a primary separation of saturated hydrocarbons together with olefins or terpenes from other, more polar, compounds in the first step of the operation. The first separation can be carried out with advantage on silica gel containing 2 – 5 % of water, using *n*-pentane as eluent, as has been said earlier. The mixture of saturated hydrocarbons can then be separated on silica gel impregnated with silver nitrate, at a sample-to-sorbent ratio of 1 : 20 to 1 : 50[5]. Saturated hydrocarbons that do not form complexes with silver ions are eluted with the front of the alkanic eluent, and olefins, or terpenes, can be eluted only with a weakly polar eluent. The elution of saturated components is carried out with *n*-pentane or cyclohexane, the elution of olefins or terpenes with a 0 – 20 % diethyl ether in *n*-pentane gradient, or 0 – 50 % of benzene in cyclohexane gradient. The eluents used may not contain sulphur-containing compounds and diethyl ether must be freed of peroxides.

The adsorption of olefins on silver nitrate impregnated adsorbents increases with the number of double bounds in the molecule. The adsorptivity of olefins also depends on the steric arrangement of the molecule. It was found that *trans*-isomers are adsorbed less strongly than the corresponding *cis*-isomers, and that these stereoisomers may be separated using this method[31]. In the case when the double bond is sterically hindered the stability of the complex decreases and thus also the retention volume of the corresponding olefin[34]. The interaction of olefins, dienes, terpenes etc. with silver ion is perfectly reversible and the substances can be easily eluted from the column. Acetylenes with a triple bond in the position 1 form silver acetylides and therefore they are firmly bound and can be easily separated from olefins.

2.2.6 PERCOLATION AND FRONTAL TECHNIQUE

Percolation and frontal analysis have a limited importance in analysis, but in common laboratory practice they are quite often employed. In both procedures mentioned the sample passes through the column packed with a sorbent without further use of mobile phase. For the performance of these procedures the sample amount should exceed the void volume of the column. In the majority of cases short columns of large diameter are used where the movement of the solution in the column takes place by the force of gravity. Percolation is used for the purification of hydrocarbon solvents or for the elimination of traces of hydrocarbons (or further organic substances) from ionic compounds soluble in water.

When hydrocarbons used as eluents in liquid chromatography and UV spectroscopy are purified, it is necessary to eliminate peroxides (for example in tetralin) or to free the solvent of water, or also to eliminate substances absorbing radiation in the measured region (for example in hexane, isooctane, etc.). For these purposes it is suitable to percolate hydrocarbons through activated polar sorbents as for example silica gel or alumina. In cases when traces of hydrocarbons have to be eliminated from aqueous solutions of ionic compounds, percolation through a layer of activated charcoal may be used, which has a strong affinity for non-polar compounds and does not retain ions.

Purification of iso-octane for UV spectroscopy

Alumina (150 g; 100—150 mesh), activated at 400 °C for 6 h is packed into a 40 × 2.5 cm I.D. column, under mild tapping, provided on the top with a container of 500 ml volume and at the bottom with a cotton plug. Iso-octane is then introduced slowly at its top into the column until the entire packing is wetted. The container is filled with 500 ml of iso-octane and the first 20 ml of the eluted solvent are returned into the column. The percolated iso-octane is dry and spectrally pure.

The *frontal technique* may be used in cases where a non-quantitative separation of the least polar component from a complex mixture is required, or when a quantitative separation (retention) of trace amounts of the most polar components from the sample is necessary which as a whole has a very weak solvent strength. When pumping the petroleum distillate through a dry column of a polar sorbent a certain amount of the eluate can be obtained, composed of saturated hydrocarbons, which moves unretained (with the solvent front). The amount of this eluate is given both by the character of the sample and the amount of adsorbent in the column. In more viscous samples the low velocity of the diffusion processes should be kept in mind and the sample pumped correspondingly slowly.

Frontal chromatography can be used with advantage in the determination of polycyclic aromates in white petroleum products[32], such as white oils and *n*-alkanes. These samples have a low elution power (they contain saturated hydrocarbons practically exclusively) and permit adsorption of trace amounts of polycyclic aromates on the column.

Isolation of polyaromatic hydrocarbons from white oils and n-alkanes

A glass tube of 100 × 1 cm I.D. dimensions, provided at its bottom with a sintered glass filter disc and a narrowed outlet and at its upper part with a stopper with an inlet for the introduction of the eluent, is packed with active silica gel for column chromatography (120—200 mesh, about 30 g), activated at 180 °C for 6 h. The homogeneity of the column bed is improved by tapping, the column is closed and connected with the pump. A solution of anthracene in the analysed sample is then prepared (about 1 000 ml of 10^{-6} g/ml concentration) which is pumped at a 100 ml/h flow rate under occasional illumination with an ultraviolet lamp (365 nm). The column is gradually saturated with the standard and the characteristic blue fluorescing zone of anthracene moves to the column outlet. When the first drops of anthracene penetrate into the eluate its volume is recorded and the experiment ended.

The column is perfectly cleaned, packed with fresh adsorbent, and the experiment with the sample is repeated under the same condition. The amount of the sample used should be 3/4 of the volume of the eluate recorded during the testing of the column with anthracene. The column is washed with 150 ml of rectified *n*-pentane, the eluate is removed and polyaromates are desorbed with 100 ml of diethyl ether of spectral purity. From the ethereal eluate the solvent is evaporated and a fraction is obtained containing all tricyclic and higher aromatic hydrocarbons in addition to a certain amount of saturated, monocyclic and bicyclic aromatic hydrocarbons or the polar substances present, of total volume of about 0.5 ml. Polycyclic aromates from 500—1 000 ml of white oils or *n*-alkanes can be concentrated on this column.

REFERENCES (2.2)

1. B. L. Karger, L. R. Snyder and C. Horwath, An Introduction to Separation Science, Wiley-Interscience, New York 1973.
2. M. Popl, V. Dolanský and J. Mostecký, *J. Chromatogr.* **91** (1974) 649.
3. L. R. Snyder and J. Kirkland, Introduction to Modern Liquid Chromatography, Wiley-Interscience, New York 1974.
4. R. P. W. Scott, Contemporary Liquid Chromatography, Wiley-Interscience, New York 1976.
5. Liquid Column Chromatography, Z. Deyl, K. Macek and J. Janák (Eds.), Elsevier Scientific Publishing Co., Amsterdam 1975.
6. L. R. Snyder, Principles of Adsorption Chromatography, Marcel Dekker Inc., New York, 1968.
7. L. R. Snyder, *J. Chromatogr.* **6** (1961) 22.

8. M. Popl, V. Dolanský and J. Mostecký, *J. Chromatogr.* **117** (1976) 117.
9. L. R. Snyder and B. E. Buell, *J. Chem. Eng. Data* **11** (1966) 545.
10. L. R. Snyder, J. Chromatogr. 25 (1966) 274.
11. M. Popl, V. Dolanský and J. Mostecký, *Coll. Czech. Chem. Commun.* **39** (1974) 1836.
12. E. Grushka, *Anal. Chem.* **46** (1974) 510A.
13. E. Grushka, L. R. Snyder and J. H. Knox, *J. Chromatogr. Sci.* **13** (1975) 25.
14. J. F. K. Huber, *J. Chromatogr. Sci.* **7** (1969) 85.
15. A. Jardy and R. Rosset, *J. Chromatogr.* **83** (1974) 195.
16. L. R. Snyder, *J. Chromatogr. Sci.* **7,** (1969) 352.
17. J. M. Miller, Separation Methods in Chemical Analysis, Wiley-Interscience, New York 1975.
18. R. E. Leitch and J. J. De Stefano, *J. Chromatogr. Sci.* **11** (1973) 105.
19. R. A. Keller and L. R. Snyder, *J. Chromatogr. Sci.* **9** (1971) 346.
20. M. Popl, V. Dolanský and J. Mostecký, *Anal. Chem.* **44** (1972) 2082.
21. L. R. Snyder and D. L. Saunders, *J. Chromatogr. Sci.* **7** (1969) 195.
22. G .C. Speers and E. V. Whitehead, in Organic Geochemistry, G. Eglinton and M. T. J. Murphy (Eds.), Springer-Verlag, Berlin 1969.
23. D. E. Anders and W. E. Robinson, *Geochim. Cosmochim. Acta* **35** (1971) 661.
24. ASTM D 2549—68 (1973), 1975 Annual Book of ASTM Standards, Part 24, Amer. Soc. for Testing and Materials, 1916 Race St., Philadelphia, Pa. 19103.
25. ASTM D 2007—75, 1975 Annual Book of ASTM Standards, Part 24, Amer. Soc. for Testing and Materials, 1916 Race St., Philadelphia, Pa. 19103.
26. D. M. Jewell, R. G. Ruberto and B. E. Davis, *Anal. Chem.* **44** (1972) 2318.
27. W. E. Haines and C. J. Thompson, Amer. Petrol. Inst. U.S., Bur. Mines, API RP 60 Publ. Rep. No. 37, July 1975.
28. ASTM D 2003—64 (1973), 1975 Annual Book of ASTM Standards, Part 24, Amer. Soc. for Testing and Materials, 1916 Race St., Philadelphia, Pa. 19103.
29. ASTM D 2002—73, 1975 Annual Book of ASTM Standards, Part 24, Amer. Soc. for Testing and Materials, 1916 Race St., Philadelphia, Pa. 19103.
30. ASTM D 936—55 (1973), 1975 Annual Book of ASTM Standards, Part 23, Amer. Soc. for Testing and Materials, 1916 Race St., Philadelphia, Pa. 19103.
31. L. R. Chapman and D. F. Kuemmel, *Anal. Chem.* **37** (1965) 1598.
32. M. Popl, M. Stejskal and J. Mostecký, *Anal. Chem.* **47** (1975) 1947.
33. H. Sawatzky, A. E. George, G. T. Smiley and D. S. Montgomery, *Fuel* **55** (1976) 16.
34. S. Lam and E. Grushka, *J. Chromatogr. Sci.* **15** (1977) 234.

2.3 Gel Chromatography

Gel permeation chromatography (GPC) is a type of liquid chromatography differing in principle from adsorption and partition chromatography. The column is packed with a fine particular material, most commonly with a *cross-linked organic gel* swollen in the eluent. For this eluent all internal spaces of the gel are freely accessible. GPC is based on the partition of the substance between the mobile part of the eluent (in the space between the particles) and its immobile part (filling the pores of the particles). In the study of GPC two theories for the separation mechanism have been proposed and elaborated: *steric exclusion* and *restricted diffusion*. According to the theory of steric exclusion the packing has such structure that only a certain part of the internal space in the pores is accessible to the sample molecules, according to their size. The pores are completely inaccessible to the largest molecules and these pass through the spaces within the particles unretained. The elution volume of these molecules is equal to the volume of the eluent in the column outside the

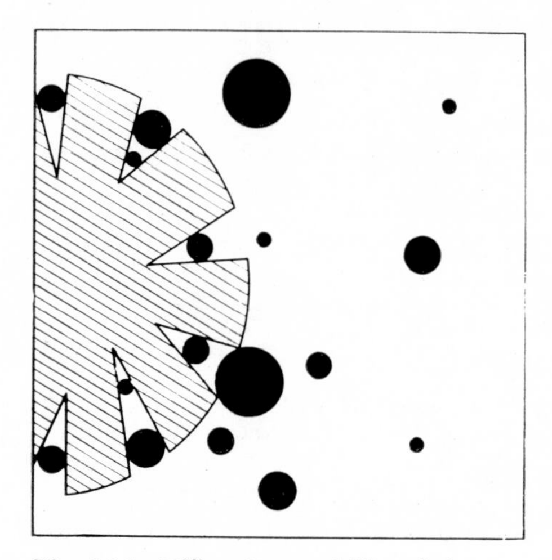

Fig. 2.3.1 Different accessibility of the pores for molecules of different size as the principle of GPC

gel particles, i.e. to the free volume of the column, V_0. In contrast to this all spaces in the pores or in the network structure of the gel are accessible to sufficiently small molecules which are then eluted with an eluent volume equivalent to the total volume of eluent within the column, i.e. to the free volume of the column V_0 and the volume of the eluent in the pores of the particles, V_i. Different accessibility of the pores for molecules of various size is shown in Fig. 2.3.1 in a simplified form.

According to the theory of restricted diffusion small molecules have higher diffusion coefficients in comparison with large ones and hence they can diffuse into the pores of the particles and stay there for a longer time. If this theory were valid, the retention volumes of substances would depend to a considerable extent on the linear velocity of the eluent, which has not been observed. According to present views the main separation mechanism is steric exclusion, but this mechanism is influenced by other phenomena, for example restricted diffusion, adsorption, etc.

The use of gels in liquid chromatography is very widespread even for separations the physical principle of which is different from GPC. In some instances it is difficult to decide whether the gel packed column operates according to the GPC principle or as a partition chromatography column. On the other hand some types of silica gel and porous glasses can be used both for adsorption and for gel permeation chromatography. As the subject of this chapter is gel chromatography of hydrocarbons, those systems will be discussed in the further text, where steric exclusion is predominantly the controlling mechanism of separation.

2.3.1 FUNDAMENTAL RELATIONSHIPS AND THE TECHNIQUE OF GPC

The fundamental relationship is expressed by an equation similar to that used in adsorption and partition chromatography (see equation 2.2.6) for the calculation of the elution volume V_R:

$$V_R = V_0 + KV_i \qquad (2.3.1)$$

where V_0 is the free volume of the column, V_i is the volume of the eluent in the pores of the packing, and K is the distribution constant of the sample between both phases, i.e. between the eluent that moves and the eluent forming the stationary phase. If the molecules are so large that the particle pores are inaccessible to them, then $K = 0$, and this extreme size of the molecules, expressed, for example, as molecular mass, is then indicated as the *exclusion limit* of the gel. All larger molecules are then in the region that is indicated as *total exclusion*. For substances with smaller molecules (with lower molecular mass) than the exclusion limit at least a part of the pores of the packing is accessible. If the molecules are sufficiently small, all pores are accessible to them, and $K = 1$. This lower limit of the operation range of gel, also called *total permeation,* cannot be determined with such accuracy as the exclusion limit. Every porous material is characterized by a certain distribution of pores and also contains a certain number of micropores which are accessible to very small molecules only. In practice calibration is carried out for every system of gel chromatography, where elution volumes of suitable standards are measured. For gels operating in the range of molecules of 100 – 1 000 mass units it is suitable to carry out the calibration of the GPC system with a series of *n*-alkanes. For gels operating in the range of larger molecules standards of polystyrene, polyethylenes, polyglycols etc. are available on the market. In the calibration graph the logarithm of the size of the molecule is plotted against its elution volume. This situation is clearly indicated in Fig. 2.3.2. From this graph V_0 and V_i may be determined, as well as the operating range of the gel and the region of linearity.

The whole problem of the processing of the GPC data and their prediction is connected with the possibility of determining the value of K. From

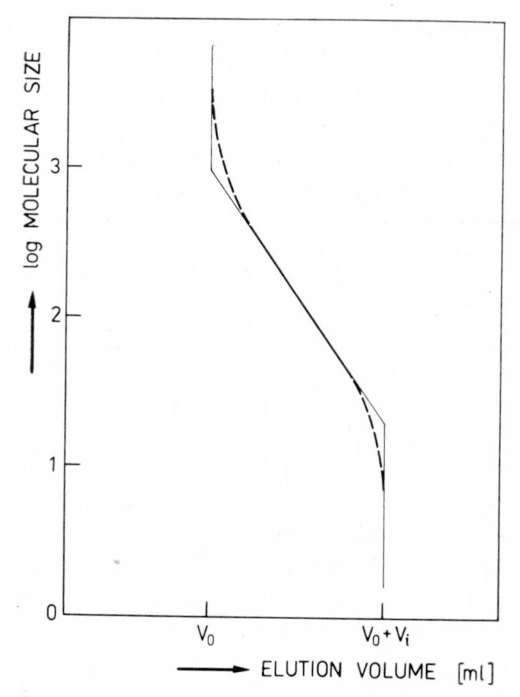

Fig. 2.3.2 Example of a GPC calibration curve
V_0 — exclusion limit, $V_0 + V_i$ — total permeation

the above principles it is evident that the elution volume depends on the molecular size of the substance analysed. The simplest way is to express the dependence of the elution volume V_R on the molecular mass[1]:

$$V_R = A - B \,.\, \log M \qquad (2.3.2)$$

where A and B are constants for a given chromatographic system. The molecular mass M is suitable in the comparison of homologues, but it does not express well the size of the molecule in substances which are structurally different. Hendrickson and Moore[2] have proposed using the effective length of the molecule, L_m. The basis of the scale is the series of n-alkanes, and from the retention volumes of other compounds equivalent numbers of carbons for various functional groups are computed. The relationship between the length of the molecule, L_m, in nanometers, and the number of carbon atoms in the molecule, C, is represented for n-alkanes by the simple equation

$$L_m = 0.25 + 0.125C \qquad (2.3.3)$$

Sometimes the hydrodynamic diameter is used to express the molecular size. The diameter is put into relation to the maximum length of the extended molecule. Smith and Kollmansberger[3] obtained better results when correlating retention volumes with the logarithms of *molar volumes.* Giddings et al.[4] proposed an average size of molecule, L, as a parameter that characterizes the GPC separation well in cases when no secondary effects interfere. In the correlation of retention data molecular mass is the least accurate parameter, the length of the molecule expresses the substances's behaviour better, but the use of the molar volume is most accurate.

It follows from these facts that GPC differs substantially from the separation on *molecular sieves,* where the minimum cross-section of the molecule, but not its length, is the decisive parameter. The fact that GPC operates with larger pores and that the process is more rapid than the separation on molecular sieves is of decisive importance. Large molecules have not enough time to assume a suitable position and penetrate into the pores that are smaller than corresponds to their maximum hydrodynamic diameter. The above discussion of the problems of GPC assumed that separation takes place on the principle of exclusion only. This applies predominantly to saturated hydrocarbons. Aromatic hydrocarbons are adsorbed on gels to a greater or lesser extent and their elution volumes are the result of simultaneous effect of adsorption and exclusion (see below).

Technique of GPC. The substantial parts of the instruments for GPC are the same as in other types of liquid chromatography: reservoirs—pump—injection port, column, detector. For some types of gels (for example of the polystyrene type) the gases dissolved in the solvent represent a serious danger, because their release in the column may easily damage the packing. Therefore a degassing device of the solvent is very often inserted in front of the pump, where the solvent is shortly heated to boiling point. An accurate measurement of the mobile phase flow is mainly in cases when molecular mass or the distribution of molecular masses has to be determined. In GPC the use of siphons placed behind the detector became common. At each emptying of the siphon, the volume of which is known, a photoelectric device records it. In older papers the retention volumes are given in counts. The differences of GPC from other types of liquid chromatography consist mainly in the type of packing and the role of solvents.

Packings for GPC. Packings for GPC may be classified according to whether they are used for work with an organic eluent or with aqueous solutions. For the separation of hydrocarbons the packings of the first type are of importance. These can be classified according to their chemical composition or mechanical properties:

According to chemical composition the following packings are known:

1. polystyrene cross-linked with divinylbenzene
2. cross-linked polyvinyl acetate
3. cross-linked alkylated polydextran
4. silica gels and glasses

According to mechanical properties the GPC packings are classified into hard, semi-hard and soft. Clearly, inorganic materials, silica gels and glasses can be classified among hard packings. Organic materials are polymers with a network structure that changes its volume in organic solvents. The degree of swelling, the gel porosity and its rigidity change with the solvent used. More swollen gels have larger pores but a lower rigidity. In some instances the producers indicate clearly that the gel is soft and that it can be deteriorated or destroyed by application of higher operation pressures. The strength of gel increases with the amount of the cross-linking agent used. Soft gels permit work with pressures of about 1 – 2 MPa and semi-hard gells with pressures 5 – 15 MPa.

Most widespread are gels based on polystyrene cross-linked with divinylbenzene, available under the commercial names Poragel, Styragel, Bio-Bead-S, Shodex, etc. They can be used with most organic eluents, with the exception of acetone, hexane and alcohol, for the separation of substances the molecular weights of which range from 10^2 to 10^4, for Styragel up to 10^8. In contrast to this the polyvinyl acetate gel represented on the market by the semihard Merckogel OR is suitable for work with polar organic solvents (acetone, alcohols). Substances within the 300-10^6 molecular weight units range can be separated on it.

Alkylated dextran gels are represented by Sephadex LH-20. This semi-hard gel is suitable both for polar organic solvents and for water. It can be used for molecular weights $10^2 - 4 \,.\, 10^4$; adsorption effects are observed in separation of aromates.

Silica gels and glasses appear on the market under the names Porasil, Spherosil, Merckogel Si, CPG Glasses, Bio-Glass. These packings are used for the separation of substances with molecular weights ranging from 10^3 to 10^6. When working with them, adsorption effects which can be considerable[5] should be taken into consideration. Further alkaline solvents should be avoided, because they dissolve the packing materials. The advantages of these packings are their hardness and the possibility of operating in aqueous and organic eluents, as well as of changing the eluents.

Similarly as in other types of liquid chromatography the development in gel chromatography also aims at the use of *microparticles.* The advantages of microparticles consist primarily in a higher mass transfer rate, which shortens the time of analysis considerably and permits the achievement of a higher resolution. Nowadays a number of packings[6] for GPC are on the market, with particle size 5 – 10 μm. Higher working pressures that are indispensable represent no problem with hard gels. In semi-hard gels of polystyrene type the maximum permissible pressures are about 15 MPa. Columns of 30 cm length packed with polystyrene gels of 10 μm particle size have an efficiency of 10 000 TP. In the case of small particles the packing procedure is carried out in suspension. The particles are suspended by an ultrasonic mixing in a liquid of such a density in which the particles float. Organic gels have a considerable electrostatic charge, they form clusters that can be eliminated by using polar liquids or aqueous solutions. Suspension technique is mainly used for the packing of semi-hard polystyrene and polyvinyl acetate gels.

Eluents. The choice of eluents does not play such an important role in GPC as in other types of liquid chromatography. The solvent does not influence the separation process directly and therefore the order of the eluted component does not usually change when the solvent is changed. Therefore stepwise or gradient elution is not used in GPC. Usually the work is done on a particular column with a single solvent during the whole operation. In the case of organic gels the change in the mobile phase can cause either a contraction or a swelling of the gel, leading to the formation of free spaces in the packing or to a mechanical damage of the gel. The pore size is also changed so that the column has to be recalibrated. A change in the order of eluted components may also take place when adsorption is involved in addition to exclusion. An eluent with a higher affinity toward the gel then restricts the adsorption effect.

General requirements put on the eluents also apply in GPC: low viscosity, good dissolving ability, suitability for a particular type of detector, low cost, and safety during operation.

The most commonly used eluent in GPC is tetrahydrofuran that has excellent dissolving properties and a low refractive index (1.407 at 20 °C). Its main drawback is the formation of peroxides. Among other eluents toluene and dichloromethane are most commonly used at room temperature. All three mentioned solvents have a low viscosity. For work at elevated temperature o-dichlorobenzene and 1,2,4-trichlorobenzene are employed. These eluents have a high refractive index and a high viscosity at room temperature. It was observed that they suppress adsorption effects in the GPC of aromates. Owing to excellent sample solubility in them, N,N-dimethylformamide and N-methylpyrrolidone are also sometimes used. These solvents are hygroscopic and water present in them damages some types of organic gels (for example Styragel).

Peak capacity and other factors. In connection with equation (2.3.1) it was mentioned that values of K can range from 0 to 1 in GPC. This also means that the elution volumes of all substances (if they are separated on the exclusion principle) range between V_0 to $V_0 + V_i$. If all substances are to be eluted within the interval V_i, then it is advantageous for the value V_i to be as high as possible. Simultaneously it is necessary to keep V_0

sufficiently low because a high value of V_0 means a prolongation of the time of analysis. Dark and Limpert[5] employ the expression V_i/V_0 as a measure of maximum capacity of the column. According to these authors the majority of gels has the value of V_i/V_0 between 0.7 and 1.3. At higher values of the V_i/V_0 ratio the volume of pores per unit of gel packing increases and the gel becomes soft or brittle. Mechanical properties of the gel are the limiting factor for further increase of the V_i/V_0 ratio.

This limitation of the operation range of the gel simultaneously limits the number of peaks which can be effectively separated by GPC on a column of a given efficiency. Adsorption and partition chromatography can elute substances in the $K = 0-10$ interval, and when gradient elution is applied this interval increases further. Giddings[7] observed that gel chromatography can resolve only one third of the peaks in comparison with other types of liquid chromatography (7 peaks on a small column of 1 000 TP efficiency). In the literature the problem is known as restricted *peak capacity*. From this it is evident that in GPC the work should be done with columns of maximum efficiency.

The introduction of the sample should not affect the viscosity of the mobile phase substantially, i.e. viscous samples have to be diluted. It is recommended that the viscosity of the injected sample should be at most double in comparison to the viscosity of the mobile phase. A dilution of the sample with the mobile phase is always recommendable, because a drastic change in the eluent composition at the place of injection may cause a change in gel volume and thus damage the packing. The amount of sample also determines the degree of separation, as shown by Pecka et al.[8] on the sample of the separation of *cis, trans* isomers of decahydronaphthalenes. The charge of about 10 mg of sample per 100 ml of total column volume or per 35 ml of dead volume V_0 seems an optimum compromise.

Temperature affects GPC far less than it does LLC. At elevated temperatures the decreased viscosity of the mobile phase improves the mass transfer, increases the column efficiency and affects the separation favourably. At elevated temperatures a low decrease in elution volumes also takes place, which is probably caused by the expansion of the gel.

2.3.2 SEPARATION OF HYDROCARBONS WITH SMALL MOLECULES

It should be said on introduction that this application of GPC is not very widespread and some authors recommend the use of gel chromatography for substances with molecular weights above 1 000 or 2 000 mass units only. However, in literature a number of papers describe the application of gel chromatography for the separation of low-molecular substances, sometimes with good success. In the case of hydrocarbons this means applying GPC for substances with molecular mass 100 – 700, which roughly corresponds to 8 – 50 carbon atoms in the molecule. Generally it applies that the use of liquid chromatography for substances with a carbon atom number lower than 15 is exceptional, because it is mostly more advantageous to apply gas chromatography.

Among the commercially available packings for example Bio-Beads S-X 12 and S-X 8 are suitable for this field of work, which have an exclusion limit at molecular mass 400 or 1 000 respectively. A disadvantage of these gels is their low rigidity. Among further materials the semi-hard Poragel A-60 A (exclusion limit 4 000), Merckogel OR-500 (exclusion limit 500), Sephadex LH-20 (exclusion limit 4 000), TSK Gel G-100 H (exclusion limit 1 000) and others are applicable.

Alkanes. From the very beginning of GPC, *n*-alkanes have been a subject of interest because endeavours were made to use them as universal calibration standards for low-molecular region. Hendrickson and Moore[2] measured elution volumes of *n*-alkanes and isoalkanes (in addition to a number of further substances) in the system polystyrene gel – tetrahydrofuran and expressed the size of the solute molecule as the length corresponding to the effective length of a particular *n*-alkane. *n*-Alkanes $C_5 - C_{36}$ and isoalkanes $C_6 - C_8$ were measured. Smith and Kollmansberger[3] measured *n*-alkanes and 2-methylalkanes in the same system and expressed the size of the molecule as its molar volume, Chang[9, 10] measured the retention volumes of *n*-alkanes $C_6 - C_{24}$ on the gel Bio-Beads S-X 8 in tetrahydrofuran and compared this system with Sephadex LH-20 in the same eluent. Thus they found that the change of the elution volume caused by one CH_2 group is about double for Sephadex LH-20. Oelert[11] investigated the properties of Merckogel OR-500 and found that the largest change of the elution volume, caused by the change of the molar volume of the substance measured, dV_R/dV_M, is obtained for *n*-alkanes when dichloromethane was used as eluent. Lower dV_R/dV_M values were found for isopropyl alcohol and the lowest for cyclohexane.

When studying isoalkanes on a polystyrene gel with tetrahydrofuran as eluent Schultz[12] proposed the parameter Z_g for the characterization of the size of the molecule. The parameter is defined as the statistical mean of the share of synclinal conformations on the steric arrangement of the molecule, determined by thermodynamic equilibrium. Together with the number of carbon atoms in the molecule this factor which includes the shape of the molecule enables an exact calculation of the elution volume of a certain isoalkane. It was shown that the elution volume increases slowly for isomers with increasing Z_g values:

$$V_R = A + B \, . \log C + D \, . \log Z_g \tag{2.3.4}$$

where A, B and D are constants for a given system and C is the number of carbon atoms in the molecule. According to Schultz[12] the constant B is about ten times larger than the constant D, and from this it is evident that the differences in elution volumes of individual isomers will not be large. The published results have shown that maximum difference in elution volumes is about 1 % between isoalkanes differing by 3 units in their Z_g values. The extreme value of Z_g is about 20. The value of Z_g increases with the refractive index, which means that of two isomeric isoalkanes the one with a higher refractive index will be eluted later.

Summing up it may be said that n-alkanes give a good linear dependence between the logarithm of the molecular size and the elution volume in a number of systems. It is irrelevant whether the molecule is characterized by the effective length of the molecule[11], molecular mass[13] or only the number of carbon atoms in the molecule. The elution of n-alkanes in a suitable GPC system is determined by the exclusion principle alone. The same is true to a large extent for isoalkanes where, however, the shape of the molecule also plays a certain role. The parameter Z_g seems so suitable[12] that a general solution is no problem in this case either.

It has already been said that the differences in elution volumes of isomeric isoalkanes are very small. The elution volumes of n-alkanes and isoalkanes with an equal number of carbon atoms in the molecule also do not differ much. In the case of isoalkanes it is not suitable to use the effective length of the molecule as a characteristic parameter. It was shown that branched isoalkanes, for example 2,2,4-trimethylpentane[2] or squalane[11], have little different elution volumes from *n*-octane or triacontane, but that their effective length of the molecule corresponds to *n*-pentane or tetracosane. For isoalkanes the dependence of the logarithm of the molecular mass on the elution volume obtained for *n*-alkanes also fits in the majority of cases. This is especially true of isoalkanes that have in their molecule 20 and less carbon atoms. Isoalkanes with a larger molecule have their elution volumes slightly higher than the corresponding *n*-alkanes.

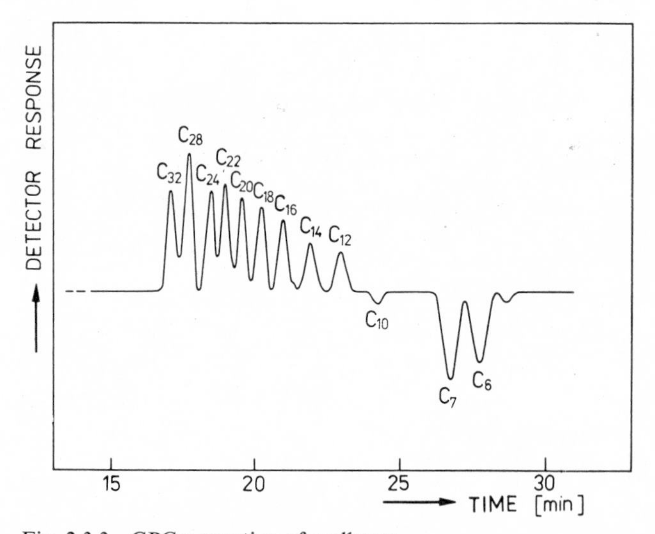

Fig. 2.3.3 GPC separation of *n*-alkanes
Two columns connected in series, 610×8 mm I.D., packed with TSK Gel (polystyrene gel); eluent: tetrahydrofuran

In GPC separation of alkanes some general problems are still unresolved: primarily the character of the dependence of the size of the molecule on the elution volume, that is not linear (see Fig. 2.3.2). From this it follows that substances with a larger molecule will be less resolved than substances with a small molecule. This also applies for the separation of alkanes when a mixture has to be separated the components of which have a constant increment of the size of molecule (CH_2 group).

A further problem is more general because in the series of *n*-alkanes and isoalkanes a suitable detector does not exist for the majority of eluents, that would record all the eluted components with a sufficient accuracy. The most often used *differential refractometer* gives inverse peaks of low-molecular *n*-alkanes if tetrahydrofuran is used as eluent. The components that are eluted within the inversion interval are not recorded. This is clearly illustrated in Fig. 2.3.3 where Majors[6] shows an example of the separation on *n*-alkanes C_6-C_{32} on polystyrene gel TSK Gel G-1000 H in tetrahydrofuran. The small particle size of the packing (10 μm) enabled the achievement of a high resolution on two short high efficiency columns (each of 610 × 8 mm dimensions).

Cycloalkanes. Gel chromatography is also a very effective method for the separation of cycloalkanes that are the most complex structural class of hydrocarbons from the point of view of the number of isomers. Cycloalkanes differ substantially in their behaviour in GPC from alkanes. Hendrickson and Moore[2] showed on the example of cyclohexane and cycloheptane that the elution volumes of these substances are substantially larger than the elution volumes of hexane and *n*-heptane, and they found that the effective length of the molecule is 3.92 or 4.20 carbons, respectively.

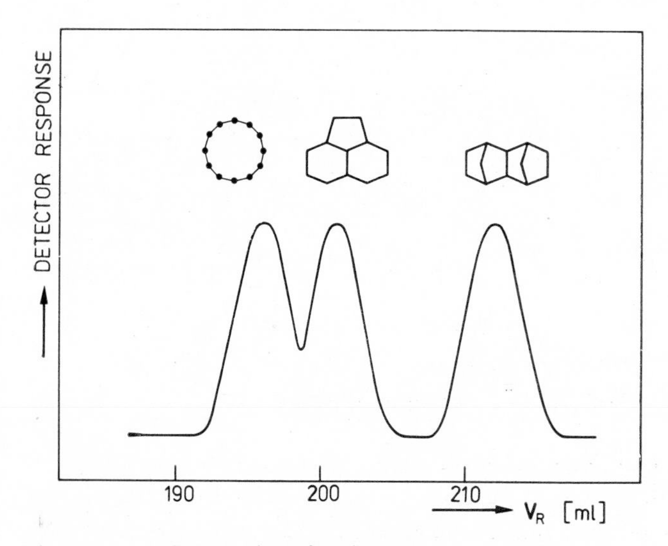

Fig. 2.3.4 GPC separation of cyclanes C_{12}
Six columns connected in series, 1 000 × 8 mm I.D., packed with S-Gel-832 (styrene-divinylbenzene); eluent: tetrahydrofuran 16 ml/h

Mair et al.[14] compared the elution volumes of some cycloalkanes and isoalkanes in the system Sephadex LH-20 – acetone.

Talarico et al.[15] also operated in the same chromatographic system and found that a mixture of cyclanic hydrocarbons can be further separated into compounds with one, two and three cycles in the molecule, and the substances are eluted from the column in this order.

Mudrovčić et al.[16,17] reproduced the preceding work and separated a model mixture *n*-hexadecane – squalane – tetralin. When separating the mixture *n*-decane – decalin – tetralin – naphthalene on Sephadex LH-20, they compared two eluents: acetone and isopropyl alcohol. A better separation of the components was achieved with isopropyl alcohol.

According to the present knowledge of the behaviour of cyclanic hydrocarbons in GPC the following can be summarized: Cyclanic hydrocarbons have substantially higher elution volumes than *n*-alkanes or isoalkanes with the same number of carbon atoms. *Unsubstituted cycloalkanes* with the same number of carbon atoms are eluted from a gel column in the order of increasing number of cycles in the molecule; for example, the order of C_{12} hydrocarbons is cyclodecane – perhydroacenaphthene – diendomethylenedecalin (see Fig. 2.3.4). Among unsubstituted cycloalkanes with the same number of cycles, and equal number of carbons, but different structure of the skeleton, the isomers with a more compact structure are retained more strongly on a gel column. For example, of the two $C_{10}H_{16}$ isomers adamantane, that has a spherical shape of the molecule, has a larger elution volume than the relatively "expanded" *exo*-trimethylenenorbornane[18].

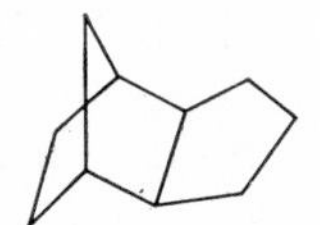

exo-trimethylenenorbornane
(V_R = 209 ml)

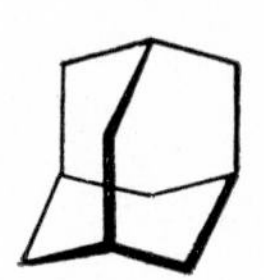

adamantane
(V_R = 214 ml)

Alkylation of cyclanic hydrocarbons decreases their elution volume. The number of substituents has a greater effect on the elution volume than the size of the substituents (two methyl groups decrease the elution volume more than one ethyl group). Alkyl carbons increase the effective dimension of the molecule much more than the same number of carbon atoms bound to the further ring. Thus 1,4-dimethylcyclohexane is eluted from a gel column before decahydronaphthalene (two carbon atoms more), and its elution volume is approximately equal to that of tricyclic perhydroanthracene that has 6 carbon atoms more in the molecule. The differences in elution volumes of alkylated derivatives caused by a homologous increase usually suffice for the separation of individual members of the homologous series.

In *isomeric cycloalkanes* with the same fundamental skeleton the narrow relationship between the molecular volume and enthalpy is reflected in some properties (density, refractive index). The isomers with a larger molecular volume have smaller non-bonding intramolecular interactions and a lower enthalpy. As the molecular volume is the directing factor in GPC separations of nonpolar substances, it may be assumed that the elution volumes are also in correlation with the enthalpies of single isomers.

Elution volumes of a number of model cycloalkanes[8] and polycycloalkanes[18] on polystyrene gel and with tetrahydrofuran as eluent were measured in the author's laboratory. Among the hydrocarbons tested with adamantane structure three groups of structural isomers occur composed of six-membered rings in chair conformation only, for which the differences in enthalpy can be calculated by simple conformation analysis[8]. The calculated enthalpy differences of the isomers and the measured GPC elution volumes are given in Table 2.3.1. In all instances the isomers with a higher content of energy have larger elution volumes. The dependence between the elution volume and $-\Delta H$ is almost linear.

The same relationship was also found in stereoisomeric cyclic hydrocarbons that differ only in *steric configuration* of the rings and in the *axial* or *equatorial* orientation of the substituents[8]. In Table 2.3.2 the data for the stereoisomers of the six hydrocarbons tested are compared. In the case of hydrocarbons with two stereoisomeric forms the thermodynamically more stable stereoisomer is eluted earlier than that with a higher heat content. For example in dimethyl- and trimethylcyclohexanes those stereoisomers are more stable that have all methyl groups equatorial, i.e. *trans*-1,4-dimethylcyclohexane and *cis,cis*-1,3,5-trimethylcyclohexane; these isomers have a lower enthalpy and also a smaller elution volume in GPC. Similarly the *trans*-isomer of decahydronaphthalene the ring configuration of which, characterized by minimum non-bonding inter-

TABLE 2.3.1

Elution volumes in GPC and the differences in the enthalpies of the isomers of alkyladamantanes

Hydrocarbon	$-\Delta H$ kJ/mol	V_R ml
$C_{14}H_{24}$		
1,3,5,7-Tetramethyladamantane	0.0*	178
1-Ethyl-3,5-dimethyladamantane	18.84	184
1,3-Diethyladamantane	37.68	189
1-(2′-Butyl-)adamantane	45.22	193
$C_{13}H_{22}$		
1,3,5-Trimethyladamantane	0.0*	186
1-Ethyl-3-methyladamantane	18.84	192
1-Isopropyladamantane	30.15	197
$C_{12}H_{20}$		
1,3-Dimethyladamantane	0.0*	194
1-Ethyladamantane	18.84	201
2-Ethyladamantane	26.38	203

* The most stable isomer to which the differences in enthalpies are referred. The elution volumes were measured under the conditions given in Fig. 2.3.4.

actions, is more advantageous than the *cis*-configuration from the energy point of view, is eluted first.

In cycloalkanes with a greater number of stereoisomers their elution volume in GPC increases linearly with increasing enthalpy. This is true both of the stereoisomers of tetramethylcyclohexane and also of perhydroanthracene and other tricyclic perhydroaromates (Table 2.3.2).

The relationship described between the elution volumes and conformational parameters affecting the total enthalpy of cyclic isomers is analogous to those observed by Schultz[12] for isoalkanes where the elution volumes of the isomers are a logarithmic function of the parameter Z_g. These correlations can be useful for the prediction of the separation effects in GPC for fractions containing isomeric cycloalkanes, and in special cases

TABLE 2.3.2

Elution volumes in GPC and the differences in the enthalpies of stereoisomers

Hydrocarbon	$-\Delta H$ kJ/mol	V_R ml
1,4-Dimethylcyclohexane		
trans	0.0*	197.1
cis	7.5	198.9
1,3,5-Trimethylcyclohexane		
cis, cis	0.0*	188.5
trans, cis	7.5	192.9
1,2,4,5-Tetramethylcyclohexane		
trans, cis, cis	7.5**	189.0
cis, trans, cis	15.1**	191.1
cis, cis, cis	23.0**	194.0
Decahydronaphthalene		
trans	0.0*	203.1
cis	11.3	207.7
Perhydroanthracene		
trans, syn, trans	0.0*	192.4
cis, syn, trans	11.3	195.8
cis, syn, cis	27.6	199.9
Perhydrophenanthrene		
trans, anti, trans	3.8***	194.5
cis, syn, trans	15.1***	197.9

* The most stable isomer to which all the differences in enthalpies are referred.

** Referred to *trans, cis, trans*-isomer.

*** Referred to *trans, syn, trans*-perhydroanthracene. The elution volumes were measured under the conditions given in Fig. 2.3.4.

the order of elution may be a guide for the determination of the structure of individual cycloalkanes.

As is evident from Table 2.3.2 the differences between the elution volumes permit the isolation of individual stereoisomers from the original mixtures in the majority of cases. Among the stereoisomers listed in the table more than half were isolated on a column 6 m long (25 600 TP) in a 91 – 99 % purity. In the majority of other cases the stereoisomers were at least substantially concentrated.

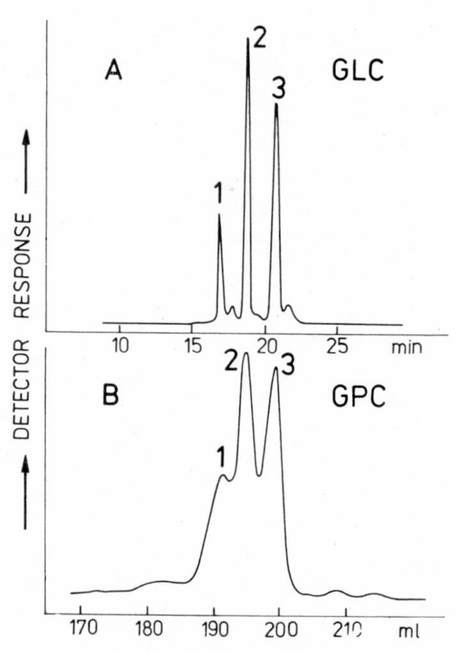

Fig. 2.3.5 GPC and GLC separation of stereoisomers of perhydroanthracene
1 — *trans, syn, trans*, 2 — *cis, syn, trans*, 3 — *cis, syn, cis*
GPC system as given for Fig. 2.3.4

The elution behaviour of stereoisomers in GPC is illustrated in Fig. 2.3.5 with a chromatogram of perhydroanthracene on which the separated fractions are indicated. For comparison the analysis of the raw material by means of GLC is also shown. The order of the stereoisomers eluted from the gel column is the same as the order in GLC on OV-225 as stationary phase. It is clear that GPC separates individual isomers with a similar efficiency as preparative gas chromatography.

Aromatic hydrocarbons. For the beginning of GPC it was observed that aromatic hydrocarbons behave in an anomalous manner which indicated that exclusion is not the sole controlling process in their separation.

A number of authors describe these deviations on polystyrene gels using tetrahydrofuran[19, 20], benzene[21] and 1,2,4-trichlorobenzene[22] as eluents. In the majority of cases it was observed that the dependence of the logarithm of the molar volume on the elution volume has an opposite slope for aromates than for alkanes, i.e. the elution volume increases with increasing molar volume. It was further established that differences exist in the behaviour of catacondensed and pericondensed aromates. When using 1,2,4-trichlorobenzene Bergman et al.[22] achieved a mild decrease of the elution volumes of aromates with increasing molar volumes.

A similar behaviour of aromates was also observed on the polydextran gel Sephadex LH-20. Wilk et al.[23] measured elution volumes of aromates on this gel using isopropyl alcohol and chloroform. In the case of chloroform the elution volumes corresponded to the exclusion principle, but in the case of isopropyl alcohol the elution volumes increased with increasing molecular weight. Oelert[24] confirmed the preceding results using the same gel and isopropyl alcohol. Quite recently Sephadex LH-20 was investigated by Rusin and Kulczycka[25] in the following solvents: cyclohexane, isopropyl alcohol, methanol, ethanol and chloroform. They found that an increase in elution volumes with increasing molecular mass of aromates took place with all solvents used, except chloroform.

Oelert[11] found, using polyvinyl acetate gel and dichloromethane and cyclohexane for elution, that the exclusion mechanism is involved in catacondensed aromates, while in pericondensed aromates it is the adsorption mechanism. In the case of isopropyl alcohol adsorption mechanism prevailed in both types of aromates. Streuli[26, 27] found that the separation course of aromates on gels is affected by the chemical nature of the gel. The strongest adsorption interaction is observed on polydextran gels and polyvinyl acetate gels. The effect of adsorption on polystyrene gels is lower and it can be completely suppressed by selecting a suitable mobile phase (1,2,4-trichlorobenzene).

The effect of *alkylation* on the behaviour of aromatic hydrocarbons was investigated by Popl et al.[28]; the measurement of the elution volumes of alkylbenzenes and benzenes with condensed cyclanic rings in the system polystyrene gel-tetrahydrofuran showed that alkyls on the benzene ring behave exactly in accordance with the exclusion principle. The contribution of the naphthenic carbon on the decrease of the elution volume is substantially smaller and it represents about 20 % of the alkyl carbon. From the data published by Hirsch et al.[21] for the system polystyrene gel – benzene a similar conclusion can be drawn for alkylbenzenes, alkylnaphthalenes and alkylphenanthrenes.

From present knowledge the following conclusions can be drawn for the behaviour of aromatic hydrocarbons in GPC: In the majority of the GPC systems used the elution of polycyclic aromates is not controlled by the exclusion principle and the elution range of these substances lies behind the operational range of the gel. This phenomenon is generally called "adsorption effect of gel". Pericondensed aromates are eluted later than catacondensed ones with the same number of carbon atoms in the molecule. The introduction of an alkyl group into the molecule of an aromatic hydrocarbon brings about important changes in the elution volume. The alkyl part of the molecule strictly follows the principle of gel separation. In the series of benzene, naphthalene and phenanthrene derivatives there are no substantial differences in the solution volumes between a particular *n*-alkane and an aromate containing in its alkyl substituent the same number of carbon atoms as *n*-alkane. It is necessary that the number of carbons in the alkyl should be larger than 5. Deviations from this rule may occur in dependence on the GPC system used, and thus also in dependence on the strength of the adsorption effect. If a polycondensed aromatic system is alkylated (for example triphenylene, coronene) the adsorption effect affecting the aromatic system is also operative to a certain extent. The resulting elution volume lies between the elution volume of the corresponding *n*-alkane (with the same number of carbon atoms as in the alkyl) and the elution volume of the parent polycyclic aromate. In the majority of cases these elution volumes lie in the working range of the gel. Alkylation with methyl groups brings about a considerable decrease in the elution volumes only up to three methyl groups on the benzene ring. The introduction of further methyl groups affects the elution volume only negligibly. *Cyclanic rings* condensed with the aromatic ring also affect the elution volume of the resulting cyclanic-aromatic hydrocarbon very little. The effect of the cyclanic carbon in the rings attached to the aromatic skeleton is 3–5 times less than the effect of the carbon of an aliphatic alkyl.

From this survey it is evident that adsorption effects are an integral part of GPC of aromates. These effects have been utilized in practice by a number of authors who used gel chromatography as a part of the separation and purification step. Klimisch[13] used the system Sephadex LH-20–isopropanol for the separation of saturated hydrocarbons from polycyclic aromates in the analysis of cigarette smoke condensate. Shook et al.[29] applied the polystyrene gel Bio-Beads SX-12 in benzene for the same purpose.

As the exploitation of the adsorption effect may become important for the analysis of polycyclic aromates, this effect was studied in the author's laboratory[30]. The average dimension of the molecule $\bar{L}$ was used[7]. In the calculation of the parameter $\bar{L}$ of aromatic hydrocarbons it was assumed that the molecules were planar and all aromatic nuclei equal. The shape of each nucleus was approximated by a toroidal surface with the radii according to van der Waals: carbon 0.17 nm, hydrogen 0.12 nm, C–C length 0.14 nm and C–H bond length 0.11 nm. The parameter $\bar{L}$ was calculated according to the equation

$$\bar{L} = C + \frac{\pi}{4}(\bar{l} + 2R - C) \qquad (2.3.5)$$

where $C/2$ and R are the radii of the toroid and $\bar{l}$ is the mean dimension of the multiangle determined by the centres of the peripheral aromatic nuclei during its rotation in the plane of the molecule. In five-membered rings all C–C bonds were considered to be 0.14 nm long, and the angle of the bonds also belonging to the mutually condensed rings is 120°. In hydrocarbons of fluorene type a regular five-membered ring is assumed with the lengths of all C–C bonds equal to 0.14 nm, and the $-CH_2-$ group was considered to be equivalent to the peripheral aromatic carbon atom.

The measurements were carried out on four columns connected in series (each 900 × 4 mm) and packed with polystyrene gel S-Gel 832 with exclusion limit about 1 000. Benzene was used as eluent. The elution volumes of the investigated polycyclic aromates are given in Table 2.3.3

TABLE 2.3.3

Experimental and theoretical elution volumes (according to eq. 2.3.6) of polycyclic aromates in the system polystyrene gel-benzene

Compound	V_R (ml)	C_A	$\bar{L}$ (nm)	$V_R^{theor} - V_R$ (ml)
Naphthalene	38.9	10	0.775	0.2
Fluorene	38.4	12	0.864	—0.5
Acenaphthylene	39.7	12	0.811	0.5
Diphenylene	38.2	12	0.845	0.5
Anthracene	39.1	14	0.897	0.0
Phenanthrene	39.8	14	0.880	0.0
Pyrene	41.4	16	0.897	0.3
Benzo(*a*)fluorene	39.2	16	0.972	—0.8
Benzo(*b*)fluorene	38.3	16	0.983	—0.4
Fluoranthene	40.2	16	0.920	0.5
Tetracene	39.7	18	1.018	—0.7
Benz(*a*)anthracene	39.5	18	0.996	0.5
Chrysene	40.0	18	0.985	0.5
Triphenylene	41.1	18	0.969	0.0
Benzo(*a*)pyrene	42.4	20	1.002	—0.1
Benzo(*e*)pyrene	42.9	20	0.985	0.2
Perylene	43.3	20	0.985	—0.5
Benzo(*k*)fluoranthene	40.5	20	1.038	0.3
Picene	41.0	22	1.090	0.1
Benzo(*b*)chrysene	40.3	22	1.101	0.3
Dibenzo(*a*, *c*)anthracene	41.6	22	1.080	—0.1
Anthanthrene	45.1	22	1.018	—0.9
Benzo(*g*, *h*, *i*)perylene	45.1	22	1.002	—0.2
o-Phenylenepyrene	42.5	22	1.040	0.8
Dibenzo(*a*, *e*)pyrene	44.2	24	1.085	—0.3
Dibenzo(*a*, *i*)pyrene	42.6	24	1.107	0.4
Coronene	47.9	24	1.018	—1.1
Dibenzo(*a*, *e*)fluoranthene	43.8	24	1.110	—1.0
Truxene	37.2	24	1.192	2.1
Ovalene	52.2	32	1.239	—4.5
Decacyclene	48.6	36	1.254	3.6

together with the number of carbons in the molecule, C_A, and calculated values of parameter $\bar{L}$.

From the measured data an equation was composed permitting the calculation of the elution volumes of basic aromatic hydrocarbons for a given system with very good accuracy. The dependence of the elution volume V_R on the parameter $\bar{L}$ and the number of aromatic carbons, C_A, can be expressed in linearized form:

$$V_R = V_B + A(C_A - 6) + B(\bar{L} - \bar{L}_B) \qquad (2.3.6)$$

where V_B is the elution volume of benzene, $\bar{L}_B = 0.654$ nm is the value system. The constant A characterizes adsorption and the constant B desorption. The V_B value is the constant for a given system. It includes both the volume of the eluent between the particles, V_0, and the volume of the eluent inside the gel matrix, V_i. This constant is independent and the absolute value of the further two constants, A and B, is linearly proportional to V_B. Equation (2.3.6) permits the prediction of the elution volumes of basic polycyclic aromates for further GPC systems too, for

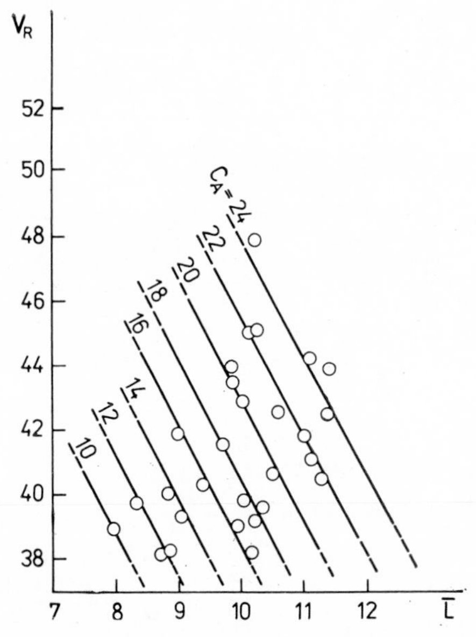

Fig. 2.3.6 Dependence of the elution volume of polycyclic aromates, V_R, on parameter $\bar{L}$
C_A — number of aromatic carbon atoms in the molecule

example polystyrene gel – cyclohexane, polystyrene gel – tetrahydrofuran, polyvinyl acetate gel – isopropyl alcohol, etc. The greatest difference of the measured data from equation (2.3.6) were found when isopropyl alcohol[30] was used. The data given in Table 2.3.3 are shown as a plot of V_R versus $\bar{L}$ in Fig. 2.3.6. From the plot it is evident that for substances with an equal number of carbon atoms, C_A, the elution volumes decrease with an increasing number of aromatic carbons. This fact indicates clearly the presence of an adsorption effect in the system polystyrene gel – benzene. In equation (2.3.6) this fact is expressed by a positive value of constant A. Of the eleven systems investigated constant A had a negative value in one single case, in the system polystyrene gel – 1,2,4-trichlorobenzene.

2.3.3 SEPARATION OF HYDROCARBON MIXTURES

In the preceding section the behaviour of various types of low- molecular hydrocarbons in GPC systems has been described. On this basis a general consideration may be formed as to the immediate composition of the eluate during the separation of any natural hydrocarbon mixture. Under the assumption that a certain *n*-alkane is eluted from the column at a certain moment it may be stated with certainty that isoalkanes with only a slightly higher (about 10 rel. %) molecular weight will be eluted simultaneously. Considerable differences exist in naphthenic hydrocarbons where components can be eluted the molecular mass of which is almost 100 rel. % higher than that in a corresponding *n*-alkane, for example *n*-heptane and perhydroanthracene[18]. The complexity further increases with aromatic hydrocarbons, especially in those pericondensed, that are often eluted only behind the working range of the gel. This means that the number of carbons in the corresponding *n*-alkane should be expressed by negative values. The part of nonalkylated polycyclic aromates in petroleums is fortunately relatively small so that this extreme usually does not come into effect in GPC separations of petroleum fractions. Relatively numerous alkylbenzenes and alkylnaphthalenes are eluted simultaneously with *n*-alkanes the molecular mass of which is always considerably smaller. Generally it can be assumed that petroleum distillates, especially those with a broader distillation range and higher boiling, will be eluted in GPC within a continuous interval without distinct peaks, and that even in systems with high efficiency. A number of published papers confirm this conclusion. If peaks appear on the chromatogram at all, they are caused by (1) substances with a molecular weight exceeding the gel's exclusion limit, forming a peak at the beginning of the elution range, (2) substances of polyaromatic character that sometimes form a sharp peak at the end of the elution range. In the first case the reason is an inadequate separation system, because a part of the substances is outside the operational range of the gel, while in the second case it is the adsorption effect of the aromates. The GPC records of petroleum fractions give relatively smooth envelope curves. In spite of this the GPC of petroleum fractions is still a subject of interest and it has two important applications: it is used for a rapid characterization of petroleums and petroleum fractions, for example as a "*fingerprint*", and for further separation of petroleum fractions that have been adjusted previously by another technique.

One of the first proposals for the use of GPC for the characterization of *crude oils* was the paper by Bombaugh et al.[31] who demonstrated the difference in the chromatograms of Kuwait crude oil before and after the elimination of light fractions. Done and Reid[32] used 600 × 9.5 mm columns packed with Styragel 60 Å of an efficiency higher than 4 000 TP for the characterization of crude oils and petroleum distillates. Elution was carried out either with tetrahydrofuran or toluene (70 ml/h) and the components were detected with an RI detector connected with an integrator. Within about 30 minutes a chromatogram was obtained that enabled the classification of various petroleums into several groups, while in the case of a smaller set of samples it was even possible to distinguish individual petroleums. To a lesser extent this also applies for petroleum fractions.

Albaugh and Talarico[33] made a further improvement of this method by using three detectors the records of which they compared. A differential refractometer gave a record corresponding to concentration and refractive index of eluted components, an UV detector recorded aromatic hydrocarbons practically exclusively, while a flame-ionization detector gave a response approximately proportional to the mass of the sample in the eluate for sample components with a sufficiently high boiling point.

The problem of detection is general and it represents one of the most important branches of LC. This problem is especially prominent in samples with a broad distillation range, such as crude oils. If the dependence of mass versus volume has to be determined at least approximately, either a detector with sample transporter must be used (as by Albaugh and Talarico[33]) or the eluate fractions must be collected, evaporated and weighed, as by Oelert et al.[34]. In both cases it means that the differences in the boiling points of the eluent must be at least 100 – 150 K. Oelert et al.[34] used two columns 150 × 1.27 cm for the characterization of crude oils. The first was packed with Poragel A-3 and the second with Poragel A-1. The samples of oils were freed from the fractions boiling up to 180 °C beforehand and then eluted with dichloromethane. The authors confirmed that in the separation of crude oils by means of GPC completely different fractions are obtained from those obtained by distillation.

Altgelt and Hirsch[35] applied GPC for the separation of heavy end oils. On a 300 × 15 cm column packed with a polystyrene gel of exclusion limit 100 000 they separated 300 g of asphalt in one run. The sample was eluted with benzene with 10 % of methanol at a 15 l/h rate. The smallest elution volume was found for asphalthenes with a molecular weight of 25 000.

The separation of petroleum fractions previously adjusted by another technique is a further field where GPC is used. A common treatment of petroleum fractions consists in the elimination of *n*-alkanes, for example with urea, and the separation of the deparaffinized sample by adsorption chromatography to a saturated, an aromatic and a polar fraction. After this procedure isoalkanes and cycloalkanes remain in the saturated fraction, and the applications of GPC are directed toward the separation of these two groups.

Mair et al.[14] who used Sephadex LH-20-acetone system demonstrated that it is possible to separate a natural mixture of isoalkanes and cycloalkanes with about 21 carbon atoms in the molecule. Talarico[15] successfully separated isoalkanes and cycloalkanes from the petroleum fraction $C_9 - C_{15}$ in the same system using a 180 × 2.54 cm column. In an attempt to separate cycloalkanes and isoalkanes from the petroleum fraction $C_{14} - C_{26}$ on a 30 × 2.54 cm column in the system Sephadex LH-20 tetrahydrofuran they were unsuccessful. Weber and Oelert[36] who attempted to separate the saturated fractions from the petroleum fraction boiling at 450 – 470 °C did not achieve a sharp separation of cycloalkanes and isoalkanes. They used Poragel A-1 as the stationary and dichloromethane as the mobile phase.

The separation of isoalkanes and cycloalkanes achieved in the author's laboratory on styrene-divinylbenzene gel with tetrahydrofuran as mobile phase confirmed that the limiting factor for a GPC separation of more complex isoalkane-cycloalkane fractions is the distillation range of the fraction under separation. In fractions with an excessively wide boiling

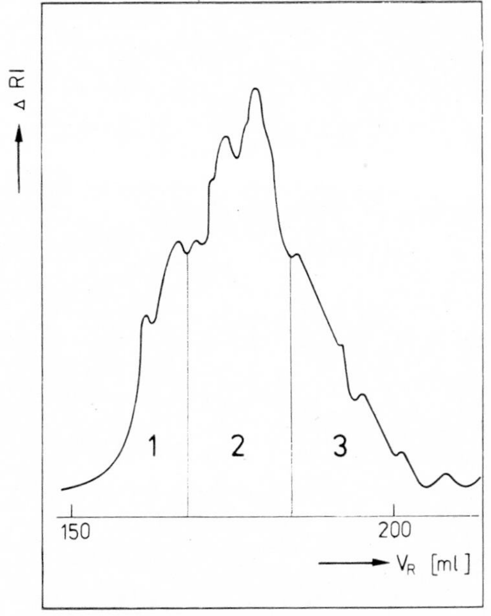

Fig. 2.3.7 GPC separation of isoalkane-cycloalkane fractions 210—250 °C
1 — isoalkanes 88 wt. % purity, 2 — isoalkanes and monocycloalkanes, 3 — dicycloalkanes 89 wt. % purity
GPC system as that given in Fig. 2.3.4

range the low-boiling isoalkanes overlap in the intermediate fraction with high-boiling monocycloalkanes that have the smallest elution volumes among cycloalkanes. This is clearly evident from the results of the separation of petroleum fractions on the mentioned gel system. In Figs. 2.3.7 and 2.3.8 the separation of an isoalkanic-cycloalkanic fraction, distilling within a 40 K interval, is compared with a similar fraction distilling in an interval of 5 K only. (The degree of separation was controlled by mass spectrometry of the isolated fractions). Using gel chromatography of a narrow fraction groups of isoalkanes, monocycloalkanes and dicycloalkanes were isolated in almost 90 % purity; on separation of a wide fraction a part of heavier isoalkanes ($C_{13}-C_{15}$) was obtained in the first concentrate, and a part of dicycloalkanes in the last fractions, in an almost 90 % purity, but two intermediate fractions contained – according to expectation – an unseparated mixture of both classes, with a prevailing

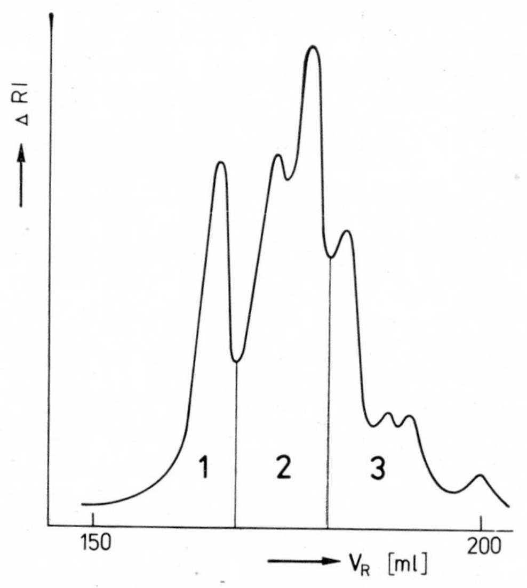

Fig. 2.3.8 GPC separation of isoalkane-cycloalkane fraction 225—230 °C
1 — isoalkanes 88 wt. % purity, 2 — monocycloalkanes 89 wt. % purity, 3 — dicycloalkanes 94 wt. % purity
GPC system as that given in Fig. 2.3.4

content of monocycloalkanes. For practical separations the suitable distillation range of the separated fractions should be found empirically for a given gel column. The mentioned results, obtained with various gel systems show that the combination of a sharp rectification (or also other preseparation methods) and GPC is so far the most efficient known means for a preparative separation of complex isoalkanic-cycloalkanic fractions.

The use of GPC for further separation of preadjusted aromatic petroleum fractions was worked out in connection with the American Petroleum Institute *Research Project No. 60*. Hirsch et al.[21] described the separation of hydrocarbons on polystyrene gel with benzene as eluent, and proposed the procedure for theoretical calculation of retention volumes. In a subsequent paper Hirsch et al.[37] described the application of the proposed method to aromatic concentrates from higher-boiling (370 – 535 °C) petroleum fractions. The GPC characterization of aromatic concentrates is a part of the separation scheme described in detail by Haines and Thompson[38]. The obtained fractions of mono-, di-, and polyaromates were separated on a 455 × 2.54 cm column, packed with polystyrene gel; its lower half contained Styragel 400 Å and its upper half Styragel 100 Å. About 1.7 g of the corresponding aromatic fraction were injected into the column, which was then eluted with benzene at a 1 ml/min rate. After the passage of 700 – 800 ml of eluate the collection of fractions (each 25 ml) was started. They were evaporated, weighed and each third was analysed by mass spectrometry or by nuclear magnetic resonance. As the analysis was carried out with previously separated aromatic groups (for example monoaromates), the separation took place within these groups on the exclusion principle, i.e. it depended on the degree and the type of alkylation. Hirsch et al.[37] measured the elution volumes of a large number of hydrocarbons and a number of sulphur compounds (sulphides, thiophenes, etc.) in the described system. On the basis of their correlation procedures the data of mass spectrometry could be made more accurate and the structural type of the hydrocarbons decided. They summarized their observations for GPC of hydrocarbons in a few conclusions:

1. For a given molecular weight the elution volume of aromates increases with the number of aromatic rings in the molecule.
2. For a given molecular weight the derivatives with cyclanic rings in the molecule are eluted in the order of increasing number of cyclanic rings.
3. One alkyl carbon atom is approximately equivalent to two or three naphthenic carbon atoms.
4. If the number of carbons in the alkyl is higher than two, all aromates are eluted with about the same elution volume, if the number of carbons in the alkyl is equal.
5. Benzene with two to three attached naphthenic rings in the molecule has the same elution volume as some alkylnaphthalene without naphthenic rings.

Point 1 comes into consideration when the groups of aromates are separated insufficiently, which is practically always the case of higher-boiling fractions. Points 2 and 3 were discussed in greater detail in the section on the GPC of cyclanic standards. Point 4 was checked mainly for alkylbenzenes, alkylnaphthalenes and alkylphenanthrenes. With a higher number of aromatic rings in the molecule the adsorption effect of this part must come into effect, and the validity of point 4 will be restricted.

For point 5 the data by Hirsch et al.[37] may be mentioned, who found that the elution volume of 1,2,3,4,5,6,7,8-octahydroanthracene is 1 480 ml and of butylnaphthalene 1475 ml. By this procedure, combining GPC and MS, it was possible to identify 30 types of substances in the group of aromates of the petroleum fractions, 28 types in the group of diaromates, and 59 types in the group of polyaromates and polar substances. Details of the correlations are given in a subsequent paper[43].

2.3.4 CHROMATOGRAPHY ON ION-EXCHANGE RESINS

Ion-exchange resins are not commonly used for the separation of hydrocarbons, but nonetheless several cases are described in literature when a retention of hydrocarbons took place on these packings. On ion exchangers with bound silver, compounds can be retained that have a double bond in the molecule. The formation of a *charge-transfer* complex between the silver ion and the molecule containing the double bond takes place in this process. In no case is the process based on ion exchange, but on adsorption, equal to that observed in the chromatography on silica gel or alumina impregnated with $AgNO_3$ (see Section 2.3.2). Silver ion exchangers can be prepared from a cation-exchange resin in H^+ or Na^+ form and an aqueous solution of a silver salt. Before use the silver ion exchangers thus prepared should be dried. Alternatively anionic resins containing amino groups can be used, because they can bind the silver ion as a complex. A silver ion exchanger of this type was used for the separation of piperylene from *n*-hexane[39]. The retention of the olefins is achieved on percolation of a mixture of hydrocarbons through a column packed with silver ion exchanger at temperatures up to 30 °C. It was observed that the capacity of these columns is very low, about 1–3 % of the theoretical capacity, referred to silver content[40]. The stability of the silver complex depends on the substitution of the double bond and on whether the hydrocarbon is a *cis*-olefin or a *trans*-olefin. Unsubstituted olefins display a higher stability of the complex and the stability of a *cis*-olefin is always higher than that of its *trans*-isomer.

Another case when hydrocarbons are retained on an ion exchanger is the adsorption of polycyclic aromates on anion-exchange resins during the elimination of acid components from petroleum fractions[21]. For this purpose Amberlyst A-29 was used, which can be used in non-aqueous medium. The anionic resin was activated with a 10 % (w/w) solution of KOH in methanol, washed with water and gradually extracted with methanol, benzene and pentane. Before use it was dried at 40 °C for 24 h. This resin, 40 g, was packed into the column, 119 × 1.4 cm, and 10–20 g of the petroleum fraction dissolved in 200 ml of pentane or cyclohexane was charged. Neutral or basic material was eluted with pentane for 42 h. The acid components were then displaced gradually with benzene, methanol and methanol saturated with CO_2. Carboxylic acids, carbazoles, phenols, amides and polycyclic aromates were then detected in the fraction obtained. When acidic components were separated in the system polystyrene gel-benzene, polycyclic aromates were eluted last. McKay and Latham[41,42] separated these polycyclic aromates further by thin-layer chromatography and identified individual components on the basis of their fluorescence spectra. The results of papers[41,42] showed that the main part is composed of pericondensed aromates that contain pyrene or perylene as the basis in their skeleton. These components have relatively small molecules and they are strongly retarded in the majority of the GPC systems. The principle of the adsorption of these polycyclic aromates on an anion exchanging resin is not yet completely elucidated. It may be assumed that it is a certain type of acido-basic interaction, similar to that observed in linearly catacondensed aromates (see Section 4.2.1).

REFERENCES (2.3)

1. E. M. W. Anderson and J. F. Stoddart, *Anal. Chim. Acta* **34** (1966) 401.
2. J. G. Hendrickson and J. C. Moore, *J. Polym. Sci.* Part A-1, **4** (1966) 167.
3. W. B. Smith and A. Kollmansberger, *J. Phys. Chem.* **69** (1965) 4157.
4. J. C. Giddings, E. Kucera, C. P. Russell and M. N. Myers, *J. Phys. Chem.* **72** (1968) 4397.
5. W. A. Dark and R. J. Limpert, *J. Chromatogr. Sci.* **11** (1973) 114.

6. R. E. Majors, *International Laboratory*, 11, November/December 1975.
7. J. C. Giddings, *Anal. Chem.* **39** (1967) 1027.
8. K. Pecka, S. Hála, J. Chlebek, M. Kuraš and B. Kremanová, *J. Chromatogr.* **104** (1975) 91.
9. T. L. Chang, *Anal. Chim. Acta* **39** (1967) 519.
10. T. L. Chang, *Anal. Chim. Acta* **42** (1968) 51.
11. H. H. Oelert, *J. Chromatogr.* **53** (1970) 241.
12. W. W. Schultz, *J. Chromatogr.* **55** (1971) 73.
13. H. J. Klimisch, *Z. Anal. Chem.* **264** (1973) 275.
14. B. J. Mair, P. T. R. H. Wang and R. G. Ruberto, *Anal. Chem.* **39** (1967) 838.
15. P. C. Talarico, E. W. Albaugh and L. R. Snyder, *Anal. Chem.* **40** (1968) 2192.
16. S. Marin-Mudrovčić, J. Mühl and M. Šateva, *Nafta (Zagreb)* **23** (1972) 593.
17. S. Marin-Mudrovčić, J. Mühl and M. Šateva, *Nafta (Zagreb)* **25** (1974) 249.
18. S. Hála, K. Pecka and B. Kremanová, *Coll. Czech. Chem. Commun.* **39** (1974) 3649.
19. R. E. Thompson, E. G. Sweeney and D. C. Ford, *J. Polym. Sci.* A **1**, 8 (1970) 1165.
20. T. Edstrom and B. A. Petro, *J. Polym. Sci.* C **21** (1968) 171.
21. D. E. Hirsch, J. E. Dooley and H. J. Coleman, U.S., Bur. Mines, Rep. Invest., 1974, RI — 7875.
22. I. G. Bergman, L. J. Duffy and R. B. Stevenson, *Anal. Chem.* **43** (1971) 131.
23. M. Wilk, J. Rochlitz and H. Bende, *J. Chromatogr.* **24** (1966) 414.
24. H. H. Oelert, *Z. Anal. Chem.* **244** (1969) 91.
25. A. Rusin and J. Kulczycka, *Chem. Anal.* **20** (1975) 361.
26. C. A. Streuli, *J. Chromatogr.* **56** (1971) 219.
27. C. A. Streuli, *J. Chromatogr.* **56** (1971) 225.
28. M. Popl, J. Čoupek and S. Pokorný, *J. Chromatogr.* **104** (1975) 135.
29. M. E. Snook, W. J. Chamberlain, R. F. Severson and O. T. Chortyk, *Anal. Chem.* **47** (1975) 1155.
30. M. Popl, J. Fähnrich and M. Stejskal, *J. Chromatogr. Sci.* **14** (1976) 537.
31. K. J. Bombaugh, W. A. Dark and R. F. Levangie, *Separ. Sci.* **3** (1968) 375.
32. J. N. Done and W. K. Reid, *Separ. Sci.* **5** (1970) 825.
33. E. W. Albaugh and P. C. Talarico, *J. Chromatogr.* **74** (1972) 233.
34. H. H. Oelert, D. R. Latham and W. E. Haines, *Separ. Sci.* **5** (1970) 657.
35. K. H. Altgelt and E. Hirsch, *Separ. Sci.* **5** (1970) 855.
36. J. H. Weber and H. H. Oelert, *Prepr. Div. Petr. Chem. A. C. S.* **15** (2) (1970) A 212.
37. D. E. Hirsch, J. E. Dooley, J. H. Coleman and C. J. Thompson, U.S., Bur. Mines, Rep. Invest., 1974, RI-7974.
38. W. E. Haines and C. J. Thompson, Amer. Petrol. Inst., U.S., Bur. Mines, API RP 60 Publ. Rep. No 37, July 1975.
39. C. L. Swarthmore, U.S. Patent 2, 865970 (1958).
40. E. T. Niles, U.S. Patent 3, 219717 (1965).
41. J. F. McKay and D. R. Latham, *Anal. Chem.* **44** (1972) 2132.
42. J. F. McKay and D. R. Latham, *Anal. Chem.* **45** (1973) 1050.
43. D. E. Hirsch, H. J. Coleman, J. E. Dooley and C. J. Thompson, Energy Res. Dev. Administration, 1977, BERC/RI-76/14.

2.4 Preparative gas chromatography

Gas chromatography – a method based on the separation of components between a mobile gas phase and a stationary liquid or solid phase – has become one of the most frequently used methods of qualitative and quantitative analysis of hydrocarbon mixtures during the last two decades. It may be said that progress in the analysis of complex hydrocarbon mixtures, from gases up to heavy fractions, has been made possible mainly owing to the development of gas chromatography (see Section 4.1).

In *analytical* gas chromatography components are separated from milligram and smaller samples, and then recorded with a detector and recorder; thus information on the relative proportion of the components and on their quality is obtained. *Preparative* gas chromatography that has developed from the same bases and simultaneously with the analytical variant, differs only in the fact that larger samples are injected and the separated components are collected in receivers.

Today preparative gas chromatography is the most efficient laboratory method for the isolation of pure substances from mixtures. It is widely used in the separation of hydrocarbon mixtures because owing to extremely small differences in the physical properties of many components, and to the absence of functional groups, other separation methods very often fail in this case. The main applications of the method in laboratory work are the following:

a) – Purification of small amounts of individual hydrocarbons for the study of their chemical and other properties and for the measurement of physical constants.

b) – Isolation of one or more components from the analysed fraction in an amount necessary for identification by spectral methods.

c) – Isolation of compounds present in the sample in trace amounts.

d) – Separation of narrow-boiling fractions from complex mixtures.

The separation effect in gas chromatography can be achieved using various techniques. The further text is devoted predominantly to the conventional *elution technique* which is commonest for laboratory preparations. Other special techniques are shortly mentioned in the last section.

Preparative gas chromatography can be classified into three different types according to the size of apparatus and its productivity:

a) – *High-capacity preparative gas chromatography.* This is developed

for pilot plant and industrial applications. Pure substances in amounts up to several kg per day can be produced even on columns of 10 cm diameter[1]. Columns of 30 cm diameter can separate up to kilogram samples in one batch[2]; still larger columns are under development[3].
b) – *Laboratory scale preparative gas chromatography.* Samples of several millilitres up to several tens of ml can be separated on columns with larger diameters (I.D. 2.5 – 10 cm). Into columns of 1 – 2 cm diameter samples of several tenths of millilitres volume are usually injected.
c) – *Micropreparation on analytical columns.* They serve for the solution of the most difficult separation problems. The mixture is injected repeatedly in µl amounts into analytical columns of several mm diameter (or less); only mg amounts of pure components are thus obtained.

In the following sections we shall discuss mainly the last two mentioned laboratory variants of preparative gas chromatography.

2.4.1 APPARATUS AND THE FUNCTION OF MAIN PARTS

For preparations on a laboratory scale a large selection of perfect commercial instruments are available[3a]. For occasional isolations of substances in mg to g amounts analytical gas chromatographs complete with accessories for preparative work are quite suitable. If preparations of larger amounts of substances (tens of g and more) are frequently necessary, preparative gas chromatographs with fully automatized cycle (injection-separation-collection) are advantageous.

A scheme of an automatic apparatus is shown in Fig. 2.4.1. A sample from a store flask is injected automatically into the *vaporizer,* according to a preset program. The sample vapours are swept with the carrier gas from the vaporizer into the *column.* The separated components leaving the column pass through a *detector* and are then introduced into a *collecting system,* where they are condensed separately. Individual receivers are opened and closed according to a selected programme on the basis of signals from the detector. The separation of the sample on the column can take place at constant temperature and at a constant flow rate or, if necessary, the change of the column temperature may be programmed, or even the change of the carrier gas velocity. In the last stage of the operation cycle the conditions are again set to the initial values. The

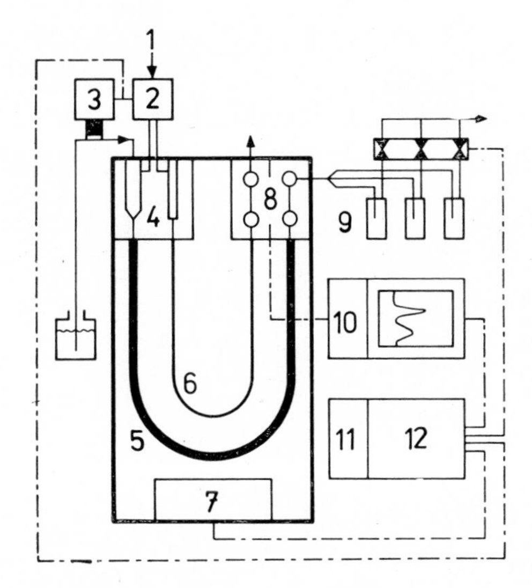

Fig. 2.4.1 Scheme of the apparatus for preparative gas chromatography
1 — carrier gas, 2 — flow programmer, 3 — automatic piston injector, 4 — vaporizer, 5 — preparative column, 6 — compensation column (analytical), 7 — heating of thermostat, 8 — detector, 9 — automatic fraction collector, 10 — recorder with amplifiers, 11 — peak selector, 12 — programming

whole cycle is repeated as often as necessary to obtain a sufficient amount of product.

The function of individual parts of the preparative chromatograph is identical in principle to analytical instruments (see Section 4.1). The differences, if any, are only a consequence of the adjustment of some parts to a larger volume of sample and to automatic operation.

Systems for sample introduction. For automatic instruments precise injection systems for gaseous, liquid and solid samples have been developed.

The simplest is the injection of compressed gaseous samples: using a time controlled magnetic valve the carrier gas flow is interrupted and the sample is introduced into the column. The amount of the sample is regulated by the pressure of the sample and the time of the opening of the sample valve.

A simple automatic system for injection of liquid samples using over-pressure is shown in Fig. 2.4.2. The container with the liquid sample is connected with the vaporizing unit by a metallic capillary tube. At the moment of the injection the three-way magnetic valve closes the carrier gas inlet into the vaporizer and passes the gas into the container. The overpressure thus created pushes a certain amount of the liquid sample onto the evaporating surface. When the injection is over the remaining sample in the capillary is expelled back into the container by the carrier gas which again can flow through the vaporizer into the column. A small back-flow of the carrier gas rinses the capillary constantly and prevents the diffusion of the sample into the vaporizer.

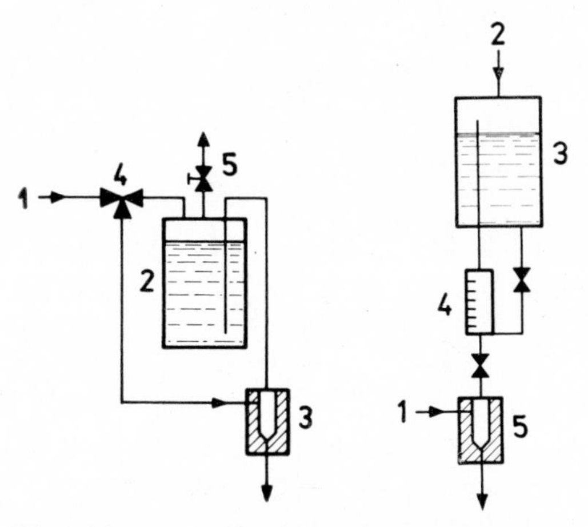

Fig. 2.4.2 Pneumatic injector of liquid samples
1 — carrier gas, 2 — store flask, 3 — vaporizer, 4 — three-way magnetic valve, 5 — needle valve for regulation of purge gas

Fig. 2.4.3 Injector of liquid samples with graduated vessel
1 — carrier gas, 2 — pressure gas, 3 — store flask, 4 — calibrated graduated vessel, 5 — vaporizer

A similar system, with a calibrated vessel, is shown in Fig. 2.4.3. Before injection the sample from the storage flask is filled into the exchangeable measuring vessel from where it is pushed into the vaporizer with an impulse of compressed gas.

A more accurate regulation of the amount of the injected sample is possible with *piston samplers*. A simple compact construction is shown in Fig. 2.4.4. During the injection the piston is pushed by a pulse of compressed air into the head, and thus the sample is injected through the valve and a small diameter stainless steel tube into the vaporizer. By the piston backward movement (under the effect of a spring) the head of the cylinder is filled with a fresh sample through the suction valve. The amount of the sample is set by micrometric regulation of the piston stroke. Various types with different ranges are available: for example, for repeated injections from 0.1 to 12 ml or for the 50 μl to 1 ml range.

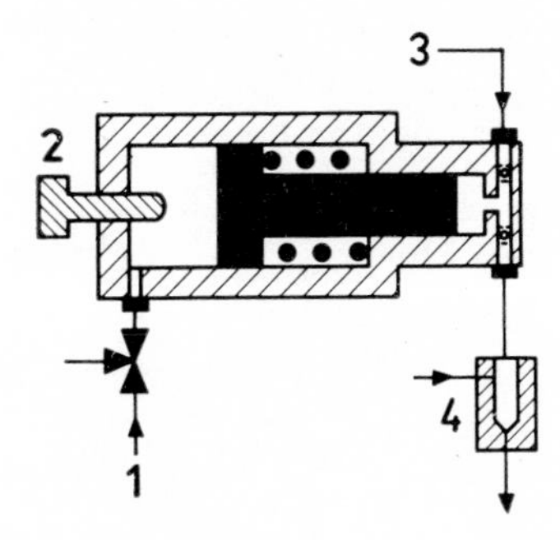

Fig. 2.4.4 Piston sampler
1 — pressure gas, 2 — micrometric regulation of the piston stroke, 3 — sample inlet from store flask, 4 — vaporizer

Another type of automatic liquid injector for volumes from 1 to 40 μl consists of a mobile holder driven by compressed air, in which a *glass syringe* of a suitable capacity is fixed. The holder is installed over the injection port and on an impulse it lowers the syringe so that the needle pierces the rubber septum. By mechanic pressing of the piston for a preset distance the required amount of sample is injected. The syringe then raises again to the initial position. Thus up to 40 samples can be injected automatically from one syringe before it has to be refilled.

Automatic injectors of *solid samples* are constructed so that the samples from a container fall at regulated time intervals through a connecting tube directly into the vaporizer. A well-tried type consists for example of a drum container with 24 holes for the samples. For the preparation of samples the solution of the substance is placed onto suitable supports (glass capillaries, porous metallic particles, etc.) and the solvent is evaporated. When the supports with the deposited dry sample are loaded into the holes, the drum container is closed under an airtight cover mounted over the vaporizer. In the case of sensitive substances the samples can be cooled. At each programmed turning step of the sample-holder drum a new sample falls into the vaporizer where the volatile substance is evaporated immediately into the stream of the carrier gas. The sample supports are collected in a glass insert in the vaporizer; when full, the insert has to be taken out and emptied.

The injection of large liquid samples into a preparative gas chromatograph is connected with the problem of *rapid* and *complete* evaporation before entrance into the column. For separation efficiency it is important that the time of injection, evaporation and inlet into the column should not exceed 10 s. The vapours of the sample entering the column should be diluted with the carrier gas as little as possible.

The vaporizers of preparative instruments are constructed as *through-flow* or *by-pass flow* heated cells, or as long *precolumns*. For samples up to several tens of ml robust throughflow cells filled with granular material of sufficient thermal capacity are suitable. The temperature of the vaporizer should be at least 50 K higher than the boiling point of the highest boiling

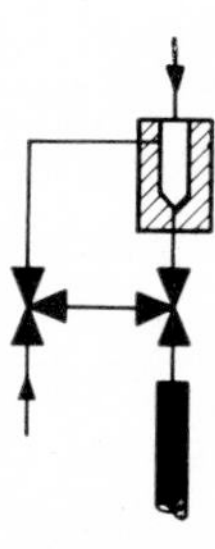

Fig. 2.4.5 Vaporizer with by-pass

component of the sample. Simple calculation shows that after the injection of 1 ml of a hydrocarbon boiling at about 220 °C into the stainless-steel vaporizer of 100 g mass, the temperature of the vaporizer drops by 20 K owing to the heat consumption. In practical cases when the temperature drop is not uniform some parts of the evaporator may be cooled still more. Hydrocarbon fractions which do not contain thermolabile components can be exposed to temperatures of up to 300 – 350 °C. Therefore, in the majority of cases it is possible to operate during the separation of hydrocarbons at a vaporizer temperature 100 K above the boiling point of the sample.

Preparative columns of large dimensions require vaporizers of special construction. Vaporizers with a carrier gas flow which can be switched over[4] (Fig. 2.4.5) are constructed for up to 500 ml of liquid sample. During the evaporation the cell is closed and the carrier gas flows into the column through a by-pass. By switching over the valves the sample vapours are then rapidly pushed into the column and the cell is closed again. The sample entering the column has an almost rectangular concentration profile, especially because a well-timed closure of the cell cuts off the tailing of the last residues of the sample.

In the case of columns of 7.5 – 10 cm diameter a by-pass-simultaneous-vaporizer proved best for injection of smaller samples, while for injection of larger samples an evaporator in the form of a long, narrow, empty precolumn is preferred[5].

Details on various systems of sample introduction for preparative instruments and on general problems connected with this may be found in the extensive monograph edited by Zlatkis and Pretorius[6] or in the book by Sakodynskii and Volkov[7].

Preparative columns. To make a column capable of separating large samples, its volume must be increased. The increase of the capacity can be achieved by increasing the diameter or the length of the column, or by connecting several columns in parallel. Usually the first way is chosen for preparative columns, i.e. columns of large diameter are used. For a first estimation the rule can be applied, that the column capacity increases with the square of the diameter; for example, in comparison with a column of I.D. 6 mm a preparative column of 1 cm diameter can be loaded with a three times larger sample (at equal time of separation) and a column of 3 cm diameter can deal with an up to 25 times larger sample.

Column efficiency in gas chromatography is mostly expressed by the *number of theoretical plates* (n). This value represents the relative broadening of the band of a component after its passage through the column; the higher the number of theoretical plates, the sharper is the band of each component and the better the separation. The number of theoretical plates is calculated from the retention and the peak width of some suitable component, using the equation

$$n = 16\left(\frac{t_R}{w}\right)^2 \tag{2.4.1}$$

where t_R is the distance of the peak maximum from the injection point and w is the peak width at its base; both values should be measured in equal units.

A correct value of the number of theoretical plates can be obtained only if the amount of the component injected for the sake of measurement is sufficiently small. In a recent paper by Albrecht and Verzele[8] it was shown that the earlier recommended extrapolation of the n value to a zero sample is not substantiated, because each column displays a maximum number of theoretical plates only at a certain *optimum size* of the sample (see Table 2.4.1). Further diminution of the sample size again worsens the separation efficiency of the column even though this deterioration is distinctly less pronounced than when the column is overloaded with a large sample.

TABLE 2.4.1

Optimum size of the sample[a] for columns 1 m long of various diameters

Column I.D. (cm)	Optimum sample size
0.2	0.1 μl
0.5	1.0 μl
1.0	10.0 μl
2.5	0.1 ml
7.5	1.0 ml
10.0	2.0 ml

[a] The listed values are only approximate; they can be changed several times in dependence on the retention time of the substance and the chromatographic system (compare Table 2.4.7).

Columns of various diameters are usually compared on the basis of their *height equivalent to a theoretical plate*, $H = L/n$, where L is the column length. The smaller the height of a theoretical plate H, the better the separation efficiency of the column. Modern efficient columns have H between 0.4 to 0.7 mm.

Columns of large diameter should theoretically have the same separation efficiency as equally long narrow columns. However, in earlier studies it was often observed that if the diameter of packed columns increased above 10 mm, a rapid drop of separation efficiency took place[9]. In addition to the common tendency to *overload* preparative columns, the main cause of this phenomenon is the *non-uniformity* of the packing. According to the method of packing and the differences in the size of the particles the bed in a broad column is more or less distributed, so that coarser particles prevail near the column walls. The nonuniform flow-rate of carrier gas in such columns leads to zone spreading and a poorer separation. Albrecht and Verzele[10] determined these irregularities in the flow profile on a broad column experimentally; as is evident from Fig. 2.4.6a the zone of the sample is already distorted on entering the column. At the exit from the column, packed by current tapping, the profile of the flow is strongly convex with visible central retardation.

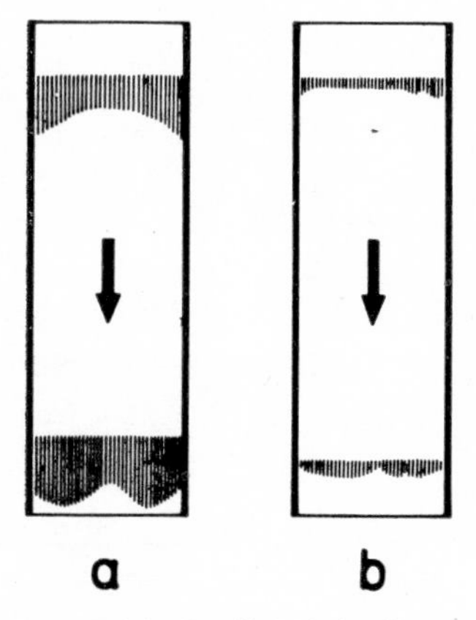

Fig. 2.4.6 Profile of the flow at the inlet and outlet of the chromatographic column of 8 cm I.D.
a — column packed by beating and vacuum suction, b — column packed by "shake and turn" method (according to Albrecht and Verzele[10])

On the basis of the recognition of the interfering factors[9,10] several methods of preparation of large diameter preparative columns have been found, the separation efficiency of which is not very different from that of analytical columns. We shall not mention details on columns with a *special geometry*[7,12,13] and on columns with inserted *flow homogenizers*[2,4,5,14,15], that have been developed for commercial instruments and industrial purposes. Mention will be only made of the principles that have to be adhered to during the packing of broader columns in the laboratory:

a) – Better results can be achieved with U-shaped columns than with spiral columns. In the case of columns with large diameter, straight tubes with two cones are preferable; the inlet cone is filled with unwetted glass beads, while the outlet cone is filled with the column packing[5].

b) – The support for the stationary phase must possess excellent mechanical strength and as much uniform particle size as possible. Most evident are special commercial supports for preparative chromatography: Chromosorb A and Anaprep. The commonly employed Chromosorb W is very inert, but it is quite friable. Chromosorb P (pink) is firm and has maximum surface area (it stands up to 30 % coating; it is insufficiently inert towards polar compounds, but it is suitable for the separation of hydrocarbons, and it gives packings with the highest efficiency. Chromosorb G is inert and has an increased strength, but it can hold only 5 % of stationary phase. The properties of individual types of supports are discussed in greater detail in Section 4.1. For short and broad columns the 60 – 80 mesh size (177 — 250 μm) is suitable, for longer columns 45 — 60 mesh (250 – 354 μm) and for very long preparative columns (above 10 m) with about 1 cm diameter where the pressure drop would be too high with smaller particles, the particles 10 – 30 mesh (595 – 2 000 μm) size is advantageous. The amount of stationary phase on the support is usually between 10 and 25 %. In large size columns the stationary phase can be deposited by *frontal technique* directly in the bed[4]; the separation efficiency of the packing is equal to when the support is coated by the conventional technique.

c) – A good filling of columns of 2.5 – 5 cm diameter may be achieved by pouring the packing material slowly into the centre of the column through a 3 mm diameter tube fixed in the column axis so, that the tube end is always closely above the surface of the packing. Only when the column is completely filled the walls are tapped.

For the packing of large diameter columns (above 5 cm) various *vibrational* and *tapping* techniques have been elaborated (cf. for example references in papers[10,17]) as well as some other procedures. The *fluidifica-*

tion technique[18] is very simple: the column is filled and the packing is fluidized for a few minutes with a strong stream of dry nitrogen introduced into the column through the bottom conical adapter. When the fluidization of the packing is steady the nitrogen stream is gradually weakened, until a firm bed is formed. The packing is thus freed of dust and it is very uniformly distributed, so that the columns have at first a good and reproducible efficiency. The disadvantage of this type of packing consists in the fact that the packing material does not fill the column space perfectly and on longer operation the packing gradually settles and thus the column

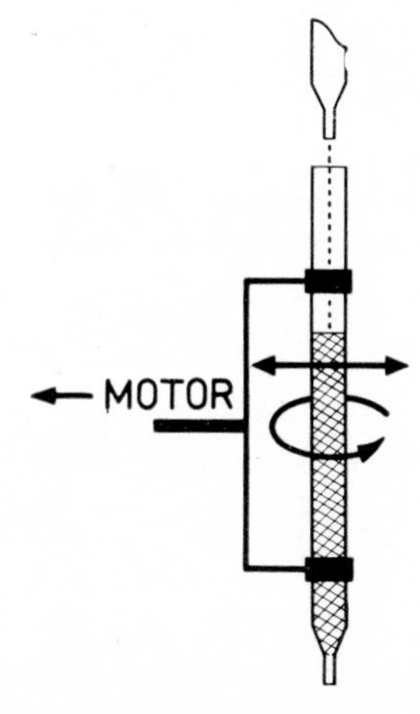

Fig. 2.4.7 Column packing by the STP-method
The pressure is applied discontinuously

efficiency decreases[10,17]. In order to achieve an equilibrium settling of the packing the friction of the particles must be overcome mechanically. Radial shaking of the column with simultaneous rotation of the column around its axis is quite efficient. In this "*Shake, Turn and Pressure*" method (STP-method) the column is shaken mechanically at a 240 strokes/min frequency and simultaneously slowly rotated[10] (Fig. 2.4.7). The pressure is applied discontinuously; after each litre of packing a pressure of 100 kPa is applied on the column for 1 min. A column of 8 cm diameter, packed by the STP-method, can deal with up to 870 plates per one meter length. As is evident from Fig. 2.4.6b the profile of the flow in the column is very flat, showing a high degree of homogeneity of the packing.

The separation efficiency of broader columns packed by various techniques is compared in Table 2.4.2. The highest known separation efficiencies in preparative columns were achieved by Resse and Grushka[17]; their new method of packing consists in the aspiration of the packing upwards into the column with a vacuum, applied at the upper end of the column. In columns 86 cm long and of 3.3 cm I.D. an H of 0.5 – 0.7 mm can be easily attained; this efficiency is equivalent to that of analytical columns. The packing of column lasts about 20 minutes and it does not require any special experience. The packing technique is described in greater detail in the subsequent instructions:

Packing of preparative columns according to Reese and Grushka[17]

A glass column, conical at both ends, is degreased, washed with acetone and dried. It is then fixed in a vertical position and the inner surface is wetted with a 1 % tetradecane solution in chloroform. The upper end of the column is closed with a stopper and chloroform is allowed to dry off slowly, so that a uniform layer of tetradecane is formed on the column walls. A cotton-wool plug is then introduced and fixed into the upper end of the column by means of a wooden stick and a vacuum is connected through the cork adapter as shown in Fig. 2.4.8. The cork adapter permits the turning over of the column and stroking it vertically on the floor during the application of the vacuum.

The packing is wetted with 10 % w/v of methanol (to prevent the effects of static electricity) and about 30 % of the packing is sucked into the column through its bottom end by a vacuum applied at the upper end. The packing is then homogenized by fluidization brought about by intermittent opening and closure of the column at its bottom end.

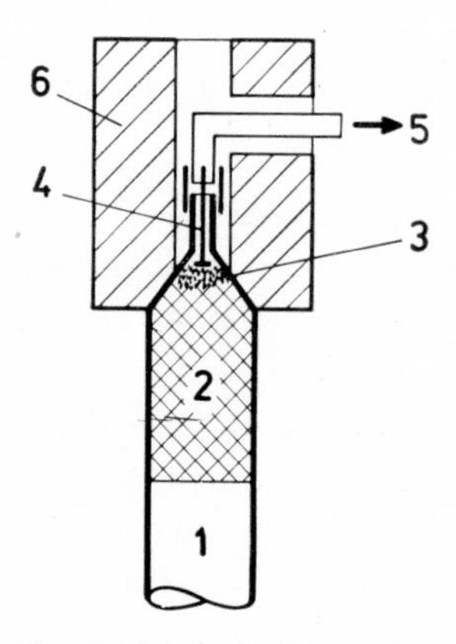

Fig. 2.4.8 Cork adapter for packing the column under vacuum
1 — column, 2 — packing, 3 — cotton wool, 4 — applicator stick, 5 — vacuum, 6 — cork ring adapter

The column is then evacuated, with the packing lying in its bottom part. Rapid opening of the column bottom end causes the packing to shift into its upper part. Careful short closures and openings of the column causes the packing in the upper part of the column to pound thoroughly and become denser. Keeping the vacuum uninterrupted the unstoppered column is taken out of the stand, then turned upside down and the packing is settled by 10 strokes on the floor (over the cork adapter). The column is then fixed in the original position and the procedure is repeated with a further 30 % of the packing. When the second portion of the packing material is consolidated by striking the floor with the column the procedure is repeated with a further 20 %, 10 % and 10 % of the packing material.

A column 86 cm long and 3.3 cm I.D. packed with acid-washed silanized Chromosorb W coated with 20 % SE-30 and tested at 170 °C and a linear velocity of nitrogen of 3.73 cm/s displays a height equivalent to a theoretical plate $H = 0.52$ cm for octane ($k' = 1.47$); for undecane ($k' = 5.7$) $H = 0.50$ mm.

Detectors. In preparative gas chromatography *thermal-conductivity detectors* with hot wires are most commonly used. This type of detector is universal and undestructive. In view of large injections and the necessity of a stable zero line during the operational cycles it is necessary to operate at an adequately low detector sensitivity, or to split off the gas stream for the detector. When nitrogen is used as carrier gas some components passing through the *katharometer* can be recorded as *inverse* peaks. In such instances nitrogen should be substituted by helium.

As a second, alternative, detection possibility in preparative instruments *flame ionization* is common. When this detector is used only a small part (0.1 – 1 %) of the carrier gas is branched off from the column outlet and led into the detector, while the main part of it is led directly into the fraction collector. The flame ionization detector (FID) is not sensitive to some uncombustible compounds; its advantage consists in the fact that cheaper nitrogen can be used in all instances as carrier gas.

Systems for collection of fractions. The carrier gas with separated components is distributed manually or automatically into individual traps according to the signals from the detector. Here two main problems arise: 1 – condensation of the components from the gas phase in maximum yield, and 2 – prevention of mutual contamination of the separated components.

For automatic control various rotating systems and systems distributing the effluent by means of pneumatic or magnetic valves are suitable. A *pneumatically* controlled system for collection of six fractions is shown in Fig. 2.4.9. The effluent from the column is distributed by a seven-

TABLE 2.4.2

Comparison of the separation efficiency of large diameter columns packed by various techniques

Packing technique	Reference	Column I.D. (cm)	H (mm)	Packing and conditions
Rocking table[a] (Method of Standing Wave)	Bayer et al.[4]	10	3.0	20 % Dinonyl phthalate on Sterchamol 0.2 to 0.3 mm; $T = 20$ °C; 5.8 l H_2/min; sample: pentane
Fluidization	Guillemin[18]	6	3.2	20 % silicone oil DC 200 on Chromosorb P, 30—40 mesh; $T = 40$ °C; 4.2 l N_2/min; sample: pentane 0.5 ml
		6	1.6	20 % silicone oil DC 200 on Chromosorb P, 400—500 μm, classified by fluidization; $T = 40$ °C; 4.2 l N_2/min; sample: pentane 0.25 ml
STP-Method	Albrecht and Verzele[8,10]	8	1.2	20 % Ucepa Pon (polymer) on Chromosorb W, 45—60 mesh; 6 l N_2/min; sample: diethyl ether 0.5 ml
Pulling the packing upward into the column with vacuum	Reese and Grushka[17]	3.3[b]	0.5	20 % SE-30 on Chromosorb W, 60—80 mesh; $T = 170$ °C, 1.9 l N_2/min; sample: undecane 1 μl

[a] packing of a column 1 m long lasts about 10 h; [b] glass column

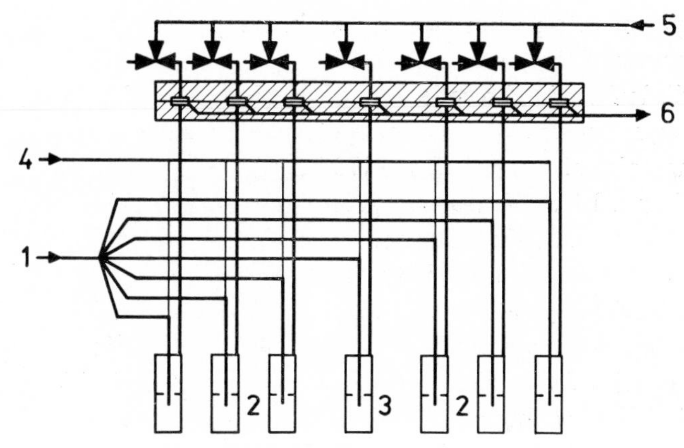

Fig. 2.4.9 Pneumatic fraction collector
1 — effluent from the column, 2 — cooled traps, 3 — discharge, 4 — auxiliary carrier gas, 5 — compressed air, 6 — vent

channel manifold which may be heated at elevated temperature to prevent vapour condensation. The effluent can stream into a cooled collecting cuvette only if the pneumatically controlled valve at the cuvette outlet is open. Except for the central waste channel through which intermediary fractions are drawn off, all the other channels are backpurged by a weak stream of scavenger gas. By this, contamination of the collected fraction with the vapours from other channels is prevented.

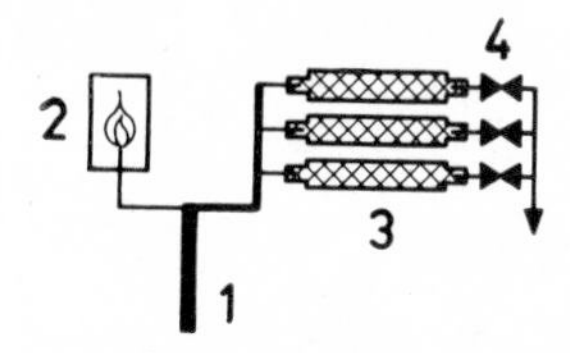

Fig. 2.4.10 Comb microcollector of fractions
1 — outlet end of the chromatographic column, 2 — FID, 3 — receiver tubes, 4 — valves

For collection of merely mg amounts of substances the simple *comb collector* shown in Fig. 2.4.10 is suitable. It is connected in parallel to the flame ionization detector. The fractions are collected into glass tubes filled with an adsorbent or with glass wool only. The distribution of the effluent into single microtraps is controlled as in the preceding case by automatic opening and closure of the outlet.

Considerable difficulties in the collection of fractions arise from components condensing as *aerosol*. This is most apt to happen with compounds the normal boiling point of which is above 170 °C, when their vapours are rapidly cooled. Very efficient *electrostatic precipitators*[19] are seldom used in laboratory instruments, so that the problem is most commonly solved by changing the temperature of the bath or by conducting the effluent over a suitable adsorbent or a material wetted with a solvent. For the collection of large fractions water-cooled precolumns of 0.6×90 cm dimensions packed with 3 mm glass beads proved good, (90 % yield of the substance). For the breaking of aerosols traps of special construction, with a temperature gradient, are also used. Traps suitable for micro-preparations (1 – 100 mg of substance) with a combined longitudinal and radial temperature gradient have been described by Magnusson[20]. A very simple microtechnique for the condensation of low-boiling compounds easily forming aerosols is used by Parliment[21]; the effluent from the distributor is led through a side tube into a flask of 5 – 10 ml volume, provided with a reflux condenser, in which 0.5 – 1 ml of a light solvent is refluxing, as for example Freon 113 (b.p. 47 °C), CCl_4 or $CDCl_3$ (for NMR spectroscopy). The collection by solvent condensation has a 95 % efficiency and in many instances the solution can be used directly for further measurements.

When *total condensation* of the effluent is used the components are obtained practically without losses. However, this technique is used only exceptionally, because the carrier gas used should be argon or some other gas which can be condensed with liquid nitrogen. The carrier gas is then slowly evaporated from the traps at low temperature, while the substance remains in them.

The automatic monitoring of the fraction collector is controlled either by a time-switch, i.e. by means of predefined "windows" for the retention times of the peaks that have to be collected or on the basis of the elution order of the peaks. The second method is more reliable because in preparative chromatography retention times of the components may be affected by the injected volume or by variation of other operational parameters. For a good yield and the purity of fractions the limits of collection are important. In simpler devices the same threshold level of detector signal is set for all selected peaks for the beginning of the collection, and another level determines the end of the collection of fractions. For more complex mixtures *discriminators of peak heights* are more convenient, in which the beginning and the end of the collection can be set for each isolated component separately. A comparison of both methods is illustrated in Fig. 2.4.11.

Further details on columns, detectors and fraction collectors used in preparative chromatography can be found in the special literature cited[6,7].

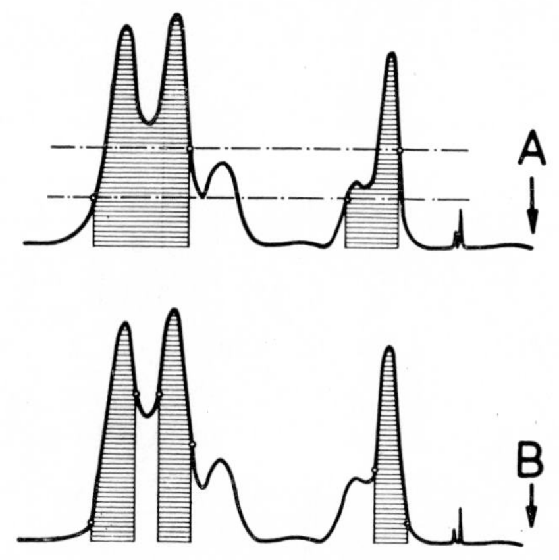

Fig. 2.4.11 Selection of limits of fraction collection
A — collection with two selectable thresholds for opening and closing the receivers,
B — collection controlled by peak selector and peak height discriminator

2.4.2 OPTIMIZATION OF OPERATION CONDITIONS

The results of chromatographic separation are dependent on many factors that must be suitably selected. Among the most important are: the type and particle size of the packing, the type and amount of the stationary phase, column geometry and the method of packing, column temperature, type and velocity of the carrier gas, size of the sample and the method of its evaporation. The effect of these factors on the separation is briefly discussed during the explanation of analytical gas chromatography, in Section 4.1, and a detailed discussion may be found in the references given there. Here we shall limit our discussion only to those relationships which immediately concern preparative work and which play a major role in the optimization of the process.

It is evident that a choice of operational conditions based on intuition is rather unreliable, owing to the opposite effect of various factors. In more complex separation problems the quickest way to find correct conditions is the exploitation of the results of chromatographic theories that quantify the relationships between the most important parameters of the chromatographic process.

Theoretical aspects of preparative chromatography. The separation of a multicomponent mixture can be solved as a problem of the two components separable with most difficulty. The degree of separation of two neighbouring chromatographic zones is expressed as a *resolution* of peaks R:

$$R = \frac{2(t_{R2} - t_{R1})}{w_1 + w_2} \tag{2.4.2}$$

where t_{R1} and t_{R2} are retention times for the first and the second component and w_1 and w_2 are peak widths at the baseline. When $R = 1.5$ both peaks are completely separated (*baseline resolution*). The value $R = 1$ represents a practically satisfactory separation, when the areas of two neighbouring peaks overlap only by about 2 % (Fig. 2.4.12).

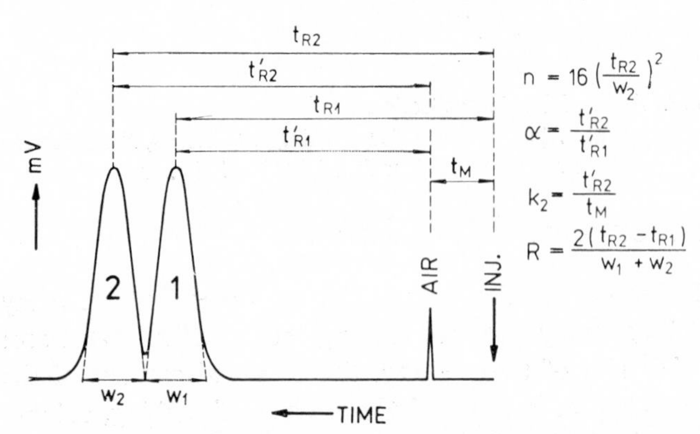

Fig. 2.4.12 Determination of retention data from chromatogram

The attainable peak resolution of the two components depends both on the column's separation efficiency and on the chromatographic behaviour of the components under consideration. The fundamental equation that expressed the interrelations of the mentioned factors has the following form:

$$R = \frac{\sqrt{n}}{4}\left(\frac{\alpha - 1}{\alpha}\right)\left(\frac{k_2'}{1 + k_2'}\right) \quad (2.4.3)$$

where n is the number of theoretical plates (eq. 2.4.1), α is relative retention of both components on the stationary phase used and k_2' is the so-called *capacity factor* of the second component. The usefulness of equation (2.4.3) consists mainly in the fact that all variables can be simply read from a single chromatogram (see Fig. 2.4.12).

Relative retention is defined as the ratio

$$\alpha = \frac{t_{R2} - t_M}{t_{R1} - t_M} = \frac{t_{R2}'}{t_{R1}'} \quad (2.4.4)$$

where t_{R1} and t_{R2} are retention times of components 1 and 2, t_M is the retention time of nonadsorbed gas and t_R' is adjusted retention time.

The capacity factor k' expresses the ratio of the amount of solute in the stationary phase and that in the gas phase; it can be read from the chromatogram for the second component as

$$k' = \frac{t_{R2} - t_M}{t_M} = \frac{t_{R2}'}{t_M} \quad (2.4.5)$$

When considering the limits within which the values of the terms of equation (2.4.3) vary in practical chromatography, the relative importance of individual factors for the resolution can be easily found.

The first term shows that an increase of R by means of an increase in the number of theoretical plates of the column is not economical. The peak resolution varies with the square root of n, so that for a doubling of the R value the column must have a fourfold number of theoretical plates, i.e. practically a fourfold length.

The *relative retention* of the chromatographed pair of components can have the value, let us say of from $\alpha = 1.01$ (a poorly separable pair) up to $\alpha = 2$ (easy separation); this means that the value of the second term on the right-hand of the equation (2.4.3) can change more than fifty times.

The *capacity factor* is a measure of retardation of the component in the chromatographic column. At $k_2' = 0$ the second component (and, of course, the first as well) passes through the column with the same velocity as the carrier gas front, and there is no separation. For a maximum separation of peaks the value of the last term should be as high as possible.

However, an increase of k' increases the resolution effectively only in the region where $k' < 5$; higher k' values just prolong the retention time of the component, but they practically no longer affect resolution. For example, for $k'_2 = 5$ the last term has the value 0.83; at a much higher capacity factor $k'_2 = 20$ the value of the last term increases only to 0.95. As in preparative chromatography such conditions are preferred for the sake of time economy at which the values of k' are between 1 and 5, the last term of eq. (2.4.3) can increase to only about double its value.

From this it follows that the resolution of peaks of a certain component pair can be most easily improved by increasing α. From Table 2.4.3, where the relative retentions of some hydrocarbon pairs are given, it may be seen that the values of α depend mainly on the selected stationary phase. Hydrocarbons with close boiling points, but differing polarity,

TABLE 2.4.3

Relative retentions of some hydrocarbon pairs on squalane and dioctyl phthalate at 25 °C[23]

Hydrocarbon pair	Boiling point °C	Relative retention α at 25 °C	
		Squalane	Dinonyl phthalate
Cyclohexane 2,4-Dimethylpentane	80.7 80.5	1.29	1.55
2,4-Dimethyl-1-pentene 2,4-Dimethylpentane	81.6 80.5	1.17	1.40
Cyclohexane Benzene	80.7 80.1	1.01	1.74[a]
trans-3-Hexene Hexane	67.1 68.7	1.06	1.17
trans-2-Heptene Heptane	98.0 98.4	1.07	1.21
3-Methyl-*trans*-2-pentene Methylcyclopentane	70.5 71.8	1.04	1.07
Cyclohexane Methylcyclopentane	80.7 71.8	1.46	1.41

[a] Reversed order of elution

have generally higher α-values on a polar stationary phase than on a non-polar one. Changes in α achieved by a change in temperature are less pronounced (Table 2.4.4). At a higher temperature cycloalkanes are somewhat more retarded than alkanes. The relative retention of hydrocarbons of the same type usually slowly decreases at higher temperatures. So, for the choice of the conditions for chromatographic separation it is best to look first for a stationary phase with a maximum selectivity for a given mixture.

TABLE 2.4.4

Change of relative retention (α) of some pairs of hydrocarbons with temperature[24] (packed column with 20% Dow Corning Silicone grease)

Pair of hydrocarbons	Boiling point (°C)	Relative retention α		
		70 °C	110 °C	150 °C
2,2-Dimethylhexane Heptane	106.8 98.4	1.20	1.17	1.15
Methylcyclohexane Heptane	100.9 98.4	1.22	1.26	1.28
Methylcyclohexane 2,2-Dimethylhexane	100.9 106.8	1.02	1.08	1.11
Nonane Ethylcyclohexane	150.8 131.8	1.09[a]	1.06	1.03

[a] at 95 °C

Even though nowadays more than 200 different stationary phases are advertized for analytical gas chromatography, the choice is limited to only several of the commonest liquids for large size preparative columns. In Table 2.4.5 some suitable phases with various degrees of polarity are listed. For chromatographic preparations phases with maximum *temperature limit* are preferred, so that the contamination of the separated substance with the evaporated stationary phase or its degradation products remains as low as possible. In view of the relatively large amount of the packing material in preparative columns, cheaper types of liquids are usually selected.

TABLE 2.4.5

Some stationary phases suitable for preparative gas chromatography

Name	Chemical composition	Temperature limit for preparative work (°C)	Average polarity expressed as distance from squalane[a]
APIEZON L	Mixture of hydrocarbons	250	67
SE-30	Dimethyl siloxane	300	100
DC-710	Phenyl methyl siloxane	250	377
QF-1	Trifluoropropyl methyl siloxane	200	709
CARBOWAX 20 M	Polyethylene glycol	200	1 052
DEGS	Diethylene glycol succinate	150	1 612

[a] On the basis of McReynolds's data[25] processed by the nearest neighbour techniqe of Leary et al.[26]. In the scale the polarity of squalane is expressed by zero and the most polar phase 1,2,3-tris (2-cyanoethoxy)propane has a distance from squalane equal to 1 885.

When the problem of the most suitable stationary phase for a preparative column is solved and the value of α for the pair of components to be separated is determined on an analytical column, the calculation of the necessary *separation efficiency* of the preparative column can be made.

The number of theoretical plates necessary for a required separation (n_{req}) can be determined using eq. (2.4.3) modified to the form:

$$n_{req} = 16R^2 \left(\frac{\alpha}{\alpha - 1} \right)^2 \left(\frac{1 + k'_2}{k'_2} \right)^2 \tag{2.4.6}$$

If the peak resolution $R = 1$ is considered sufficient, and that the value of k'_2 will be about 4 for the majority of separations, then equation (2.4.6) becomes simplified to the form:

$$n_{req} = 25 \left(\frac{\alpha}{\alpha - 1} \right)^2 \tag{2.4.7}$$

Hence, by means of the value of α determined on an analytical column the efficiency of a preparative column can be calculated using the simplified equation (2.4.7), that is necessary for a complete separation of components. For orientation some calculated values are listed in Table 2.4.6. It is evident that for α close to 1 the number of plates increases rapidly. On the other hand this means that for poorly separable mixtures even a slight increase of relative retention is reflected in a substantial decrease of the requirements of the column efficiency (length). From Tables 2.4.3 and 2.4.6 it can be seen, for example, that a mixture of *trans*-2-heptene and *n*-heptane, that requires on *squalane* 5 850 theoretical plates, can be separated on a *dioctyl phthalate* column with a mere 900 theoretical plates.

Table 2.4.6

Number of the theoretical plates (n_{req}) necessary for the separation of components with various relative values of retention α
(calculated from equation (2.4.6))

α	n_{req}	α	n_{req}	α	n_{req}	α	n_{req}
1.01	225 000	1.06	7 800	1.15	1 480	1.40	308
1.02	65 000	1.07	5 850	1.20	900	1.50	225
1.03	29 400	1.08	4 560	1.25	625	1.60	183
1.04	16 900	1.09	3 660	1.30	463	1.80	133
1.05	11 000	1.10	3 030	1.35	380	2.00	100

For practical preparative work the following should be born in mind: a preparative column constructed for a separation efficiency equal to the calculated n_{req} value will separate the mixture under consideration in the expected manner only if the amount of the injected sample is close to the *optimum size* of the sample (see Table 2.4.1). This, of course, is a less frequent case; in the case of easily separable mixtures as a rule the injected samples are much larger than the optimum of the given column, which leads to an *overloading* of the column and a drop in its separation efficiency. The phenomena connected with the overloading of chromatographic columns are so typical of preparative work that they must be discussed in greater detail.

Overloading of the column. In the preceding text it was mentioned that the number of theoretical plates of a column decreases if excessively large samples are injected. This phenomenon is caused by the simultaneous effect of several factors that are commonly called column overloading.

One of the main reasons of column overloading is the excessively large *volume* of the sample vapours entering the column. If the sample – either pure or diluted with the carrier gas – occupies too large a zone at the beginning of the column, additional band spreading at the outlet will take place. In this case the column separates the sample with a smaller number of theoretical plates than when a sample of optimum size is injected.

The investigation of the problem of the *permissible sample size* was of great importance for the development of the theory as well as the practical application of gas chromatography. Van Deemter et al.[22] and Purnell[9] derived an equation for the determination of the permissible sample volume:

$$V_F \leqq \frac{V_R}{2\sqrt{n}} \qquad (2.4.8)$$

where V_F is the volume of the vapours of the sample at column temperature that does not worsen the separation efficiency of the column by more than 2 %. V_R is the retention volume of the sample and n is the number of theoretical plates of the column.

Equation (2.4.8) shows that the permissible sample size may be the larger the larger the retention volume of the component, i.e. the larger the volume of the carrier gas necessary for the elution of the component from the column. The second important fact is that the permissible sample size decreases with increasing column efficiency.

In Table 2.4.1 optimum sample sizes were given for columns of various diameters; the values were obtained partly from various experimental data and partly by extrapolation, and therefore they have only approximate validity. From equation (2.4.8) the limit sizes of the samples may be computed more exactly: For example, for a column 2 m long and of 1 cm I.D., having 1 200 theoretical plates, we can calculate the limit volume of the injection for heptane that has a retention volume of 400 ml on a column at 100 °C, in the following manner:

$$V_F = \frac{400}{2\sqrt{1200}} = 5.77 \text{ ml of vapour}$$

As 1 mg of heptane has a vapour volume of about 0.3 ml at 100 °C, 5.77 ml correspond to 19 mg of heptane. An injection of 19 mg of heptane, corresponding to 28 μl of liquid heptane, will not appreciably worsen the separation efficiency of the column. In the same manner limit sample sizes are calculated in Table 2.4.7 for columns with various efficiencies and for various retention volumes.

TABLE 2.4.7

Limit size of sample (in μl of liquid), calculated from the equation $V_F = V_R/2\sqrt{n}$ for heptane and column temperature of 100 °C

Retention volume V_R' in ml	Number of theoretical plates of the column			
	1 000	2 000	4 000	8 000
100	8	6	4	3
200	16	11	8	6
400	31	22	16	11
800	62	44	31	22
1 600	124	88	62	44
3 200	248	176	124	88

A further reason for decreasing column efficiency on overloading is the high *concentration* of the sample in the gas phase. The distribution isotherm, expressing the solute concentration in the stationary phase in dependence on partial pressure of the solute in the gas phase, is usually linear at the beginning, but in the region of high concentrations it deviates

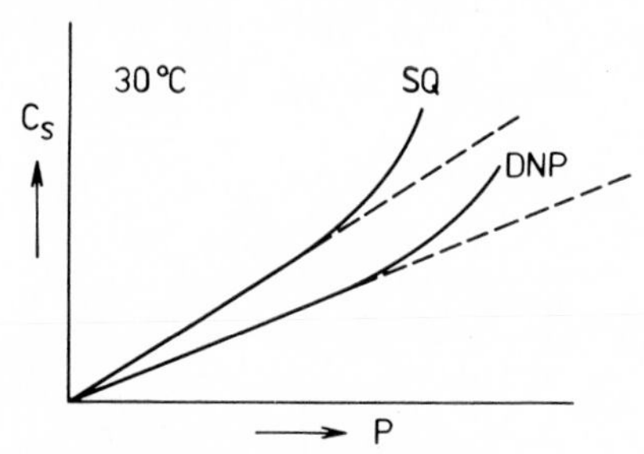

Fig. 2.4.13 Distribution isotherm for dissolution of cyclohexane vapours in squalane (SQ) and dinonyl phthalate (DNP)
C_s — cyclohexane concentration in stationary phase, P — partial pressure of cyclohexane in carrier gas

from linearity as a rule. In Fig. 2.4.13 the isotherms for the dissolution of cyclohexane vapours in squalane and dinonyl phthalate[27] are shown as an example; in both instances the isotherm is curved in the direction of the axis of concentrations in the stationary phase at a high vapour pressure of cyclohexane in carrier gas. In the case of an isotherm of *anti-Langmuir* type (as in Fig. 2.4.13), that is commonest in partition GLC, a part of the zone of the sample of maximum concentration moves through the column more slowly than the peripheral parts, where the concentrations of the solute are lower. On overloading the peak of the component at the column outlet is asymmetrical, with a slow leading edge and a steep trailing edge, as illustrated in Fig. 2.4.14b. The retention time of the peak maximum is prolonged with increasing sample size.

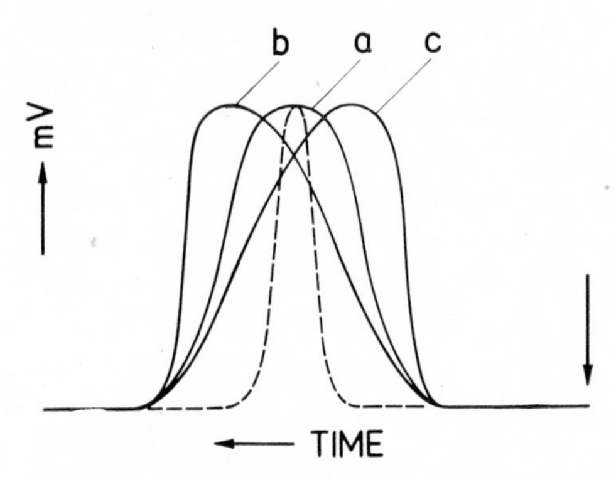

Fig. 2.4.14 Shapes of peaks on column overloading
The peak indicated with a broken line corresponds to the optimum size of the sample
a — peak broadening due to overloading with a large volume of sample vapours,
b — asymmetry of the peak caused by concentration overloading of the column (fronting),
c — asymmetry of the peak caused by enthalpic overloading of the column (tailing)

Both these reasons act together, and when large samples are injected the band broadening can be eliminated only with difficulty. The contradictory character of both phenomena prevents it: the sample concentration can be decreased, for example, by slow injection, during which the sample is diluted with the carrier gas, but on the other hand the volume occupied by the sample in the gas phase increases, which contributes to the band broadening. In the opposite case the rapid injection of the undiluted sample shifts the distribution into the region of the non-linear isotherm from which peak deformation results, accompanied by a decrease in efficiency.

In preparative columns of large diameter in addition to the above-mentioned effects *enthalpic overloading* also plays a role. When the sample zone is passing through the column a considerable amount of heat is developed during the dissolution of the solute in the stationary phase. In broad columns the heat of dissolution cannot be abducted (the column packing is a bad conductor) and it increases the temperature of the stationary phase locally. The region of maximum concentration in the sample zone is thus chromatographed at a higher temperature than peripheral regions; this results in the deformation of the peak, characterized by a steep leading edge and an extended trailing edge; increasing of the sample size leads in this case to a shortening of the retention time (Fig. 2.4.14c).

Enthalpic overloading may cause a distinct change in relative retention of a pair of components, eluted one after the other. In the frontal half of the zone of the first component the heat of dissolution is released, a part of which is abducted and the other part increases the temperature of the stationary phase. In the rear, desorption part of the zone the same amount of heat is again consumed. In view of the non-adiabatic conditions the temperature of the stationary phase is decreased after the passage of the desorption part of the zone. If the second component is eluted immediately after the desorption zone of the first component when the column is enthalpically overloaded, it is actually chromatographed at a lower temperature and the difference between the retention times increases.

Another reason for the peak deformation at overloading may be the *condensation* of the sample at the beginning of the column. This occurs if the sample concentration of the gas phase is high and the temperature is below the saturation point of the vapours.

In addition to these factors the separation efficiency of a preparative column can also deteriorate under the effect of a radial temperature gradient in the column, under the effect of the changes in activity coefficients at high sample concentration, and under further effects that are difficult to explain completely.

Only some of the phenomena accompanying overloading can be

restricted to a certain extent; for example, the consequences of enthalpic overloading are not so striking at a low flow rate of the carrier gas, when the sample dissolution approaches isothermal conditions; however, a high carrier gas flow rate is usually selected, in order to shorten the time of the cycle as much as possible. The condensation of vapours at the beginning

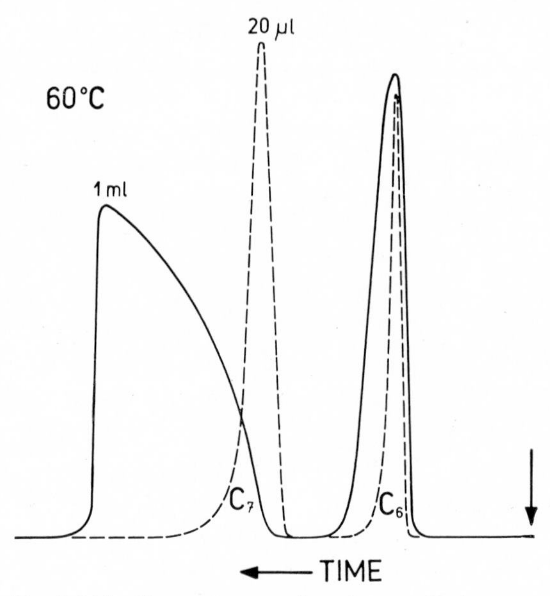

Fig. 2.4.15 Chromatogram of a mixture of *n*-hexane and *n*-heptane on a preparative column of 1 cm I.D., packed with Chromosorb W (30—60 mesh) with 20% of Apiezon L, Helium 60 ml/min, column temperature 60 °C, broken line—injection of 20 µl of mixture, full line—injection of 1 ml of mixture
The heptane peak displays fronting and a prolongation of retention time, while relative retention increased

of the column can be prevented by increasing the temperature of the column. In order to minimize the effect of overloading it is recommended to inject the sample rapidly, so that the zone occupied by the vapours should be as short as possible, while the column temperature should be slightly above the boiling point of the sample. Even correctly selected conditions can, however, lead to partial improvement only, and generally when large samples are injected into a preparative column, a decrease of its efficiency should be reckoned with.

Peak deformation is a regular phenomenon in preparative work. If the components are well separated the asymmetrical shapes of the peaks do not impair the collection of pure components. Changes in the shape of peaks in dependence on the increase in sample size and column temperature are shown in Fig. 2.4.15 and 2.4.16 for a mixture of *n*-hexane and *n*-heptane on a preparative column with Apiezon L as stationary phase.

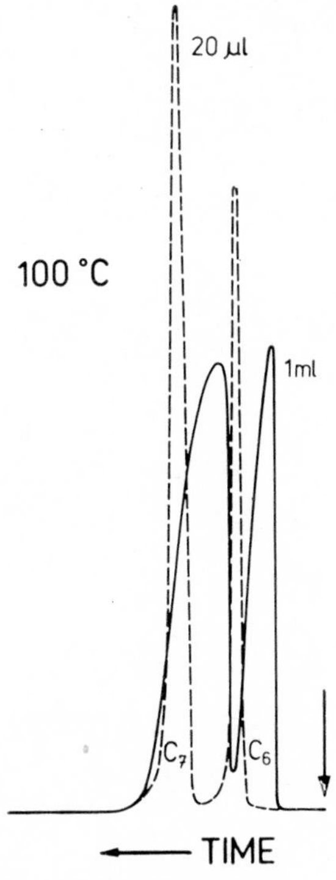

Fig. 2.4.16 Chromatogram of a mixture of *n*-hexane and *n*-heptane obtained on a preparative column (described in Fig. 2.4.15) at 100 °C
broken line—injection of 20 µl of sample, full line—injection of 1 ml of mixture
Both peaks display tailing and shortening of retention time, relative retention decreased a little

Preparations on an overloaded column are undertaken in those frequent cases when the mixture is easily separable and the main problem is the isolation of a substance on a larger scale. For this purpose it is necessary to determine the dependence of the number of plates on the sample size experimentally for the preparative columns available, using suitable model components. For two columns of different diameters the course of this dependence is shown in Fig. 2.4.17. With increasing sample size the apparent number of theoretical plates (n') decreases steeply at first, and later slowly. In the case of broader columns the decrease in efficiency is usually slower than in narrower columns. As evident from the graph, for separations requiring less than 350 theoretical plates a less efficient 10 cm column is more convenient, because it can take a higher load.

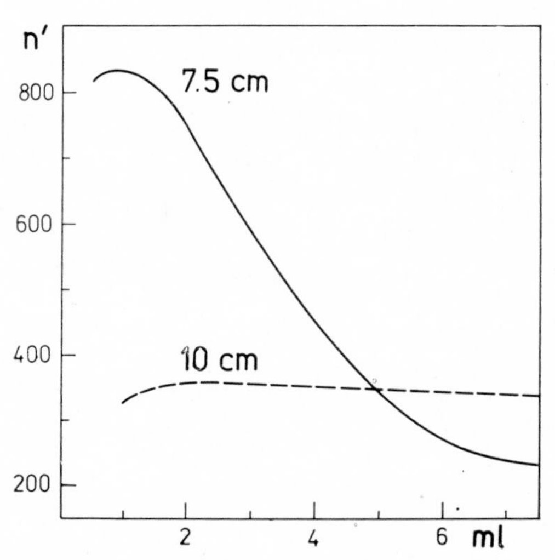

Fig. 2.4.17 Apparent number of theoretical plates (n') in dependence on the size of the injected sample (in ml of liquid) for columns 1 ml long and with I.D. 7.5 cm and 10 cm (according to ref.[4,8])

For the selection of conditions the following course is taken: the necessary number of plates is calculated on the basis of the relative retention α of the separated components using equation (2.4.7). The size of the sample which can be injected without impairing the necessary separation efficiency of the overloaded column (i.e. when $n' = n_{req}$) is read from the corresponding graph of n' *versus* sample size. The value of sample size obtained in this manner applies, of course, to a single component only. The maximum permissible sample size for the whole mixture is then calculated by multiplying the value read from the graph by 100 and dividing by the percentual content of the required component in the mixture.

Recommendation for the choice of working conditions. During laboratory preparative chromatography the operational parameters are usually chosen with respect to the maximum *throughput*; only in exceptional cases of especially complex mixtures the aim is to achieve a maximum *separation efficiency*. In this the laboratory technique differs from high-capacity production chromatography where the *minimum cost* of the product is the main criterion[3].

The main principles for the selection of working conditions in laboratory preparations can be summarized under the following points:

a) – *Type and velocity of the carrier gas*. The separation is only slightly affected by the type of the carrier gas; mostly nitrogen, helium and argon are employed. The use of hydrogen is dangerous because even for 1 cm columns flow-rates in the range of 1 litre per minute are necessary. It is indispensable that the carrier gas should not contain even traces of oxygen which could oxidize the stationary liquid. In contrast to analytical columns the carrier gas flow-rate in preparative work is selected higher than optimum; the ensuing low drop in column efficiency is of no importance in comparison with the loss of efficiency caused by column overloading. On the other hand a high flow-rate of the carrier gas is advantageous because it shortens the time of the cycle and permits work at a lower temperature.

b) – *Type and amount of stationary phase*. The type of stationary phase is selected according to the results of the separation experiments on analytical columns. The results can be easily compared because the retention volume on analytical columns and on broad preparative columns with the same phase are in good accord. The main criterion for the selection of phase is maximum relative retention α of the components under separation. The value α depends much more on the type of phase than on temperature. The selected liquid must be completely stable and have a negligible vapour pressure at column temperature. The temperature limits given for analytical columns are decreased for preparative work by about 50 K.

The support of the stationary liquid should be little friable and it should have an inert surface that does not catalyze the thermal decomposition of the stationary phase. In preparative columns a relatively high percentage of impregnation is preferred. It is true that a large amount of liquid on the support increases the retention times, but it permits larger injections of the sample. For a selected stationary liquid the amount of impregnation affects the capacity ratio k' which is further only dependent on the column temperature.

c) – *Column temperature.* At a high column temperature the solubility of the components in the stationary liquid improves (k' increases and the retention times are shortened). At a high temperature the volatility and the decomposition of the stationary phase also increase, thus causing a decrease in the purity of the product. At a low temperature retention times are prolonged and the danger that the sample will be condensed at the beginning of the column increases. In view of these contradictory effects of temperature a convenient compromise should be selected for each case separately. It is best if the column temperature is slightly above the boiling point of the component separated.

2.4.3 TECHNIQUES USED FOR THE SEPARATION OF HYDROCARBONS

In the separation of hydrocarbon fractions by preparative gas chromatography the selection of conditions depends on whether the mixture is well or badly separable and whether the required amount of the separated components is large (in g) or small (in mg).

Small amounts of well separable components can be obtained relatively rapidly on short preparative columns of 1 – 2 cm diameter. If large amounts are required, columns of larger diameters are more suitable, working under the conditions of overloading.

The greatest problem is, of course, to obtain large amounts of poorly separable components; for this purpose either 10 – 20 m long preparative columns with a diameter of about 1 cm[28] are used, or also several metre long columns with 3 cm diameter, packed so that their H equals that of analytical columns[17]. In both cases the separation must be repeated many times with a sample of optimum size, using an automatic arrangement. For the separation of poorly separable components in microamounts efficient analytical columns can also be used[29,30].

Table 2.4.8 gives examples of practical separations of hydrocarbon mixtures on columns of various diameters taken from literature.

For preparative chromatography of complex mixtures various auxiliary

TABLE 2.4.8

Some examples of separations of hydrocarbons on preparative columns of various diameters

Dimensions of columns in cm		Packing	Conditions	Separated mixture	Product	Reference
I.D.	*L*					
0.6	610	20 % SE-30 on Chromosorb W 60/80 mesh	$T = 210$ °C, hydrogen, sample size: 1 ml	Kerosene fraction	6 fractions within the range of 100 index. units	Mitra et al.,[31]
1.1	244	1.8 % OV-17 on Gas-Chrom P 60/80 mesh	$T = 240$ °C, nitrogen 250 ml/min, sample size: 1 mg	Mixture of steroids	Coprostane 5-Cholestene Coprostanol Cholestanol	Vandenheuvel and Kuron[32]
1.8	320	Wideporous silica gel, surface area 50 m^2/g	$T = 70$ °C sample size: 6 ml	Mixture of 3-methyl-pentane and 3-methyl-2-pentene	3-Methylpentane of 99.9 % purity	Sakodynskii and Volkov[7]
6.4	203	20 % Apiezon L on Chromosorb P 10/60 mesh	$T = 30$ °C, helium 2 l/min, sample size: 20 ml	Commercial cyclopentene 97 %	Cyclopentene purity 99.99 %	Pausch[15]
10.1	203	20 % Carbowax 20M on Chromosorb W 30/60 mesh	$T = 125$ °C, helium, 12.5 l/min, sample size: 60 ml	Mixture of α- and β-pinene	β-Pinene purity 98.6 %	Craven[1]

techniques have been developed by which some parameters of the process are improved. For hydrocarbon fractions the techniques of the acceleration of the process, high purity separations and effective microtechniques are especially important.

Shortening of the separation cycle. As shown, in preparative chromatography a larger amount of the separated component can be obtained either to the detriment of the separation efficiency of the column or at the cost of a considerable prolongation of the time of the experiment. Techniques that lead to an acceleration of the separation are therefore of great practical importance; the following are most often employed:

a) – *Narrow boiling sample.* This is the simplest and also one of the most efficient methods of shortening the separation cycle. The injected sample should be adjusted so that the components to be isolated should be present in as high a concentration as possible. Most important of all is the elimination of higher boiling ballasts, because even a small amount of impurities with a high boiling point which are eluted slowly from the column prolong the time of the cycle excessively. If a sufficient amount of sample is available it is best to separate the sample to narrow-boiling fractions on an efficient distillation column first. For this purpose spinning band columns are most convenient; for small samples miniature columns are available on which fractions can be distilled from only several ml batches (see Section 2.1.2). For preparative chromatography only those fractions are selected that contain the required components in maximum concentration. The time spent on the preliminary distillation is usually more than compensated for by the shortening of the time necessary for a complete elution of the sample; the higher the number of cycles in preparative chromatography, the greater is the time economy achieved.

b) – *Fast cycling.* When the separation cycle is repeated, it is not always necessary to wait with the injection of the subsequent sample until the first sample is completely eluted from the column. The highest output is obtained if the next sample is injected before the preceding cycle is finished. In the techniques of fast cycling[33] the point of injection is chosen so that the first separated component is eluted shortly after the last component of the preceding sample. It does not matter if an overlapping of the components which are not of particular interest takes place. Of course, it is important for the purity of the isolated compounds that the high-boiling components should be cut out sharply from the sample. Therefore it is useful to combine the fast cycling with the above-mentioned preadjustment of the sample. Conventional separations with subsequent cycles and the separation with the use of fast cycling are compared in Fig. 2.4.18.

c) – *Back-flushing*. When samples are separated that have a broader boiling range, the main time loss is caused by the fact that even the heaviest components have to pass through the whole length of the column.

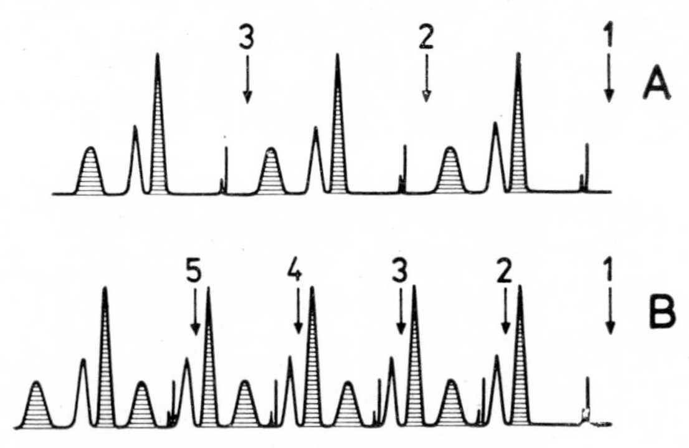

Fig. 2.4.18 Sample injection on consecutive cycles (*A*) and fast cycling (*B*)
Hatched areas indicate the collected components

The advantages ensuing from the preliminary cutting off of the sample by distillation have already been mentioned. The technique of back-flushing has substantially the same goal, but the higher boiling components that are not of interest are separated from the sample directly in the chromatographic apparatus. A short precolumn serves this purpose, in which a rough separation of the sample to a volatile and a less volatile part takes place. The lighter fractions that contain components to be separated are let into the main column, while the heavy ballasts are flushed out from the precolumn by a reversed gas flow.

Schomburg, Kötter and Hack[34] used a precolumn in preparative gas chromatography, in connection with the back-flushing technique according to Deans and co-workers[35], and they mention a shortening of the time of the cycle by more than two thirds. The arrangement of the columns is drawn in Fig. 2.4.19. The carrier gas flow is switched by two pairs of simultaneous switch solenoid valves that in no case come into contact with the sample.

d) – *Programming of the column temperature and of the carrier gas flow.* The advantages of the temperature programming in analytical gas chromatography of broadly boiling fractions are quite evident; by increasing column temperature during the analysis each component pair can be eluted at optimum temperature, so that the separation of the components in the whole boiling range is approximately equally effective.

The use of programmed temperature in preparative gas chromatography is more problematical, however; the broader the columns used, the more easily can the negative consequences of the temperature increase outweigh positive effects.

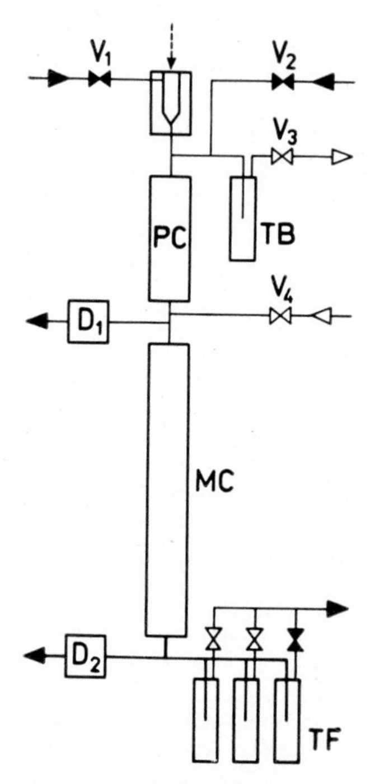

Fig. 2.4.19 Back-flushing
V_1, V_2 — solenoid valves switched over simultaneously; they remain open until the required cut from the precolumn (PC) comes into the main column (MC), V_3, V_4 — the second pair of simultaneously switched over solenoid valves; after their opening the cut of light components is eluted through the main column and the heavy ends are back-flushed into the receiver TB, D_1 — detector for switching over the columns, D_2 — detector for switching over the receivers TF

The advantages of temperature programming include the fact that the components are eluted from the column in the form of narrow zones. The higher concentration of the substance in the zone, and a higher symmetry of the peaks facilitates the collection of pure components. The column can also be loaded with a larger sample. Equation (2.4.8) for the permissible volume of the vapours of the sample (V_F) that can be injected without a deterioration of column efficiency has the following form for programmed temperature[36]:

$$V_F = \frac{V_{R(T_0)}}{2\sqrt{n}} \tag{2.4.9}$$

where $V_{R(T_0)}$ is the isothermal retention volume, which the sample would have at the starting temperature of the programme. Hence, permissible sample volume depends merely on the starting temperature of the column and not on the true retention volume. As $V_R(T_0)$ is practically always higher than V_R under isothermal conditions, larger samples may be tolerated at programmed temperature. The lower the starting temperature of the programme, the larger the sample which may be injected.

The advantages of low-temperature injection can be illustrated by the example of the separation of C_4 and higher hydrocarbons from natural gas: in preparation with programmed temperature[36] up to 500 ml of natural gas would be injected into the column used without causing band spreading of the separated hydrocarbons.

It is characteristic of the movement of the components that during the starting phases of the programme the heavy components move only negligibly and remain at the beginning of the column. With each increase of temperature by 30 K the distance covered by the component in the column is doubled; this means that only during the last 30 K before the elution temperature has the component passed through the whole second half of the column.

Among the advantages of programmed temperature the fact should be mentioned that both the column and the stationary liquid are exposed to high temperatures for a short time only.

On the other hand the negative accompanying effect of temperature programme also should be considered. The time saving achieved by the shortening of the retention times of high-boiling fractions is usually lost during numerous repetitions of the cycles; this is due to the fact that after the termination of each cycle a cooling period for the column and an interval for the re-establishment of the starting conditions have to be included.

Another serious obstacle is the fact that in preparative columns with a diameter higher than 1 – 2 cm a distinct radial temperature gradient is formed in consequence of the poor thermal conductivity of packing, even at a mild increase in temperature. The sample then moves more rapidly near the column walls than in its centre, which of course, leads to zone spreading and a decrease in separation efficiency.

Even when the temperature gradient in broad columns can be lowered to an acceptable value, for example by metallic ribs reaching into the packing and by other special shapes of the column[6,7], it is generally recommended to use the temperature programme only in unavoidable cases, and that mainly in the case of long and narrow columns. An improved separation can be expected only for the lightest components. Samples with a broad boiling range should be preseparated preferably by distillation or applying the technique of back-flushing in the precolumn, rather than by using a temperature programme.

A more detailed discussion of the general problems of temperature programming in preparative chromatography is given for example by Kaiser[37] and Harris and Habgood[38].

A similar effect as in temperature programming is observed when the velocity of the carrier gas is gradually increased: heavy components are eluted from the column more rapidly in the form of symmetrical, narrow peaks. As the carrier gas velocity can be easily regulated by a change in the inlet pressure, the flow programming is relatively simple and it can be additionally connected to any instrument. In contrast to the indirect logarithmic dependence of the retention time on temperature, the retention times are shortened linearly when the carrier gas flow velocity is increased. This means that for the same shortening of the retention time much larger changes of flow-rate are necessary than of temperature changes. On the other hand, the gas flow in the column can be changed rapidly and easily within much broader limits than temperature.

Although the operation is carried out outside the region of optimum linear velocity of the carrier gas during the flow programming (see Section 4.1), the losses in the separation efficiency of the column are surprisingly low. In any case they are compensated for by the possibility of a rapid elution of high-boiling components at a relatively low temperature of the column. The advantages of low-temperature elution is the fact that the relative retention α of the components of the same type usually increases at lower temperatures, and also that the strain put on the stationary liquid is much lower: the rate of volatilization of the stationary liquid increases exponentially with the increase in temperature, but only linearly with the increase in carrier gas flow-rate.

Theoretical aspects and the applications of programmed flow-rate in preparative gas chromatography are discussed in detail in a number of papers[29,37,39–41]. Generally, the programming of the flow-rate has several advantages over temperature programming. For work with repeated cycles the cooling period falls off, and after the end of the cycle the original flow conditions are reestablished in a very short time. The best prospect seems to be the combination of the flow programming with the switching column and back-flushing[42].

High purity separations. If preparative gas chromatography is led so that a maximum amount of product is obtained in a very short time, the purity of the isolated components only rarely exceeds 90–95 %. For many purposes this suffices. In other instances, however, a much higher purity is required. The preparation of pure individual compounds necessary for the measurement of some chemical or physical properties or for the identification of unknown components by spectral methods are typical cases when the separation conditions are adjusted rather with respect to the purity of the product than to the rapidity of the cycle or yield.

The purity of the isolated components is limited both by the properties of the mixture, and by the properties of the chromatographic system used. The main sources of impurities are the close proximity of neighbouring components, i. e. a low relative retention of the components of the mixture, and then the volatility of the stationary phase. An increase in the purity of the product can be achieved, in principle, in three ways: adjustment of the sample before separation, a suitable choice of the chromatographic system and the conditions of separation, and a terminal adjustment of the product. The most common ways used in practice are the following: a) – *Adjustment of sample by distillation.* By preliminary distillation of the sample the concentration of impurities eluted immediately before and after the separated component can be decreased. The product isolated by preparative chromatography (at identical collection range) is then

less contaminated by neighbouring components than when an unprepared sample is chromatographed. The same effect is, of course, obtained when the separation is repeated on the same chromatographic column.
b) – *Improved resolution of components.* The main condition for the high purity separations is a minimum overlapping of the required component by neighbouring ones. The resolution of the peak can be improved either by increasing the selectivity of the stationary phase, or by an increase in the separation efficiency of the column. Both methods are indicated in Fig. 2.4.20. Even though a large number of stationary phases are known,

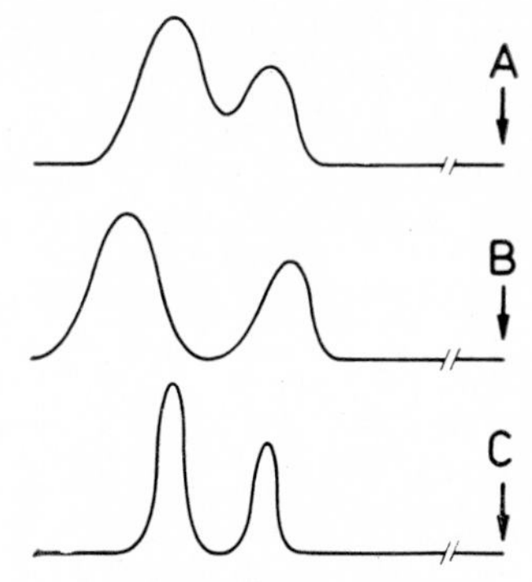

Fig. 2.4.20 Possibilities of improving the peak resolution
A — chromatogram of an incompletely separated binary mixture, B — chromatogram of the same mixture on a stationary phase with a higher selectivity (increased relative retention α of the components), *C* — chromatogram on the same phase as in the case of *A*, but the column has a higher number of theoretical plates n

the use of a liquid with maximum selectivity for the separated mixture is not always feasible, because thermally labile and easily volatile phases must be excluded. However, even a low increase of α has a large effect on the purity of the product; using common preparative columns a purity of the product exceeding 99.9 % can be attained at $\alpha = 1.1$, and at $\alpha = 1.3$ compounds can be isolated the purity of which is higher than 99.99 %[7].

If the most suitable stationary phase is used the resolution can be further improved by increasing the number of theoretical plates of the column. As already said, the efficiency of a given column can be exploited only if a sample of optimum size is injected. A higher number of theoretical plates can then be achieved only if the effect of the packing is improved (a more suitable support, better packing technique) or by prolongation of the column. Even though columns up to 75 m long[43] have been described, work with long columns is not without complications. The high pressure drop of the column and the long time of the cycle are the main disadvantages. Therefore the techniques of column packing, giving a high efficiency (H) are still in the forefront of interest. There is still a large reserve in the quality and the homogeneity of the support particles.

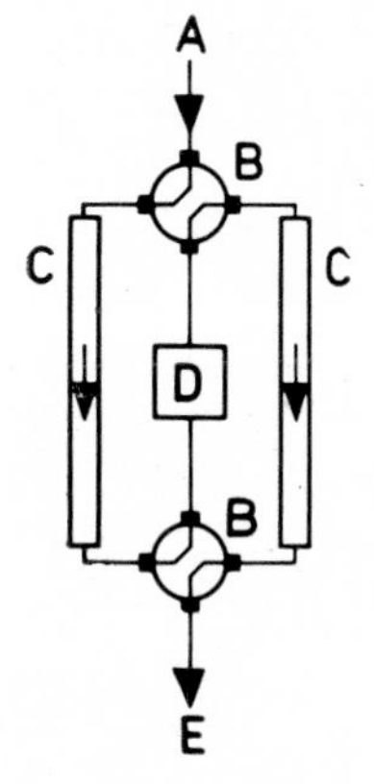

Fig. 2.4.21 Scheme of circulation chromatography
A — carrier gas, B — four-way valves, C — columns, D — detector, E — to fraction collector

Another solution is offered by so-called *circulation chromatography*[7]. The principle is shown in Fig. 2.4.21. Two relatively short chromatographic columns are connected with two synchronous valves to a system in which the separated mixture circulates as long as the required peak resolution is achieved. Both valves are always switched over at the moment when the last component of the mixture has passed the detector and entered the column. Although the efficiency achieved is high and enables, for example, the separation of durene from isodurene and the separation of other couples of isomers with a value of α up to 1.02, circulation chromatography is not yet widely used. In a recent paper[44] circulation chromatography was compared with a long packed column and it was shown that the main disadvantage of the circulation system is its lower productivity.

c) – *Chromatography repeated on different phases.* The most reliable method for the isolation of pure components even from complex hydrocarbon fractions is double chromatography on stationary phases of different polarity. Complex fractions as a rule contain hydrocarbons of various structural classes, that have identical or very close boiling points. In chromatography on a non-polar phase the required hydrocarbon can be cut off sharply from the lower- and higher-boiling components, but the contamination of the product with a hydrocarbon of different structure, having the same retention time on a non-polar phase, cannot be prevented. However, even a small difference in the polarity of the contaminating component and the prepared hydrocarbon, due to their different structure, usually suffices for the separation of both types on repeated chromatography on a polar phase, and for obtaining a pure product. The procedure is illustrated in Fig. 2.4.22 for a hypothetical case of the separation of the alkanic hydrocarbons *P* from a mixture containing three further cycloalkanic hydrocarbons N_1—N_3. It should be noted that a direct separation of the starting mixture on a polar phase would not be success-

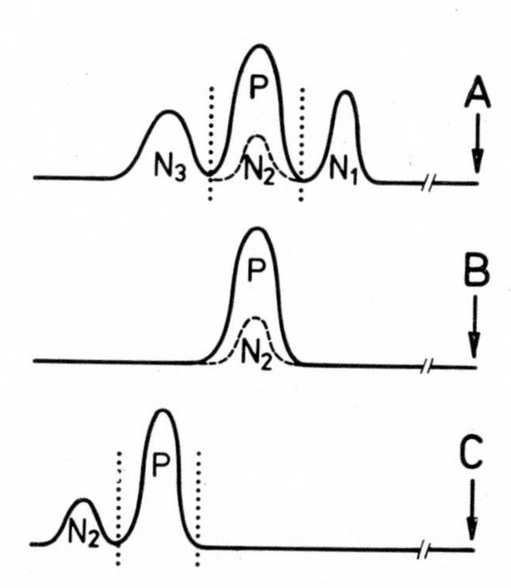

Fig. 2.4.22 Separation on two stationary phases of different polarity
A — chromatogram of a mixture of alkane hydrocarbon *P* and three cycloalkane hydrocarbons N_1 — N_3 on a nonpolar column; dotted lines indicate the limits of fraction collection, B — chromatogram of a cut out fraction (on a non-polar column); the isolated alkane is contaminated with cycloalkane N_2, *C* — separation of fraction *B* on a column with a polar stationary phase; pure alkane *P* is obtained in the collected fraction

ful; in view of the greater retardation of cycloalkanes, the hydrocarbon P would in such a case be eluted together with the cyclic component N_1.

Sequential chromatography on a non-polar and a polar phase can be applied successfully for various types of hydrocarbon fractions. The procedure can be used for example, in the separation of petroleum isoalkane–cycloalkane fractions: the narrow cut prepared on the non-polar column can be separated easily on a polar phase to an isoalkane fraction and a cycloalkane fraction with a longer retention time. In the case of high boiling fractions of this type the cut from the non-polar column can be separated even to several peaks that contain isoalkanes, monocycloalkanes, dicycloalkanes and possibly higher polycycloalkanes. However, the overlapping of peaks increases with increasing boiling point of fraction.

Another example from the field of synthetic mixtures may be the separation of isomeric diisobutenes, which was investigated some time ago in our laboratory[45]. A mixture of olefins formed by oligomerization of the butane–butene fraction contains six main trimethylpentenes of different structures. On a column with Apiezon L 2,3,3-trimethyl-1-pentene is eluted in a common peak with 3,4,4-trimethyl-2-pentene; on a column with 7,8-benzoquinoline as stationary phase this couple is indeed separated, but 2,3,3-trimethyl-1-pentene is again eluted together with another isomer – 2,3,4-trimethyl-1-pentene. For the isolation of pure 2,3,3-trimethyl-1-pentene it is therefore useful to repeat the preparative gas chromatography on two different stationary phases.

The separation on two stationary phases of different polarity was used by Mair and Barnewal[46] in the *API Research Project 6* for the identification of individual aromatic and cycloalkanic hydrocarbons in the fractions of light gas oil. They operated with Apiezon L as the non-polar phase and diethylene glycol succinate as the polar one.

The separation of the kerosene fraction by combination of a non-polar and a polar preparative column has been described recently by Mitra and co-workers[31].

In simpler cases the first chromatography on a non-polar phase can be replaced by rectification on a highly efficient column. Best results are achieved if only narrow-range – almost constantly boiling – fractions are submitted to separation on a polar phase.

d) – *Narrowing of the collected fraction.* Even with an incomplete peak resolution of the required compound from neighbouring components a very pure preparation can be obtained if only a narrow zone around the peak top is collected. This procedure is one of the simplest; the high purity of the preparation is connected, of course, with a steep drop in the productivity of the separation.

e) – *Decrease of contamination of the product with stationary phase.* The volatilization of the stationary phase, or of its decomposition products (*column bleed*) is a serious problem from the point of view of the purity

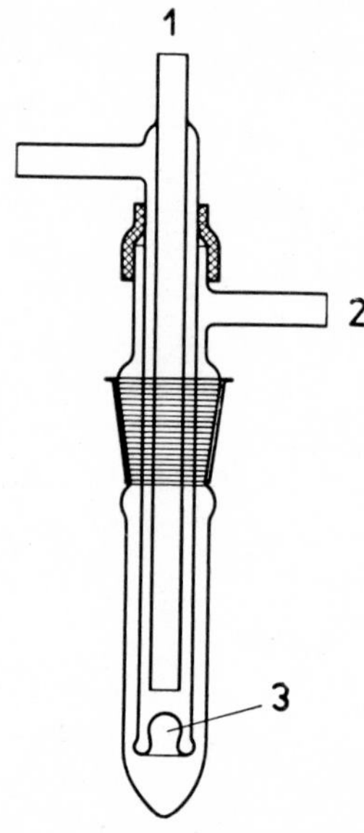

Fig. 2.4.23 Device for microdistillation
1 — cooling water inlet, 2 — connection to vacuum, 3 — cavity of several mm diameter in which the distillate is held by capillary force

of the collected preparations. In preparative chromatography generally columns with a high content of stationary liquid are used, which are operated with high flow-rates of the carrier gas. This means that a considerable amount of the phase can be retained in the traps even at temperatures far below the indicated limit of the phase. The thermal and the chemical stability of the selected stationary phase and the operating temperature of the column are of prime importance for the purity of the product. As at lower temperatures retention times are prolonged the operating temperature is selected by compromise. For high purity separations the requirement of as low a temperature as possible should be especially respected. It should be born in mind that with current control of the purity of the product by analytical gas chromatography the contamination of the product by the stationary phase will not be detected; however, in a spectroscopic analysis of hydrocarbons and during the measurement of other properties the content of the stationary phase can interfere very much.

Mere choice of the separation conditions can seldom prevent the contamination of the collected hydrocarbon with the stationary phase completely. When highly pure substances are prepared it is therefore indispensable to complete the purification of the product *afterwards*. In the majority of cases the isolated hydrocarbon can be freed from the traces of the stationary phase by distillation. Various microdistillation apparatuses are used for this purpose. For example, for the distillation of a few drops of the product the apparatus shown in Fig. 2.4.23 is suitable. Non-polar hydrocarbons containing traces of polar phase can be purified by liquid chromatography on a column of adsorbent.

These problems with the volatility of the stationary liquid are completely absent in *adsorption gas chromatography*. Gas chromatography on adsorbents (GSC) is used in practice mainly for the separation of gas mixtures; however, in principle it is also suitable for preparations in the domain of high purities, just because the columns do not contain any volatile organic material and because they are stable even at elevated temperatures. The columns packed with an adsorbent cannot be loaded as much as in GLC. This disadvantage is partly compensated for by shorter retention times and the possibility of shortening the cycles. Modified *silica gels* and *aluminas* can be considered in the first place as packing material for the columns, then *zeolites* and various forms of *carbon*. Chromatography on adsorbents is suitable mainly for the preparation of light hydrocarbons and other non-polar compounds that show only a low tendency to *tailing*. Highest purities can be achieved in those cases when the prepared compound is eluted before the impurities. Sakodynskii and Volkov[7] describe, for example, the preparation of 99.9 % 3-methylpentane from a mixture containing an olefin, on a column packed with wide-pore silica gel at 70 °C (see Table 2.4.8).

Micropreparation on analytical columns. Preparative gas chromatography on analytical columns is today the most perfect microtechnique for the separation of complex organic mixtures. Samples weighing only several mg or fractions of mg can be separated without difficulty with the efficiency of several thousands of plates (chromatographic), while 10 and more components can be collected separately.

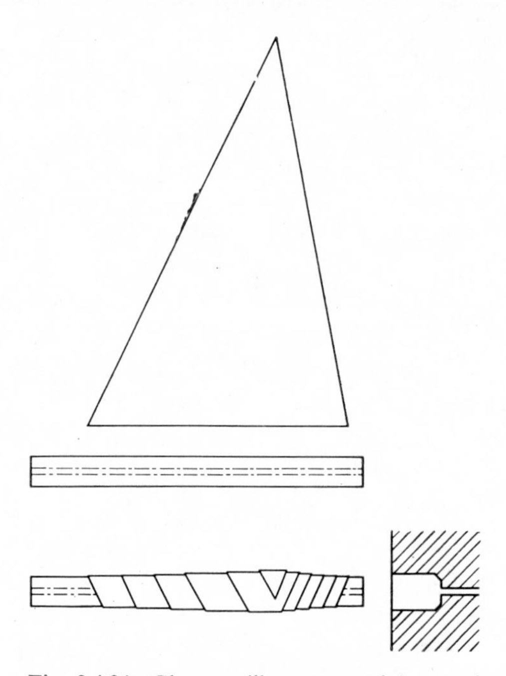

Fig. 2.4.24 Glass capillary wrapped in an aluminum foil, serving as a receiver in micropreparations

In cases when only a few sample injections suffice and when only one or two components are collected, current analytical instruments without automation can be employed. Straight glass capillaries (for m. p. determination) are commonly used as receivers. They are inserted by hand into the outlet opening of the detector at the moment of the elution of the required component. The glass capillary is wrapped with a triangular piece of aluminum foil, as shown in Fig. 2.4.24. The foil wrapping serves both to tighten the capillary in the outlet of the detector and to conduct the heat from the detector along the capillary. The vapours of the component condense in the wrapped capillary in high yield, because the longitudinal temperature gradient restricts the formation of aerosol effectively. The condensed component can be washed out from the capillary with a few drops of solvent.

In preparations on analytical columns the decrease in the separation efficiency with an increasing amount of the injected sample is much steeper than in the case of preparative columns. To preserve the high efficiency the injected sample must be kept at low level: for current packed columns of I.D. 4 – 5 mm, about 0.1 μl per peak. The isolation of pure components in mg amounts therefore requires many tens of cycles, and in such instances, naturally, it is more convenient to operate with instruments with automatic collection of fractions.

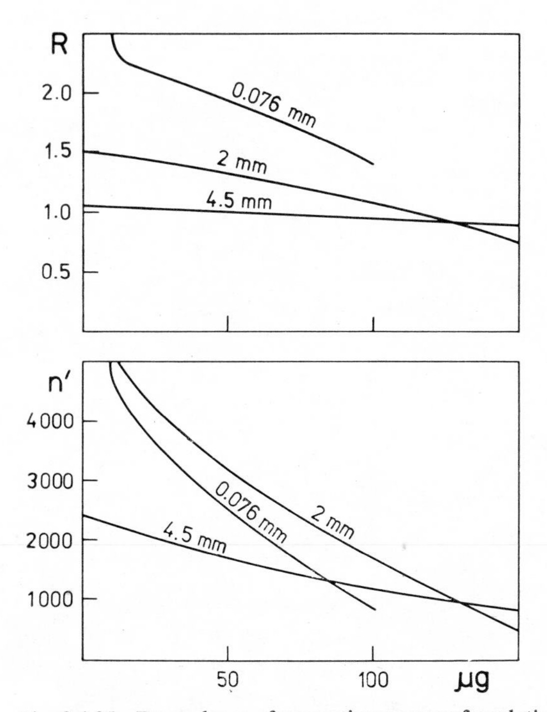

Fig. 2.4.25 Dependence of separation power of analytical columns of various diameters (given in Table 2.4.9) on the sample size (in μg/peak)
R is peak resolution and *n′* is the apparent number of theoretical plates of the column

The advantages of preparative chromatography of hydrocarbon mixtures with analytical columns were already stressed by Scott[29]. When separating a seven-component mixture of gasoline hydrocarbons C_6—C_8 on a 1.5 m long column of 4 mm I.D. packed with Celite with 25 % of squalane, he pointed out the positive effect of the programming of the carrier gas flow-rate on the shape of the separated peaks.

The suitability of the analytical packed and capillary columns for preparative work was also investigated by Gelpi and co-workers[30] in connection with the isolation of steranes and triterpanes from Oil Shale (Green River). The parameters of several analytical columns of various diameters, operating with sample injections of optimum size, are shown in Table 2.4.9. The resolution of the peaks, R, is measured for neighbouring peaks of 5α- and 5β-cholestane. Fig. 2.4.25 shows the effect of the size of the injected sample on the decrease in separation efficiency of columns of various diameters. Whether the separation power is expressed by the number of theoretical plates n or by the peak resolution R,

TABLE 2.4.9

Parameters of some analytical columns tested for micropreparations (Gelpi et al.[30])

Dimensions of columns		Type of column and packing	Optimum velocity of carrier gas (cm/s)	Number of theoretical plates (n)	Resolution[a] (R)
I.D. (mm)	L (m)				
0.076	30	Capillary coated with 7-rings polyphenyl ether	9.8	18 278	4.8
1	3	Packed; 5 % SE-30 on 100/120 Varaport-30	8.1	6 789	1.86
2	3	Packed; 5 % SE-30 on 100/120 Varaport-30	4—16	5 996	1.50
4.5	3	Packed; 3 % SE-30 on 80/100 Aeropack-30	3.5	2 828	1.08

[a] For the separation of 5α- and 5β-cholestane at 240 °C for the capillary column, and at 270 °C for packed columns.

more efficient columns of smaller diameter are generally more sensitive to overloading, and their efficiency decreases more steeply with increasing sample size This property of the chromatographic columns was already pointed out by Van den Heuvel and Kuron[32] who investigated the separation of steroids on analytical columns of 4 and 11 mm I. D. The same tendency is also observed in preparative columns of larger diameters, as was shown in Fig. 2.4.17.

On the basis of preliminary studies Gelpi et al.[30] selected columns 6 m long and 4.5 and 2 mm I. D. for the isolation of the main components of isoalkane-cycloalkane fractions of shale-oil. Applying automatic preparative gas chromatography they isolated mg amounts of the main components from a mixture of branched and cyclic hydrocarbons. Typical operation conditions in the isolation of steranes and triterpanes were the following: Column: 6 m, 4.5 mm I. D., 5 % OV-1 on Supelcoport 60/80, column temperature 270 °C, carrier gas flow 163.5 ml of helium per min. Resolution for the 5α- and 5β-cholestane mixture was $R = 1.7$ at an injection 100 μg per peak. Average volume of the injected sample was 17.4 μl, i.e. about 79 μg per 5α-cholestane. The time of one cycle was 59 min, total number of peaks was 42.

Under the given conditions the main components were isolated in high purity; for example 5α-cholestane and 5α-ergostane were purer than 99 %.

Preparative gas chromatography on *capillary columns* is more problematic. The original excellent separation efficiency of capillary columns decreases steeply if the volume of the injected sample exceeds 10 μg. As is evident from Fig. 2.4.25, in the region of utilizable sample volumes (about 100 μg) the capillary column (I.D. 0.076 mm) displays a lower number of theoretical plates than the tested packed columns. On the other hand even a strongly overloaded capillary column keeps a relatively high *resolution* ability. As shown in the upper part of Fig. 2.4.25 a capillary column gives satisfactory resolution of cholestane isomers, with an R value between 1.8 and 1.3, even when the sample weighs 70–100 μg. Although at such an overload the number of theoretical plates is very small (2 500–850) and the peak shapes are strongly distorted, the distances between the peaks permit a practically complete separation of both isomers. In extreme cases, requiring a high peak resolution, capillary columns can be used for preparative purposes and it is possible to isolate up to mg amounts of preparations by repeated injections. Gelpi et al.[20]

describe a practical example, where the isolation of 1 mg of 5α-cholestane of a purity higher than 99.7 %, from a mixture with 5β-cholestane, requires 14 injections of 70 μg per each peak on a capillary column with polyphenyl ether (mentioned in Table 2.4.9), at 240 °C; total time for the 14 cycles was about 12 h.

For micropreparative work the use of broader packed capillaries (*SCOT-columns* described in Section 4.1) seems to have good prospects. These columns have a relatively high efficiency (≃ 50 000 TP) while standing the injection of the sample directly into the column, without a splitter.

If merely a characterization of several components of the mixture by mass spectrometry is required, it is better to replace the time-consuming micropreparation by direct connection of the column outlet with a mass spectrometer. The GC – MS tandem enables the measurement of the mass spectra of a large number of separated components in one injection without its being necessary to collect them. Analogously the IR spectra of components of a mixture can be measured directly on a GC – IR tandem instrument. These techniques are discussed in greater detail in Section 4.1.

2.4.4 CONTINUOUS GAS CHROMATOGRAPHY AND OTHER TECHNIQUES

A logical way to increase the capacity of chromatographic devices is the continuous arrangement of the process. In spite of the evident advantage of continuous chromatography in principle, continuous instruments described so far are more complex from the constructional point of view and less versatile than chromatographs operating by pulse technique.

As the continuous instruments are developed predominantly for production scale gas chromatography, and not for various laboratory purposes, we shall restrict ourselves to a short characterization of the developmental trends only. More detailed data on production chromatography can be found in the reviews by Conder[3, 47]. Continuous chromatography is treated in special chapters of the monographs by Zlatkis and Pretorius[6] and Sakodynskii and Volkov[7], and in more recent articles by Sussman and Rathore[48] and Barker and Deeble[49].

The equipment for continuous chromatography can be classified according to the arrangement of the movement of the solid and the gas phase into systems with a countercurrent flow, with a transverse flow, or with alternating flow. From the constructional point of view the following systems can be considered as main developmental trends:

a) – *Moving bed.* The sample is fed into the central part of a vertical column in which the packing coated with the liquid phase falls against the ascending flow of an inert gas. The lighter component (or the lighter fraction of the sample) is transported by the gas into the column top, while the heavier component (which has a higher affinity for the stationary phase) is expelled by heat from the packing in the lower stripping section

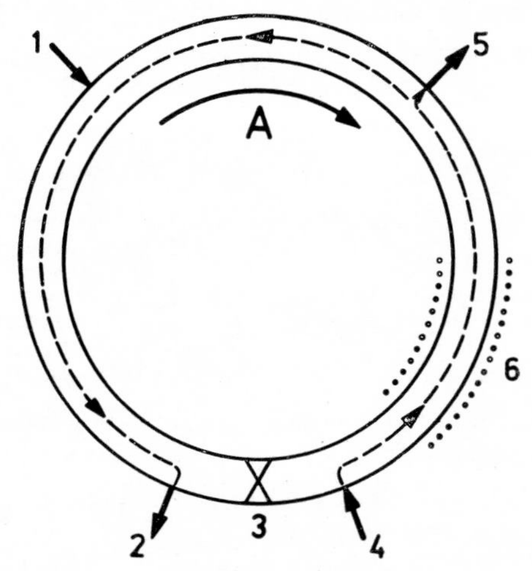

Fig. 2.4.26 Continuous chromatography on a circular column[50]
The column with the packing is rotated (1—10 rph) in the direction of the arrow *A*; the opposite flow of the carrier gas is indicated by a broken line. The valves are stationary: 1 — feed, 2 — light fraction withdrawal, 3 — gas lock, 4 — carrier gas inlet, 5 — heavy fraction withdrawal, 6 — heating of the stripping section

of the column. The stripped packing is again recycled to the top of the column by a gas lift. The arrangement is analogous to the well known process – *Hypersorption*, no longer used today.

b) – *Moving column.* A countercurrent movement of the carrier gas and the packing can be achieved by rotating the column bent into a ring and fixed on the circumference of a slowly rotating disc. The mixture, the carrier gas, and the purge gas are introduced through the stationary valves located on the circumference of the column, and the light and heavy fractions are led off. Several different systems with a circular column have been described. Fig. 2.4.26 shows the scheme developed by Barker[50]. The instrument separates the mixture into two fractions: the heavy one is carried by the packing in the direction of the rotation of the column, and the light one is eluted by the carrier gas in the opposite direction. In view of constructional difficulties with the tightening of the stationary valves on the rotating column, the instruments based on this principle did not pass the experimental stage. Later on Barker improved the construction by dividing the column into several short, straight sections that are placed cylindrically and fastened between two rings. The whole cylindrical bundle of columns rotates slowly; four stationary valves are located on the circumference of the upper ring and the valves forming gas locks are located on the lower ring. For example, on this compact *circular chromatographic machine* highly pure cyclopentane (99.999 %) was prepared from 97 % cyclopentane (containing 2,2-dimethylbutane) at a feed rate of 154 ml/h in almost quantitative yield.

Another apparatus with a moving column has been described by Dinelli et al.[51]. Their multicolumn unit (*Roto-Prep*) is composed of 100 straight parallel tubes (6 mm I.D., 1.2 m long) arranged in rows at the circumference of a vertical cylinder, rotating slowly around its axis. The fronts of the cylinder are stationary. Carrier gas is distributed into the columns through the upper front. The sample is injected into each column that passes under the stationary inlet valve during the rotation. The outlet ends of the columns communicate with the receivers via the bottom stationary front. The movement of the carrier gas and the packing creates a system of transversal flow here: the gas flows through the column downwards (vertically) while the packing moves in a horizontal direction. The result is that each component of the injected mixture moves along its own helical path to the outlet point. The rotation of the cylinder with the columns and the velocity of the carrier gas must be adjusted so that the same component separated on different columns always comes into the same receiver. At a 200 ml/h injection rate the apparatus separates the mixture benzene-cyclohexane or heptane-toluene to components of 99.9 %. If the injection rate is increased to 1000 ml/h, the separated components are still highly pure, i.e. 98.5 – 99.6 %.

c) – *Moving ports.* The developmental work of Barker and co-workers eventually led to the construction of a *sequential chromatographic separator*

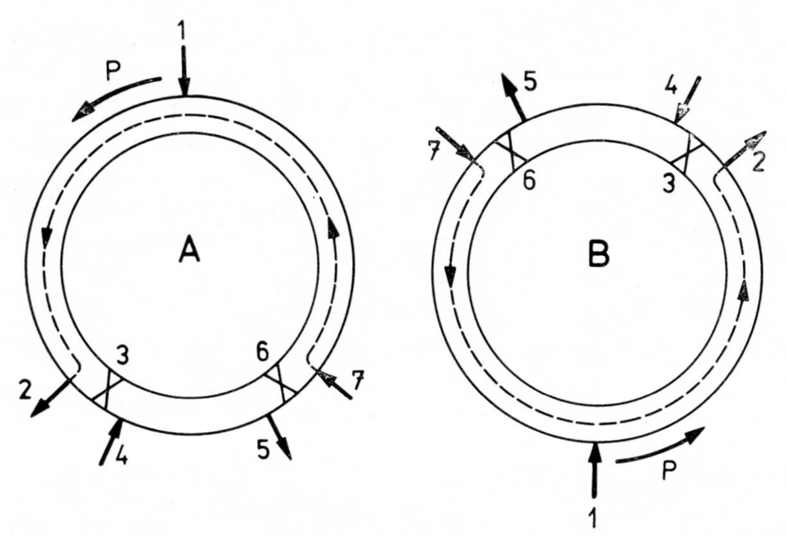

Fig. 2.4.27 Continuous chromatography on sequential chromatographic separator[52] (moving port system)
The column is stationary; the carrier gas flows in the direction of the broken line. Arrow *P* indicates the direction of the rotation of seven valves; 1 — feed, 2 — light fraction withdrawal, 3, 6 — gas locks, 4 — inlet of purge gas, 5 — heavy fraction withdrawal, 7 — inlet of carrier gas, *A* — function of the valves in the initial position, *B* — function of the valves at half of the circulation

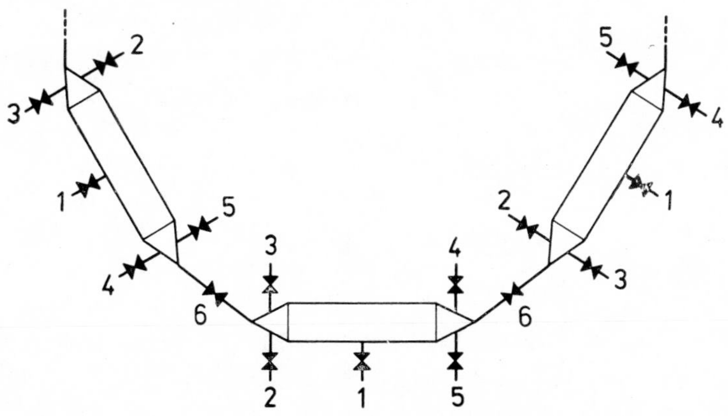

Fig. 2.4.28 Location of solenoid valves on three sections of the circular column of the sequential chromatographic separator[52]
1 — feed valve, 2 — carrier gas inlet, 3 — purge gas inlet, 4 — carrier product outlet (light fraction), 5 — purge product outlet (heavy fraction), 6 — transfer valve

operating continually, without the tightening or the column having to move. The countercurrent movement of the carrier gas and the packing is simulated by gradual switching over of the solenoid valves located at regular distances along the fixed circular column. The scheme of this "moving port" system is shown in Fig. 2.4.27. The experimental prototype described by Barker and Deeble[49, 52] has a column of 7.6 cm diameter. The column is divided into 12 sections, each 61 cm long, connected circularly. Each section is fitted with valves for the introduction of the mixture, inlet and outlet of the carrier gas, inlet and outlet of the purge gas, and gas locks are put between the sections. The location of the valves on the

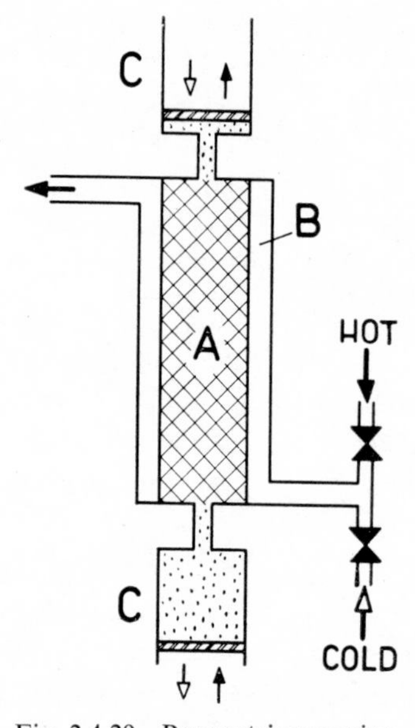

Fig. 2.4.29 Parametric pumping
A — chromatographic column with packing, *B* — heating and cooling jacket, *C* — pumps for the reversal of the fluid flow

sections is shown in Fig. 2.4.28. When in operation the valves of individual sections are gradually switched on and off according to a preset program. The rate of procedure of the switching over of the valves around the column is lower than the rate of the light component carried by the gas through the column packing, but higher than that of the heavy component. In contrast to earlier continuous systems the sequential chromatograph is constructionally simpler because there are no problems with the tightening of the apparatus. At injection of 500 ml/h the binary testing mixtures separate to products of 99.9 % purity.

d) – *Parametric pumping*. Continuous separation is also possible with chromatographic systems in which the direction of the fluid flow is changed *periodically*. Wilhelm et al.[53] have elaborated a technique called Parametric pumping in which the separation takes place on a bed of solid adsor-

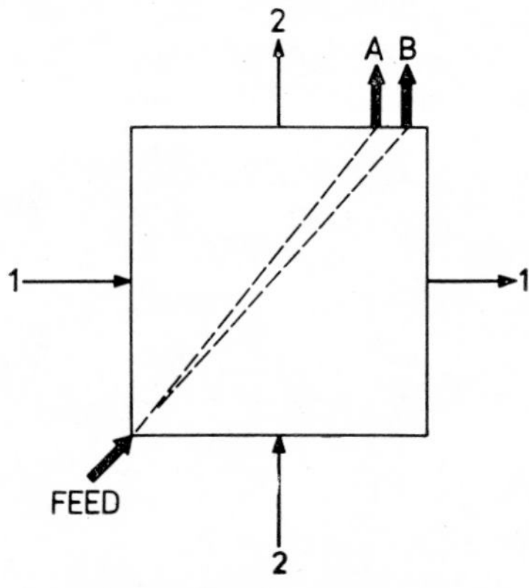

Fig. 2.4.30 Scheme of continuous chromatographic separation on a thin slab of adsorbent with a rectangularly alternating fluid flow[55]
1 — direction of fluid flow at temperature T_1, 2 — direction of fluid flow at temperature T_2, A, B — outlet points of the light and the heavy component

bent particles in consequence of periodical changes in the direction of the flow of the fluid, that are synchronized with the changes in column temperature. The principle of the function of this column is illustrated by the scheme in Fig. 2.4.29. The switching over of the heating and the cooling medium into the column jacket, and the reversing of the direction of the fluid flow takes place with the same frequency and in the same phase. A survey of the methods utilizing parametric pumping for continuous separations is given by Sweed[54].

The concept of continuous chromatography, described by Tuthil[55] is close to the above described parametric pumping. In this system the flow of the fluid is changed to a transverse direction synchronously with the oscillation of the thermal field. The separation unit can have the form of a right-angle thin slab of adsorbent, schematically shown in Fig. 2.4.30. The mixture is introduced continuously into one corner of the slab. The direction of the fluid flow alternates at a right angle simultaneously with the temperature changes of the whole system. As the ratio of the rate of elution of the components (v_A, v_B) on the chromatographic packing changes with temperature,

$$(v_A/v_B)_{T1} \neq (v_A/v_B)_{T2}$$

the paths of the light and the heavy component split and the components can be collected separately at the opposite corner of the slab. In principle even multicomponent mixtures can be fractionated on the slab unit.

In spite of recent distinct progress in the development of continuous gas chromatographs it is still not clear whether they are more suitable for production purposes than the apparatuses operating by pulse technique.

Other techniques. For chromatographic separation of pure substances *frontal* and *displacement* technique can also be used in addition to the most usual elution technique.

Extensive theoretical and experimental elaboration of the problems of frontal chromatography has been given by Krige and Pretorius[56]. As a preparative method frontal chromatography is utilizable mainly for the separation of gases. The mixture is introduced continuously into the chromatographic column with the adsorbent. The pure, least sorbed component of the mixture is then eluted from the column for a certain time, then further more sorbed components penetrate gradually into the effluent, and eventually the mixture of the original composition is eluted. It is evident that the most suitable application of the method is in the purification of some gases from small amounts of more strongly sorbed contaminants. Thus, for example, on a column with active charcoal pure methane can be obtained from natural gas in a yield of about 50 g methane/100 g of sorbent. The column must then be regenerated at 180–200 °C. Similarly ethane can be purified from ethylene and C_3 hydrocarbons on a column with a *molecular sieve*[7]. The advantage of these applications is that the collected product is not diluted with the carrier gas.

Displacement chromatography, developed mainly in the Soviet Union[7,57,58] is one of the variants of preparative gas chromatography without a carrier gas. Applications are similar to the preceding case.

For example, on a column with a molecular sieve 5A (0.5 – 1 mm) and using CO_2 as the displacing agent, impurities were eliminated from ethylene in the fore and the rear fractions, while the product was obtained in 60 % yield and 99.999 % purity[59]. Columns packed with adsorbent are used mainly for the preparation of pure hydrocarbon gases, but when using the so-called "*pseudo-displacers*" even high-boiling compounds can be separated by the GLC technique on columns containing a stationary liquid[58].

The method of displacement of the compounds from a chromatographic column by a thermal field, called "*Chromathermography*" (thermochromatography) has been developed predominantly by the Soviet teams Zhukhovitskii et al.[60] and Roginskii et al.[61]. The method exists in several variants and it is suitable for the separation of gaseous hydrocarbons and other mixtures[7]. A comparison of chromathermography with isothermal GLC has been given by Ohline and DeFord[62]. The combination of thermal desorption with the technique of frontal analysis is also advantageous. Recently light hydrocarbons were separated in this manner preparatively in 99.6 – 99.9 % purities[63].

Description of some further, less widespread techniques of preparative gas chromatography can be found in monographs[6, 7].

REFERENCES (2.4)

1. D. A. Craven, *J. Chromatogr. Sci.* **8** (1970) 540.
2. A. B. Carel, R. E. Clement and G. Perkins, *J. Chromatogr. Sci.* **7** (1969) 218; *Anal. Chim. Acta* **34** (1966) 83.
3. J. R. Conder, *Chromatographia* **8** (1975) 60.

3a. H. M. McNair, *J. Chromatogr. Sci.* **16** (1978) 578.

4. E. Bayer, K. P. Hupe and H. Mack, *Anal. Chem.* **35** (1963) 492; *Angew. Chem.* **73** (1961) 525.
5. J. Albrecht and M. Verzele, *J. Chromatogr. Sci.* **9** (1971) 745.
6. A. Zlatkis and V. Pretorius (Eds.) Preparative Gas Chromatography, Wiley-Interscience, New York (1971).
7. K. I. Sakodynskii and S. A. Volkov, Preparativnaya gazovaya khromatografiya, Khimiya, Moscow (1972).
8. J. Albrecht and M. Verzele, *Chromatographia* **4** (1971) 419.
9. H. Purnell, Gas Chromatography, J. Wiley, New York (1962).

10. J. Albrecht and M. Verzele, *J. Chromatogr. Sci.* **8** (1970) 586.
11. J. C. Giddings, *J. Gas Chromatogr.* **1** (1963) 12: **2** (1964) 290; R. P. W. Scott *Anal. Chem.* **35** (1963) 481.
12. B. M. Mitzner and W. V. Jones, *J. Gas Chromatogr.* **3** (1965) 294.
13. C. Erba, Short Notes 2—66, Milano (1966).
14. N. B. Diximer, B. Roz and J. Guichon, *Anal. Chim. Acta* **38** (1967) 73.
15. J. B. Pausch, *Chromatographia* **8** (1975) 80.
16. G. M. Higgins and J. F. Smith in Gas Chromatography—1964, A. Goldup (Ed.), Institute of Petroleum, London (1965).
17. C. E. Reese and E. Grushka, *Chromatographia* **8** (1975) 85.
18. C. L. Guillemin, *J. Chromatogr.* **12** (1963) 163; **30** (1967) 222; *J. Gas Chromatogr.* **4** (1966) 104.
19. D. W. Fisch and D. G. Crosby, *J. Chromatogr.* **37** (1968) 307.
20. G. Magnusson, *J. Chromatogr.* **109** (1975) 393.
21. T. H. Parliment, *Anal. Chem.* **45** (1973) 1792.
22. J. J. Vandeemter, F. J. Ziderweg and A. Klinkenberg, *Chem. Eng. Sci.* **5** (1956) 271.
23. B. Smith, R. Ohlson and G. Larson, *Acta Chem. Scand.* **17** (1963) 436.
24. J. Bricteux and G. Duyckaerts, *J. Chromatogr.* **22** (1966) 221.
25. W. O. McReynolds, *J. Chromatogr.* **8** (1970) 685.
26. J. J. Leary, J. B. Justice, S. Tsuge, S. R. Lowry and T. L. Isenhour, *J. Chromatogr. Sci.* **11** (1973) 201.
27. G. F. Freeguard and R. Stock, *Nature* **192** (1961) 257.
28. M. Verzele and M. Verstappe, *J. Chromatogr.* **26** (1967) 485.
29. R. P. W. Scott, *Nature* **198** (1963) 782.
30. E. Gelpi, P. C. Wszolek, E. Yang and A. L. Burlingame, *Anal. Chem.* **43** (1971) 864; *J. Chromatogr. Sci.* **9** (1971) 147.
31. G. D. Mitra, G. Mohan and A. Sinha, *J. Chromatogr.* **99** (1974) 215; *J. Chromatogr. Sci.* **11** (1973) 419.
32. W. J. A. Vandenheuvel and G. W. Kuron, *J. Chromatogr. Sci.* **7** (1969) 651.
33. J. R. Conder and M. K. Shingari, *J. Chromatogr. Sci.* **11** (1973) 525.
34. G. Schomburg, H. Kötter and F. Hack, *Anal. Chem.* **45** (1973) 1236.
35. D. R. Deans, M. T. Huckle and R. M. Peterson, *Chromatographia* **4** (1971) 279.
36. W. E. Harris, *J. Chromatogr. Sci.* 11 (1973) 184; *Can. J. Chem.* **43** (1965) 1560.
37. R. Kaiser, Ref. 6, Chapter 6.
38. W. E. Harris and H. W. Habgood, Programmed Temperature Gas Chromatography, J. Wiley, New York (1966).
39. J. D. Kelley, J. Q. Walker, *J. Chromatogr. Sci.* **7** (1969) 117.
40. L. S. Ettre, L. Mázor and J. Takáct in Advances in Chromatography, Vol. 8, J. C. Giddings and R. A. Keller (Eds.), Dekker, New York (1969).
41. L. Mázor and J. Takács, *J. Chromatogr.* **29** (1967) 24; *J. Gas Chromatogr.* **6** (1968) 58.
42. J. Gyimesi and G. Guiochon, *J. Chromatogr.* **86** (1973) 25.
43. M. Verzele and M. Verstappe, *J. Chromatogr.* **19** (1965) 504.

44. V. P. Khizhkov, G. A. Yushina, L. A. Sintzina and B. A. Rudenko, *J. Chromatogr.* **120** (1976) 35.
45. L. Markovec, S. Hála and S. Landa, *Neftekhimiya* **5** (1965) 835.
46. B. J. Mair and J. M. Barnewal, *J. Chem. Eng. Data* **9** (1964) 282; **12** (1967) 432.
47. J. R. Conder, Production Scale Chromatography in New Developments in Gas Chromatography, p. 137—186, H. Purnell (Ed.), J. Wiley, New York (1973).
48. M. V. Sussman and R. N. S. Rathore, *Chromatographia* **8** (1975) 55.
49. P. E. Barker and R. E. Deeble, *Chromatographia* **8** (1975) 67.
50. P. E. Barker, Ref. 6, Chapter 10.
51. D. Dinelli, S. Polezzo and M. Taramasso, *J. Chromatogr.* **7** (1962) 477; 11 (1963) 19; *J. Gas Chromatogr.* **2** (1964) 150.
52. P. E. Barker and R. E. Deeble, *Anal. Chem.* **45** (1973) 1121.
53. R. H. Wilhelm, A. W. Rice, R. W. Rolke and N. E. Sweed, *Ind. Eng. Chem Fundamentals* **7** (1968) 337; **5** (1966) 141; *Science* **159** (1968) 522.
54. N. H. Sweed in Progress in Separation and Purification, E. S. Perry and C. J. Van OSS (Eds.), Vol. 4, Wiley-Interscience, New York (1971).
55. E. J. Tuthill, *J. Chromatogr. Sci.* **8** (1970) 285.
56. G. J. Krige and V. Pretorius, *Anal. Chem.* **37** (1965) 1186, 1191, 1195, 1202.
57. A. A. Zhukhovitskii, N. D. Turkeľtaub, L. A. Malyasova, A. F. Shlyakhov, V. V. Naumova and T. I. Pogrebnaya, *Zav. lab.* 29/10 (1963) 1162.
58. O. V. Altshuler, O. D. Vinogradova, S. Z. Roginskii and J. N. Khirkov, *Dokl. Akad. Nauk SSSR*, **152** (1963) 892.
59. V. C. Mirzayanov, V. G. Berezkin, E. G. Proskurneva and V. P. Pakhomov, *Khim. i tekhnol. topliv i masel* 9/9 (1964) 66.
60. A. A. Zhukhovitskii, O. V. Solotareva, V. A. Sokolov and N. M. Turkeľtaub, *Dokl. Akad. Nauk SSSR*, **77** (1951) 435; **88** (1953) 859; **92** (1953) 987; **94** (1954) 77; **102** (1955) 771.
61. O. V. Altshuler, O. D. Vinogradova, S. Z. Roginskii and M. N. Yanovskii, *Dokl. Akad. Nauk SSSR*, **140** (1961) 1307.
62. R. W. Ohline and D. D. DeFord, *Anal. Chem.* **35** (1963) 227.
63. A. F. Shlyakhov, A. A. Zhukhovitskii, O. A. Kancheeva, L. N. Ryabchuk and M. L. Sazonov, *Neftekhimiya* **13** (1973) 300.

2.5. Adductive crystallization

The possibilities of separation and analysis of hydrocarbon mixtures are considerably extended by the methods based of the so-called *molecular-sieving effect*. This name indicates the separation selectivity to molecules of a certain shape and size in which the boiling point, chemical structure, and other properties of molecules play a secondary role. The result of the process — classification of the molecules the size of which is within a certain defined limit — is analogous to the classification of loose materials according to particle size, on mechanical sieves.

Adductive crystallization is one of the important methods with shape selectivity used in the separation of hydrocarbon mixtures. It is based on the ability of some compounds to give relatively stable crystalline complexes with hydrocarbons of certain molecular dimensions. The procedure excels by its simplicity: a solution of a complex-forming reagent is mixed with the hydrocarbon fraction; hydrocarbons of suitable dimensions form a solid complex which is then filtered off and decomposed. Thus the trapped hydrocarbons are again set free and can be isolated. Adductive crystallization is widely used in laboratory separations of hydrocarbon mixtures, because various complex-forming reagents with special selectivity are known. Some processes of adductive crystallization have already been introduced in industry[1].

Attractive forces between hydrocarbon molecules and the molecules of the complexing agent are of various nature, according to the type of complex[2]. The majority of hydrocarbon complexes cannot be counted as true chemical compounds; intermolecular adhesive forces are considerably weaker than in covalent or ionic bonds, but they are stronger than those involved in simple adsorption on the surface of common adsorbents.

The composition of hydrocarbon complexes can be expressed by the general formula:

$$(\text{reagent})_a \,.\, (\text{hydrocarbon})_b;$$

according to the mole ratio of the components, three types of these complexes are known:

Molecular compounds – their a:b ratio is stable and corresponding to small whole numbers.

Inclusion compounds – their reagent to hydrocarbon ratio is non-stoichiometric. Two types of inclusion compounds are known: *adducts*, the mole ratio a:b of which is not expressed by small whole numbers, but which is stable; the second type are *clathrates* that can contain a changeable amount of hydrocarbon, up to a certain maximum value, depending on the conditions of preparation. In inclusion compounds the molecules of hydrocarbon (guest) are trapped inside the spaces of the crystal lattice of the complexing agent (host). The size and the shape of these spaces determines the selectivity of the host for hydrocarbon molecules.

For practical separations of hydrocarbon mixtures adducts with urea are of main importance, then thiourea adducts and some types of clathrates and molecular compounds containing aromatic hydrocarbons.

2.5.1 ADDUCTS WITH UREA

From a saturated solution, aqueous or methanolic, pure urea crystallizes in *tetragonal* form. If *n*-alkanes are present in the solution, *hexagonal* crystals of the adduct precipitate from the solution much more easily in characteristic needles. During the crystallization the molecules of urea settle around each linear molecule of *n*-alkane in the form of a tight helix which is stabilized in this hexagonal crystalline structure by intermolecular

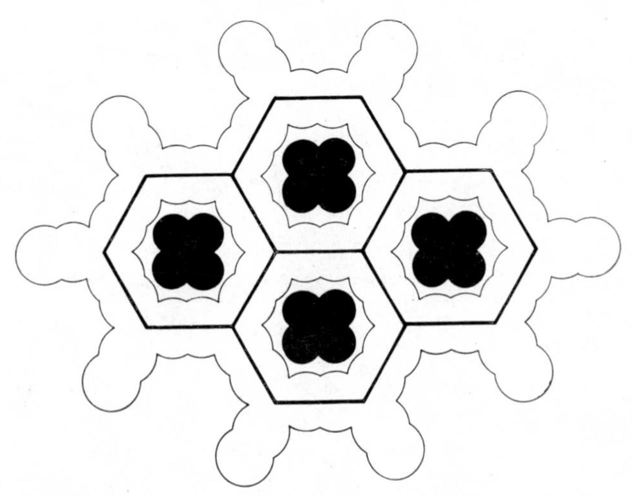

Fig. 2.5.1 Deposition of molecules of *n*-alkanes (black areas) in the channels of the urea adduct

hydrogen bonds between the oxygen atom and amino groups. The chains of *n*-alkanes are thus gradually withdrawn from the liquid phase and built into the crystal lattice of the host where they assume positions in regular cylindrical channels penetrating the whole crystal structure. The deposition of *n*-alkane molecules in the crystal structure of the adduct is evident from Fig. 2.5.1. The diameter of the channels is 0.5 – 0.6 nm.

The molecules of n-alkanes with the critical diameter 0.49 nm are bound in the channels by weak van der Waals forces only, and they possess a certain measure of freedom for rotations and oscillations[3]. The chains of n-alkanes are stretched to a maximum in the adduct. According to X-ray analysis[4] the molecule of n-alkane occupies a length in the channel, corresponding exactly to the calculated theoretical length of the molecule. Individual molecules of n-alkane lie closely behind each other in the channel. The distances between the terminal carbons (centres of atoms) of neighbouring molecules are only 0.348 nm long. Without the presence of hydrocarbon molecules the structure of the adduct is not stable. The linear molecules of the guest form the scaffolding that supports the whole crystalline structure, and if the hydrocarbon molecules are eliminated from the adduct, for example by distillation, the crystal lattice collapses.

The composition and the stability of the adducts urea—n-alkanes. The formation of the adduct is an exothermic reaction that can be expressed by equation

$$a \,.\, \text{urea} + \text{hydrocarbon} \quad \rightleftarrows \quad \text{adduct} + \Delta H \qquad (2.5.1)$$

The lowest n-alkane that forms an adduct with urea at room temperature is n-hexane. The longer the n-alkane chain, the more easily is the adduct formed and the more stable it is (the higher the heat of formation). The upper limit of the length of the adducting alkane is given practically only by the solubility of the hydrocarbon in the reaction mixture. (Urea adducts of extreme stability were prepared[5] with linear polymers of molecular weight $10^5 - 10^6$). The mole ratio of the components in the adduct is different for each n-alkane because the number of urea molecules surrounding one molecule of alkane depends on the length of its chain. In Table 2.5.1 the mole and weight ratios of the components in the adducts of individual n-alkanes and their heats of formation[6-9] are given. The number of moles of urea (a) in the adduct per one mole of n-alkane increases linearly in dependence on the number of alkanic carbon atoms (C). This dependence can be expressed for all n-alkanes by the simple equation:

$$a = 0.64C + 1.5 \qquad (2.5.2)$$

The weight ratio is approximately constant. From Table 2.5.1 it is evident that for a complete conversion of any n-alkane into the adduct at least a 3.5-fold amount of urea is necessary.

TABLE 2.5.1

Composition and heats of formation of urea adducts of *n*-alkanes

Hydrocarbon	Moles of urea per 1 mole of hydrocarbon	Grams of urea per 1 g of hydrocarbon	Heat of formation in kJ/mol
n-Heptane	6.1	3.7	30.6
n-Octane	6.7	3.5	40.6
n-Nonane	7.4	3.5	49.4
n-Decane	8.3	3.5	54.9
n-Undecane	9.1	3.5	61.1
n-Dodecane	9.3	3.3	67.4
n-Hexadecane	11.8	3.1	87.9
n-Octadecane	14.0	3.3	—
n-Tetracosane	17.8	3.2	—
n-Octacosane	21.6	3.3	—
n-Dotriacontane	23.3	3.1	—

Reaction conditions. A great number of procedures for the isolation of *n*-alkanes by adductive crystallization have been described in literature. They differ in various reaction conditions, depending on whether the main stress is put on the purity of the adduct or on the quantitative yield of *n*-alkane, or else, whether the procedure is intended for the processing of a large volume of a hydrocarbon fraction. The majority of the procedures described, however, use one of the following techniques:

a) – *Solution technique.* The hydrocarbon fraction or its solution is added to a concentrated or saturated solution of urea. After 1 – 3 h of contact (stirring, shaking) the precipitated adduct is filtered off, washed and dried.

For urea water, alcohols, methyl ethyl ketone or combined solvents are used. Less viscous hydrocarbon fractions are added undiluted; the heavier fractions and solid hydrocarbons are dissolved in benzene, tetrachloromethane or petroleum ether.

Depending on the nature of the hydrocarbon fraction or on the stability of the adduct either a solution of urea saturated at room temperature or a solution saturated at the solvent's boiling point are used. In the latter case the hydrocarbon fraction is added to a hot solution of urea

(or a refluxing one) and the adduct is separated during the cooling of the reaction mixture.
b) – *Slurry technique*. Powdered urea is added to a hydrocarbon fraction or to a solution of the fraction in an inert solvent. The finer the urea powder, the more rapidly the reaction takes place. Instead of the solvent only a small amount of methanol is added which activates the adducting and the mixture is stirred vigorously for several hours. The microcrystalline slurry containing the adduct is filtered off and washed and dried in the conventional manner.

The filtered off dry adducts of higher *n*-alkanes (from C_{10} up) are very stable and they can be stored for a long time. On heating their appearance does not change up to 132.7 °C (melting point of urea), when they melt. However, at a certain temperature below the melting point the adducts dissociate by heat, which can be seen under the microscope[10, 11] or recorded by *differential thermal analysis* (DTA)[11, 12]. The decomposition (dissociation) temperatures of some adducts are given in Table 2.5.2.

The isolation of *n*-alkanes from the adducts is the same for both va-

TABLE 2.5.2

Dissociation temperatures of urea adducts of *n*-alkanes (determined by DTA)

Number of carbons in *n*-alkane	Dissociation temperature of urea adduct in °C	
	McAdie[11]	Topchiev et al.[12]
10	82.4	
12	92.5	
14	102.4	
16	108.4	106.4—108.5
18	114.2	115.8—117.1
20	117.9	120.0—122.4
22	123.2	120.6—123.1
24	125.3	122.3—124.2
26	130.0	
28	131.2	125.6—128.6
30		129.5—131.3
32		130.0—131.9

riants of adductive crystallization: the isolated adduct, or the mixture of the adduct and crystalline urea, is decomposed in hot water at 70–90 °C. Urea is dissolved and the layer of hydrocarbon set free is separated or extracted with a suitable solvent.

Unreactive hydrocarbons are isolated from the filtrate after the separation of the adduct. The dissolved urea and the polar solvent are eliminated from the filtrate by extraction with water. In cases when a dilute fraction is worked up, the solvent used is distilled off from the residue of the non-adducting hydrocarbons. That part of hydrocarbons that was isolated *via* the adduct (mostly *n*-alkanes) is usually called the *extract*; the remaining unreactive hydrocarbons pass into the filtrate and they are called the *raffinate*.

In addition to these techniques a number of other methods of contacting *n*-alkanes with urea have been described[2, 9, 13], which lead to the formation of adducts. For example DTA shows that an adduct is formed even when crystalline urea is ground with a solid hydrocarbon in a mortar[12].

Adducts are also easily formed if sublimating urea (at 120 °C and 10 Pa) is condensed on a surface coated with a film of a reactive hydrocarbon. The process can be put into effect as a microtechnique in a common laboratory sublimator[14].

Kisielow[15] found that the adducts are also formed if a solution of *n*-alkanes is *percolated* through a solid bed of crystalline urea. Later some other authors[16, 17] used a column technique analogous to *liquid chromatography* for the preparation of the adducts.

In equation (2.5.1) it is shown that the formation of an adduct is an equilibrium reaction. At high concentration of reacting components the equilibrium is shifted in favour of the adduct, and *vice versa,* on addition of solvent (for urea or for *n*-alkane) the adduct can be dissociated back to the original components. A complete conversion of *n*-alkanes to adducts is furthered by the following factors:

a) A high as possible a concentration of urea in the solution. It is best when during the whole course of the precipitation of adduct crystals the solution is constantly saturated with urea. (A saturated solution of urea in methanol contains 30–35 wt. % of urea at about 50 °C and 20 wt. % of urea at 25 °C).

b) – A system of solvents that brings both *n*-alkanes and urea into a common liquid phase.

c) – A suitably selected temperature. Low temperature suppresses the dissociation of the adduct. At elevated temperature the solubility of *n*-alkanes is better and the reaction rate of adducting is higher. The adducts of *n*-alkanes having 16 and more carbon atoms in the chain are already so stable that they can be isolated from the hydrocarbon fraction with urea without difficulty, in a quantitative yield even at 25 °C. Quantitative isolation of *n*-alkanes from C_{12} up can be achieved by precipitating the adduct at 0 °C. For the conversion of lower *n*-alkanes (C_6—C_{12}) to adducts temperatures of 0° to −10 °C are used, but if current procedure for the reaction is used, quantitative yields cannot be achieved.
d) – Sufficient reaction time. When stirring or shaking the reaction mixtures the formation of the adduct is over within 1–3 h. For a quantitative yield some procedures recommended leaving the reaction mixture with the precipitated adduct to stand overnight.

Among the main factors that decrease the yield of the adduct, and that should be suppressed as much as possible in quantitative work, are these:

a) – Low concentration of *n*-alkanes in the reaction mixture. For a better contact with the urea solution the viscous hydrocarbon fractions (and solid hydrocarbons as well) are diluted with an inert solvent. Excessively high dilution of the reaction mixture with an inert solvent however furthers the dissociation of the adduct.
b) – Washing of the adduct. The crystalline adduct should be washed after filtering off, because it is voluminous and keeps a certain amount of unreacted hydrocarbons (mainly aromatic ones) and polar substances on its surface. No solvent is known that dissolves the adhering hydrocarbons well without decomposing the adduct. Mostly the adducts are washed with low-boiling hydrocarbons – isopentane, cyclopentane, petroleum ether, or with saturated solution of urea in methanol or methyl ethyl ketone. In any case the adduct should be washed with a minimum amount of solvent, and at low temperature.
c) – Presence of *inhibitors*. Pure *n*-alkanes form adducts even in contact with aqueous solutions of urea. The hydrocarbon fractions from petroleum, tars, etc. contain inhibitors, however, that retard or prevent the precipitation of the adduct[6] when reacting with an aqueous solution of urea or with powdered urea. The addition of a small amount of methanol or some other polar activator (alcohols, ketones) destroys the effect of the

inhibitor. Of course, when working with a methanolic solution of urea the effect of inhibitors is not evident.

Selectivity. The formation of adducts with urea is not limited to *n*-alkanes only. A number of organic compounds have such molecular dimensions that they can stay in the channels of the crystal lattice of the adduct. These are primarily the compounds that carry a smaller substituent at the end of the long polymethylene chain, such as for example chlorine, the primary alcoholic group, the carbonyl group, the ether group, etc. The stability of these adducts is generally lower than in the case of equally long *n*-alkanes.

Among hydrocarbons the straight-chain unsaturated hydrocarbons also have suitable molecular dimensions for adducting. The lowest olefin that can form an adduct under current conditions is 1-octene. From C_{16} up, the adducts of 1-alkenes are even more stable than the adducts with corresponding *n*-alkanes[18]. Among other long-chain olefins *trans*-olefins generally form more stable adducts than the *cis*-isomers, the chains of which deviate more strongly from linearity.

Among alkynes those with 9 and more carbon atoms[19] form adducts with urea. The relative stability of the adducts varies with the position of the triple bond. For example, 1-, 2- and 3-nonyne form adducts, while the isomers with the triple bond in the middle of the molecule – such as 4-nonyne and 5-decyne – do not give an adduct. Similarly, none of the tested C_6—C_9 diynes gave an adduct.

The selectivity of the separation of *n*-alkanes from hydrocarbon fractions by means of urea is most strongly disturbed by isoalkanes with a side-methyl in the extreme position of the linear chain, and by cyclic hydrocarbons substituted with a long linear alkyl.

A condition for the formation of adducts in monomethylalkanes is a sufficiently long unbranched part of the molecule. The lowest members of the homologous series of monomethylalkanes capable of forming adducts with urea are 2-methylundecane, 3-methylundecane, 4-methylpentadecane and 5-methylhexadecane. It is evident that the linear part of the molecule must be the longer the farther away from the edge of the chain the side-methyl is bound[18]. Still longer must the linear chain be in dimethylalkanes; the lowest homologues capable of forming an adduct are 2,3-dimethylheptadecane and 2,4-dimethyloctadecane.

Cyclopentane, cyclohexane or benzene nucleus can be "pulled" into the adduct only if it is bound to a linear chain with at least twenty carbon atoms. Sometimes only small differences in the structure decide whether the hydrocarbon can enter into the adduct or not. For example, of the subsequent pair of hydrocarbons

$$\begin{array}{l} C-C \\ |\quad\;\; \rangle C-C_{21} \\ C-C \end{array} \qquad \begin{array}{l} C-C \\ \quad\;\; \rangle C-C_{21} \\ C-C \end{array}$$

cyclopentyleicosane forms an adduct with urea, while the freely rotating ethyl group in 2-ethyltetracosane increases the cross-section of the molecule to such an extent that the adduct is not formed.

Some isoalkanes and cyclic hydrocarbons that, when in pure state, do not form an adduct, can be "*inductively*" pulled into the adduct[20] in the presence of *n*-alkanes.

In spite of the mentioned possibilities of contamination of the extract with various non-linear structures, the adductive crystallization with urea is very well utilizable for the separations of *n*-alkanes from hydrocarbon fractions. It is advantageous that the reaction rates and the stabilities of the adducts of branched and cyclic hydrocarbons are generally much lower than of the corresponding *n*-alkanes. Therefore, choosing suitable reaction conditions a quantitative isolation of higher *n*-alkanes can be achieved, and at the same time it is possible to keep the content of nonlinear hydrocarbons at a low level. In any case it is more accurate to call the hydrocarbons present in the extract not *n*-alkanes, but adducting or complexing hydrocarbons. Fig. 2.5.2 shows gas chromatographic analysis of an extract isolated with urea from a *diesel fuel* from paraffinic petroleum as a typical example. Non-linear hydrocarbons that have passed into the adduct form characteristic multiplets on chromatograms that are repeated between the dominant peaks of *n*-alkanes C_{10}—C_{20}. In the analysed sample as well as in other lighter and middle petroleum distillates the hydrocarbons that are eluted in these multiplets are mostly methylated alkanes. In each multiplet the component with the longest retention time (on the stationary phase OV-225) is 3-methylalkane (*d*), and the most abundant component (*c*) eluted before it is 2-methylalkane (in the common peak with 2-methylalkane 4- and 5-methylalkanes can also be eluted). In bands with the shortest retention times (*a*, *b*) methylalkanes substituted further from the chain end are eluted. From this point of view the extracts from various natural hydrocarbon mixtures are very similar to each other; for example, the extract isolated recently in our laboratory from the oil fraction (C_{13}—C_{22}) of *ozokerite*[21] has almost the same character. The extract isolated with urea from heavy oils and especially from petroleum waxes usually contain cyclic hydrocarbons in addition to *n*-alkanes and isoalkanes[9, 22].

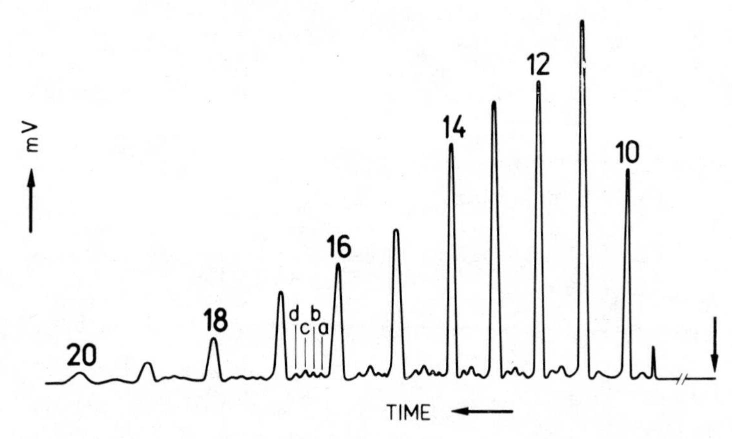

Fig. 2.5.2 Gas chromatographic analysis of the extract isolated from Diesel fuel with urea
Dominant peaks are *n*-alkanes C_{10}—C_{20}. *a*, *b* — methylhexadecanes with the methyl group farther from the chain end; *c* — 2-methylhexadecane; *d* — 3-methylhexadecane

Applications. In the analysis of hydrocarbon mixtures the urea adducts are applied mainly for
a) the purification of *n*-alkanes
b) isolation of *n*-alkanes and other adducting hydrocarbons
c) quantitative determination of *n*-alkane content in fractions.

For the purification of synthetic or isolated individual *n*-alkanes the procedures are best suitable during which the adduct precipitates slowly in well developed crystals from a large volume of a homogeneous solution. In the authors' laboratory, for example, the following procedure[23] proved successful for the purification of C_{16}—C_{20} *n*-alkanes.

Purification of n-alkanes by means of urea adducts

Crude *n*-alkane (1 g), urea (6 g, recrystallized from methanol), acetone (65 ml) and methanol (35 ml) were put into a flask provided with a reflux condenser and the mixture refluxed on a water bath. When the hydrocarbon and urea are dissolved to an homogeneous solution the flask is allowed to cool overnight. The precipitated long needles of the adduct are filtered off, washed with a minimum amount of acetone, dried in air and decomposed in hot water. The yield of *n*-alkane is 70—75 wt.%. The second product (11—15 %) can be isolated from the filtrate by adding 2.5 g of urea; the solution is shortly refluxed and the procedure is repeated.

In the investigation of the composition of hydrocarbon materials of various origins the separation of the group of *n*-alkanes usually means a considerable simplification. The isolation of hydrocarbons adducting with urea is therefore often included as one of the first steps of the separation scheme. This is advantageous because the process is extraordinarily simple and enables a laboratory work-up of large amounts (tens of kg) of the starting fraction. However, adductive crystallization can be applied equally well as a microtechnique.

For the isolation of adducting hydrocarbons both previously mentioned techniques are used: for the working up of gram amounts of a fraction preference is given to the solution technique, while for kilogram amounts of a fraction the use of the slurry technique is more suitable. Conditions of several proved procedures are given in Table 2.5.3.

For analytical purposes the formation of adducts with urea is made use of in various ways. After an experimental check, some isolation procedures can even be used for quantitative determination of *n*-alkanes. For example, among eight different procedures of the determination of *n*-alkanes in solid paraffin by means of urea, checked by Kisielow and Kajdas[34], the procedure *E* given in Table 2.5.3 gives results that are very close to the true content of *n*-alkanes, determined by means of *molecular sieves* (see Section 2.6.1).

According to another method the content of complexing hydrocarbons in a fraction can be determined from the increase in weight of urea. The original procedure elaborated by Veselov[35] also utilizes slurry technique: the fraction (10 g) and urea (30 g) are weighed into a weighed flask and 100 ml of petroleum ether and 6 ml of methanol as activator are added. The slurry formed is stirred for 1 h at room temperature. The liquid part is then filtered off from the flask with a filter stick. The remaining solid complex in the flask is washed with five portions of petroleum ether (50 ml each) – a filter stick is again used for sucking off the filtrate – and the remains of the solvent are distilled off on a water bath at 2.5 kPa (100 min.). The vacuum dried complex is weighed together with the flask. The increase in weight corresponds to the content of complexed hydrocarbons in the fraction.

Best results are obtained with the mentioned urea method with fractions distilling within the 240 – 360 °C interval. In these fractions *n*-alkanes are converted to adducts quantitatively and they are accompanied by only a small amount of non-linear hydrocarbons. The results also agree well with the determination of *n*-alkanes by means of molecular sieves[32,34]. The urea method does not require previous dearomatization of the fractions, and it is more advantageous than molecular sieves for the boiling range indicated. For lower boiling fractions (below 240 °C) the urea method does not give quantitative results because a part of *n*-alkanes remains in the raffinate. On the other hand, in the case of heavy fractions boiling above 360 °C a large amount of branched and cyclic hydrocarbons also pass into the adduct, and in comparison with the molecular sieve method the urea method gives excessively high results.

The procedure worked out by Veselov is used in a modified form for middle petroleum fractions as a standard GOST method[36].

Determination of the content of complexing hydrocarbons in fractions 240 – 360 °C

About 15 g of fraction (with an accuracy of 0.01 g) is weighed into a 500—750 ml flask, and 250 ml of petroleum ether (40—70 °C), 15 ml of ethanol and 60 g of ground urea, dried previously at 50 °C to constant weight, are added to it. The mixture in the flask is stirred mechanically at 20—25 °C for 30 min. The adduct formed is transferred quantitatively onto a Büchner funnel and filtered off using a weighed filter. The adduct is then washed four to five times with petroleum ether (each time with 40—50 ml). The first portion is used for rinsing the flask. The washed adduct is transferred together with the filter onto a weighed filter paper (30 × 30 cm) where it is spread and dried in a vacuum drying oven at 50 °C and 5 kPa to constant weight. The increase in weight corresponds to the content of complexing hydrocarbons in the fraction.

Seleznev et al.[37] have accelerated the standard method by mixing the fraction with urea directly in a weighed fritted-glass filtration funnel (the stem is closed with a stopper). Thus the transfer of the adduct from the flask to the filter is avoided.

For the determination of the content of *n*-alkanes in lighter fractions

TABLE 2.5.3

Reaction conditions for some procedures of adductive crystallizations with urea

Method and ref.	Ratio of components in the mixture				Activator	Reaction time and temperature	Washing	Method's application
	Hydro-carbon fraction	Diluting solvent	Urea	Polar solvent				
A[24]	2 g	12—25 ml of gasoline 60—70 °C	12 g	35 ml methanol	—	1 h[a] 50 °C to 25 °C	20 ml of gasoline fraction 60—70 °C	Pure hydrocarbons
B[25]	10 g	700 ml CCl_4	70 g	700 ml C_2H_5OH	—	2.5 h under heating	C_2H_5OH	Paraffin wax[b] ceresin
C[26,27]	9.3 g	—	93 g	760 ml C_3H_7OH[c]	—	Reflux[d]	C_3H_7OH 2× 100 ml at 0 °C	Natural waxes bitumens crude oils[e,29]
D[30]	50 g	50 ml benzene	300 g	670 ml methanol	—	40 min[f] 50 to 25 °C	250 ml benzene	Paraffin wax
E[31]	10 g	12 ml benzene	30 g	18 ml MEK	0.5 ml of water or methanol	1.5 h 64 °C	mixture of benzene + MEK	Paraffin wax

F[32]	50 g	100—200 g petroleum-ether	100 g	—	15 g methanol	25 °C	petroleum ether	Viscous petroleum fractions
G[33]	1 kg	$CHCl_3$	2—3 kg	—	200 g CH_3OH or C_2H_5OH	1 h 10—15 °C	$CHCl_3$	Middle and heavy petroleum fractions
H[6]	3 kg	—	1 kg	—	200 g methanol	1 h[g] 25 °C	1.5 l iso-octane	Middle petroleum fractions (C_{13} — C_{18})
J[23]	4.7 kg	3 l hexane	3 kg	—	1 litre methanol	2 h 25 °C	5 l hexane	Neutral oil C_{16} — C_{20} from low-temperature tar

[a] A solution of hydrocarbon is added dropwise and under stirring to a hot solution of urea.

[b] For a more accurate separation of *n*-alkanes from isoalkanes the procedure is repeated three times.

[c] The amount of *n*-propanol depends on the mol. weight of the hydrocarbon; usually a 50—80 fold amount of the mass of hydrocarbon.

[d] Enough *n*-propanol is added under reflux just to dissolve the hydrocarbon and the urea, and the solution is then allowed to cool slowly in a Dewar flask for 25—30 h.

[e] Fractions boiling up to 200 °C and polar substances are previously separated from the crude oil (the last by chromatography on silica gel).

[f] An urea solution is added slowly at 50 °C and under stirring to a 50 °C warm solution of hydrocarbon.

[g] For heavy petroleum fractions (C_{17} — C_{42}) optimum reaction temperature is 30—40 °C.

Orság and Báthory[38] have described a method based on *refractometric* measurement of the loss of urea from the solution after adductive crystallization. The method gives good results with gasolines and kerosenes, but it fails with heavier oils.

The most reliable results are obtained with the urea method in combination with *gas chromatography*. In this manner the content of *n*-alkanes even in heavy petroleum fractions and dearomatized residues can be determined accurately. The method combines the rapid and quantitative isolation of high-boiling *n*-alkanes with urea with the chromatographic determination of the content of individual *n*-alkanes in the extract. Thus the decreased selectivity of the urea method for high-boiling hydrocarbons is mostly eliminated. In gas chromatography the pulled-in isoalkanes are mostly well separated from the main peaks of *n*-alkanes (see Fig. 2.5.2), so that both types of hydrocarbons can be determined simultaneously by integration of the areas below the peaks. For an accurate determination the areas of *n*-alkanes can be referred to the area of the added weighed *internal standard*; the results naturally also give the distribution of *n*-alkanes according to the number of carbon atoms. The procedure worked out by Marquart et al.[39], which was checked on *heavy gas oils* and heavier fractions, is the following:

Determination of the content of n-alkanes in middle and heavy hydrocarbon fractions by adductive crystallization of urea and gas chromatography

The hydrocarbon fraction is weighed accurately in a small flask, so that the content of *n*-alkanes is about 0.3 g, and 8 g of urea and 25 ml of methanol added to it. The content is then heated on a water bath at 55—60 °C and shaken at this temperature for 30 min. The water bath is then withdrawn and the shaking is continued for another one hour. During this time the contents cool to 25 °C. The separated crystals are filtered off using a Büchner funnel and washed with four portions (20 ml each) of methyl ethyl ketone saturated with urea. The adduct is transferred from the filter quantitatively into a volumetric flask where it is decomposed with water. About 2 ml of *n*-decane (as solvent) are added to the liberated hydrocarbon layer in the narrow neck of the flask, and *squalane* is weighed accurately and added into the flask as internal standard (it is eluted between C_{26} and C_{27} *n*-alkanes). The well stirred *n*-decane solution is analyzed by gas chromatography, for example using a column packed with 2 % of Carbowax 1 000 on Chromosorb W, at a temperature gradient of 75—250 °C (6 °C per minute).

Under the given conditions *n*-alkanes from C_{16} up are isolated quantitatively and determined. If the procedure is adjusted so that the adduct is separated at 0 °C, even *n*-alkanes above C_{12} can be determined.

Method[40] is based on the same principle which is used for the isolation of *n*-alkanes from saturated fractions with boiling point above 320 °C, obtained by silica gel separation of crude oil bottoms. The method is arranged as a microtechnique:

Isolation of n-alkanes from heavy saturated hydrocarbon fractions by adductive crystallization with urea

About 50 mg of hydrocarbon fraction is put into a test tube, 5 ml of a benzene-methanol mixture (3 : 1 v/v) are added and the tube is closed and shaken until the fraction is dissolved. Then 2 ml of a saturated urea solution in methanol are added and the mixture is thoroughly mixed. The precipitated mixture of the adduct and the crystalline urea is allowed to stand overnight. The solvent is then withdrawn from the test tube with a syringe with a long needle and the adduct is allowed to dry for several hours in the open test tube. The adduct is kept for gas chromatographic analysis in a closed tube, for example for the determination of the distribution of *n*-alkanes according to the number of carbon atoms.

Special use of urea adducts. Both main variants of adductive crystallization – solution technique and slurry technique – can be easily adjusted for the *fractionation* of mixtures of alkanes. Small portions of urea are added gradually to the hydrocarbon fraction thus forming the adduct with only a part of the alkanes present at a time. Then individual fractions of the extract are obtained from the separated adduct by decomposition. The procedure can be repeated as many times as necessary for the isolation of all adducting hydrocarbons. In consequence of a higher reaction rate and higher stability of the adducts higher *n*-alkanes are concentrated in first fractions.

The fractionation can also be carried out by chromatographic technique on a column of urea. A solution of *n*-alkanes is allowed to flow through a glass column packed with fine crystalline urea (wetted with an activator). The highest *n*-alkanes contained in the sample are retained in the upper part of the column, while the lower homologues are displaced into the lower parts of the column, or they are eluted. After washing the column and cutting it, fractions with different content of individual *n*-alkanes can be isolated from individual sections of the column. The main difficulty in this technique consists in the fact that the penetrability of the column decreases on adduct formation. A certain improvement can be achieved by packing the column with a mixture of urea and an inert material. Karr and Comberiati[17], for example, separated *n*-alkanes from *pitch oils* on a column packed with 40 parts (by weight) of urea and 60 parts of 20 – 100 mesh silica gel.

Probably the best method of fractionation on a urea column has been elaborated by Bolshova and Starobinets[41]. They arranged the procedure analogously to partition chromatography:

Separation of n-alkanes on a column of urea

A column of 1.5 cm diameter, containing 62 g of urea (height of the column 73 cm) is washed with a mixture of petroleum ether (25 %) and acetone (75 %) until equilibrium is attained. A sample containing *n*-alkanes C_{23}—C_{27} (50 mg of each) is dissolved in 1—1.5 ml of benzene and the solution is allowed to soak into the column within 3—4 h. The components are then eluted with a mixture of petroleum ether and acetone (the content of acetone is decreased from 50 % to 20 % during the operation) at a 1 ml per 15 min rate. From individual fractions of the eluate completely separated individual *n*-alkanes (see Fig. 2.5.3) can be isolated by evaporation in quantitative yield.

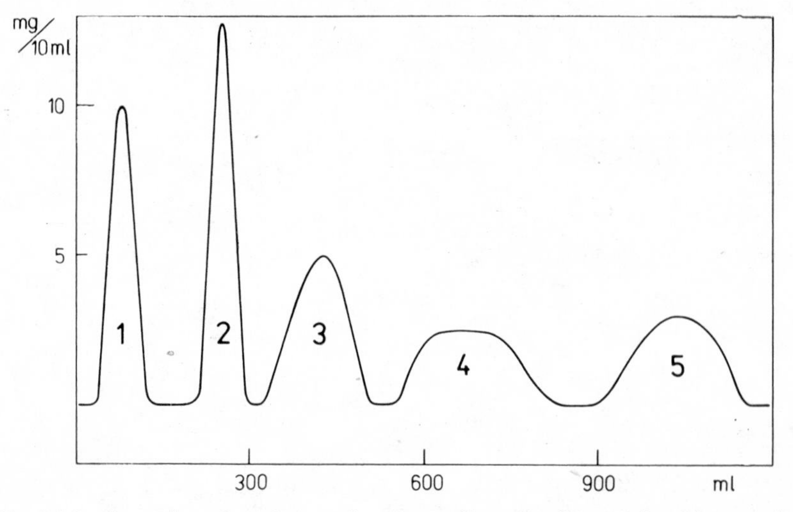

Fig. 2.5.3 Separation of a mixture of *n*-alkanes C_{23}—C_{27} (50 mg of each) on a column of urea according to Bolshov and Starobinets.[41] The yields are given in mg of hydrocarbon per 10 ml of eluent, 1 — $C_{23}H_{48}$; 2 — $C_{24}H_{50}$; 3 — $C_{25}H_{52}$; 4 — $C_{26}H_{54}$; 5 — $C_{27}H_{56}$

The possibilities of the separation of straight-chain compounds on urea using the technique of thin-layer chromatography and paper chromatography were demonstrated by Bhatnagar and Liberti[42]. The properties of urea adducts as the stationary phase in gas chromatography were investigated by Mařík and Smolková[43]. The separation of hydrocarbon mixtures in the gas phase on the adduct takes place with the same selectivity as in the liquid phase.

Schlenk discovered a remarkable ability of the urea adduct to split the *racemates* of *chiral* adducting substances to optical *antipodes,* and he then further developed this technique with his collaborators[44]. As has been said, the chains of *n*-alkanes in the crystal lattice of the adduct are wrapped with regular helix of urea molecules. From this it necessarily follows that the crystal lattice can exist in two *enantiomeric* forms – one with plus *helicity* (*P*) and the second with minus helicity (*M*) – that are mirror images of each other. In Fig. 2.5.4 both antipode forms of the adduct are represented schematically.

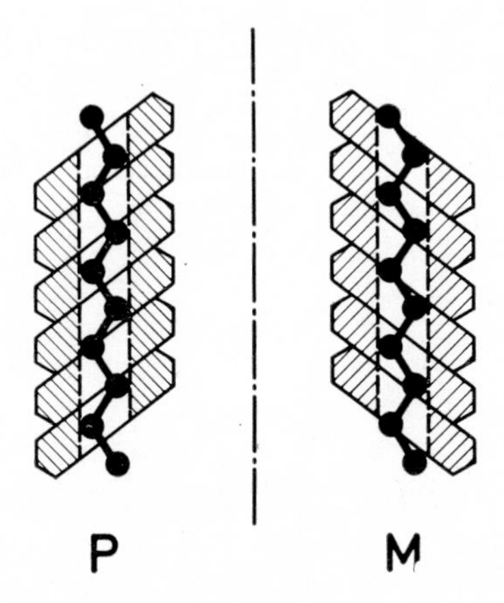

Fig. 2.5.4 Chirality of the urea adduct. The chains of *n*-alkanes are wrapped round by the helix of urea molecules, that has either helicity plus (*P*) or its mirror image minus (*M*)

Measurements have actually shown that the monocrystals of the urea adduct containing optically inactive *n*-alkane rotate the plane of polarized light[45]. In this the length of the guest alkane molecule has no measurable effect on the rotation of the crystal. The monocrystals of the adduct with the helicity *P* rotate the plane of the polarized light to the left, while the adduct with the helicity *M* turns it to the right.

When the guest molecule is chiral itself, four theoretical combinations can take place. For the adduct of urea with chiral 3-methylalkane these combinations are represented schematically in Fig. 2.5.5. If the chirality of the hydrocarbon is indicated according to the *R*/*S*-system – and the helicity of the urea helix with *P* and *M*, then four combinations are obtained: *S*–*P*, *R*–*M*, *R*–*P* and *S*–*M*. In the figure the indicated pairs are antipodes and form racemates. However, for example, the adducts of the type *S*–*P* and *R*–*P* represent a new form of *diastereoisomery*. On the basis of the differing properties of these adducts (stability, solu-

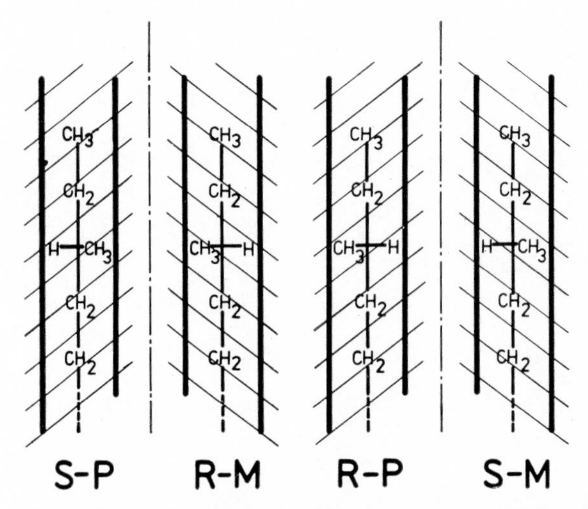

Fig. 2.5.5 Diastereoisomery of the urea adducts with 3-methylalkanes
R, *S* — indicate absolute configuration of hydrocarbons; *M*, *P* — indicate the helicity of urea molecules in the channel. Diastereoisomeric configurations, for example *S*—*P* and *R*—*P*, differ in physical properties

bility, etc.) chiral hydrocarbons (and other adducting compounds) can, in principle be separated from racemates. A condition for this separation is that the adduct with one type of helicity should crystallize from the solution containing the racemate. Often on slow crystallization of pure solutions one or the other mirror form of the lattice is formed spontaneously. This was how the separation of racemate by adductive crystallization was carried out for the first time[46]. Nowadays several methods of controlled crystallizations are known, for example, by inoculating the solution with an optically active monocrystal of the adduct of the required helicity, by crystallization from optically active solvent, etc[44].

A series of experiments carried out by Schlenk and co-workers confirmed that a certain enantiomer of the guest compound crystallizes, as a rule, more easily in one of the mirror images of the lattice than in the other. In the narrow channel with the helical relief on the inner side one antipode is evidently more oppressed, while the other fits into this space more "comfortably", similarly as a hand can slip into one glove of a pair more easily than into the other. Of the group of chiral methylalkanes, for example, all members of the homologous series (+) (*S*)-5-methylalkanes C_{11}—C_{24} crystallize preferentially in the adduct with the *P* lattice. In other homologous series the correlation between the configuration of the enantiomer and the helicity of the lattice is not so clear. In the series of (+) (*S*)-3-methylalkanes the hydrocarbons with 11, 13, 16, 18, 19, 21 and more carbon atoms prefer the lattice with the helicity *P*, while (*S*)-antipodes with 12, 14, 15, 17 and 20 carbon atoms crystallize more easily in the adduct with the lattice *M*.

When enumerating special applications of the adducts *rhomboedric* adducts of urea[47] should also be mentioned for the sake of completeness. In addition to hexagonal adducts urea forms in some exceptional cases a rhomboedric adduct (space group D_{3d}^6—$R\bar{3}c$) with the guest substance. The crystal lattice also has a channel structure. However, the urea molecules are not arranged on the walls of the channels in helical form, but in the form of regularly repeated cuffs. In contrast to the homogeneous force field inside the channels of the hexagonal adducts, in the channels of the rhomboedric adducts different force fields are repeated regularly at distances of 0.55 nm. It is interesting that the rhomboedric adduct is formed just by isoalkanes (and some other compounds) the molecular length of which is approximately equal to the multiple of the indicated period 0.55 nm, i.e. 1.1, 1.65, 2.20, 2.75 nm etc. For example the rhomboedric adduct is formed by 2,15-dimethylhexadecane (length of the molecule 2.26 nm) or 2,19-dimethyleicosane (length of the molecule 2.78 nm). The mole ratio of urea and hydrocarbon is stoichiometric in these adducts and it is in all instances a multiple of 3 : 1; for the first hydrocarbon it is 12 : 1, for the second it is 15 : 1. However, with *n*-alkanes only the hexagonal adduct is always formed even when the length of the molecule would suit the rhomboedric adduct.

From the above survey of the applications of adductive crystallization with urea it is evident that the method is useful for isolations, purifications and separations of straight-chain hydrocarbons and for their analytical determination. The method is still developing and it will evidently become a valuable acquisition for the most subtle separation techniques in the field of stereoisomers and optical antipodes.

No less important are the applications of urea adducts for synthetic chemistry. The guest molecules in the adduct can *polymerize,* which ensures their unambiguous orientation. In the channels of the urea aduct *stereospecific polymerization* of 1,3-butadiene[48], 1,3-pentadiene[49] and other monomers was carried out, for example, under the effect of gamma-radiation (^{60}Co). However, these applications are outside the scope of this monograph.

2.5.2 THIOUREA ADDUCTS

The channel adducts of *thiourea* with hydrocarbons have a *rhomboedric* lattice and are similar to the above mentioned rhomboedric adducts with urea. The molecules of thiourea forming the walls of the channels are not developed as helices (as in the urea hexagonal adducts), but lie there in regular layers at a distance of 0.625 nm[47]. The bulkier sulphur atom in the thiourea molecule causes the cylindrical channels in the adduct to have a larger diameter (about 0.7 nm) than in the urea adducts. From this a wider choice of guest hydrocarbons also ensues. Tens of aliphatic and cyclic hydrocarbons are already known which give crystalline adducts with thiourea. Hence, the selectivity of thiourea is lower in comparison with urea. The decisive factor is the *critical diameter* of the molecule, which can be calculated from van der Waals's atomic radii, and which is determined – for the hydrocarbon chain – by the diameter of the largest circle circumscribing the molecule in the plane perpendicular to the chain length; *n*-alkanes have a critical diameter of the molecule 0.49 nm, monomethylalkanes 0.56, etc. Stable adducts are formed by hydrocarbons the molecules of which fill the cylindrical channel in the adducts tightly. Hydrocarbons the diameter of which is larger do not form adducts. A greater tolerance is observed for smaller molecules. The chains of *n*-alkanes have a considerable freedom in the channel of the thiourea adduct, and therefore lower homologues do not form adducts. With higher *n*-alkanes (C_{14}—C_{19}) it was possible to form adducts [50,55], but they are not very stable. It is thought that the longer chains of *n*-alkanes are to a certain extent coiled in the adduct, so that they fill the channel better.

So far the most reliable method for predicting whether some hydrocarbon will form an adduct with thiourea or not is the method proposed by Schiessler and Flitter[52]. Three dimensions of a molecule are measured on its space-filling model (1 cm = 0.1 nm) as shown in Fig. 2.5.6. Hydro-

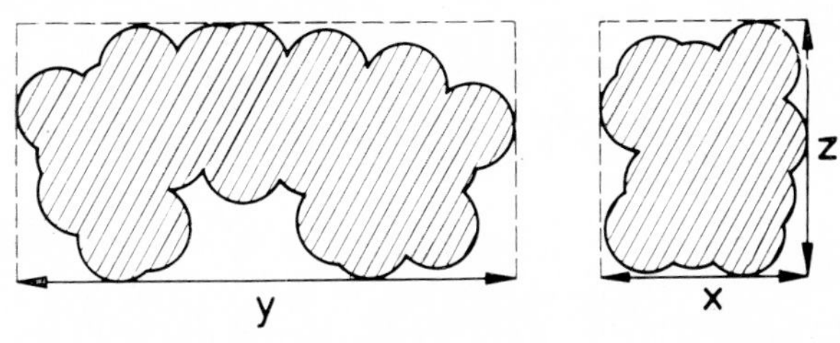

Fig. 2.5.6 Measurement of the dimensions of molecular models according to Schiessler and Flitter[52]
y — length of the molecule; x, z — dimensions of the critical cross-section of the molecule

carbons with a cross-sectional dimension $X = 0.58 \pm 0.08$ and $Z = 0.68 \pm 0.06$ nm have the required conditions for the formation of an adduct. Benzene and its lower homologues are an exception, because though having the required dimensions, they do not form adducts.

According to *X-ray* measurements[51] the channels of thiourea adducts are occupied with less regularity than those of urea. The hydrocarbon molecules are also more compressed in the channel.

With the exception of cases which will be mentioned later, the mole ratio of the components in the adduct is non-stoichiometric and it depends on the structure of the guest hydrocarbon. In Table 2.5.4 the composition of the adducts of several C_{10} hydrocarbons of different structure is given. It may be seen that the mole ratio decreases in the order isoalkanes – monocycloalkanes – dicycloalkanes – tricycloalkanes. The more compact the structure of isoalkanes, or else the more condensed the rings of cycloalkanes, the "more economically" the space of the channels in the adduct is utilized, and the lower is the number of thiourea molecules necessary to shut one hydrocarbon molecule into the lattice. The variation of the mole ratio in dependence on the structure and the number of carbon atoms in the guest hydrocarbon is shown in Fig. 2.5.7. For comparison mole ratios of urea adducts with *n*-alkanes are also shown in the graph.

For the choice of reaction conditions the weight ratio is important, which is between 2 and 3 g of thiourea per 1 g of hydrocarbon for the majority of adducting hydrocarbons.

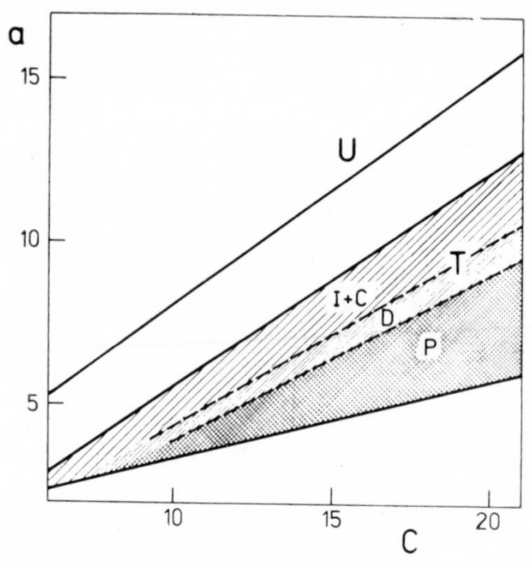

Fig. 2.5.7 Dependence of the molar ratio of the adducts on the number of carbon atoms (*C*) and the structure of hydrocarbon
a — number of moles of thiourea (urea) per 1 mole of hydrocarbon; *U* — adducts of urea with *n*-alkanes; *T* — thiourea adducts; *I* + *C* — isoalkanes + cycloalkanes; *D* — dicycloalkanes; *P* — polycycloalkanes

The relative stability of thiourea adducts can be compared according to the measured *heats of formation* or *equilibrium constants*[1,9]. More recently the ability of a hydrocarbon to form an adduct has been judged according to the *heat of dissociation* of the adduct, measured by DTA[53], and according to the so-called *individual equilibrium concentration* (C_e)[54]. The last value is measured as the residual concentration of the hydrocarbon in the inert solvent after the formation of adduct in the presence of a large excess of thiourea; in other words, it is the minimum concentration, at which the hydrocarbon must be present, if the adduct is to be formed. Hydrocarbons with a low equilibrium concentration form adducts with

TABLE 2.5.4

Composition of some thiourea adducts with C_{10} hydrocarbons

Hydrocarbon	Composition	Ratio of thiourea to hydrocarbon	
		molar	weight
2,7-Dimethyloctane	$C_{10}H_{22}$	5.9	3.2
2,2,3,5-Tetramethylhexane	$C_{10}H_{22}$	5.1	2.7
Isobutylcyclohexane	$C_{10}H_{20}$	4.9	2.7
1,2,4,5-Tetramethylcyclohexane	$C_{10}H_{20}$	4.3	2.3
trans-Decalin	$C_{10}H_{18}$	4.6	2.5
cis-Decalin	$C_{10}H_{18}$	4.0	2.2
Perhydrodicyclopentadiene	$C_{10}H_{16}$	3.7	2.1
Adamantane	$C_{10}H_{16}$	3.4	1.9

a high dissociation temperature and *vice versa.* In hydrocarbon mixtures of known composition the composition of the raffinate (mother liquor) after adductive crystallization can be calculated from individual equilibrium concentrations C_e of adducting hydrocarbons. The concentration of the adducting hydrocarbon A, in the raffinate is given by the relationship:

$$C'_{eA} = \frac{w_A}{\Sigma w} \cdot C_{eA} \qquad (2.5.1)$$

where C'_{eA} is the residual concentration of the adducting hydrocarbon A in the raffinate and C_{eA} is the individual equilibrium concentration of hydrocarbon A (both in wt. %), w_A is the wt. % of hydrocarbon A in the original mixture and Σw is the sum of the concentrations of all adducting hydrocarbons in the mixture (in wt. %). The composition of the raffinate calculated according to equation (2.5.1) is in good agreement with experimental data[54].

The structures of adducting hydrocarbons. Among aliphatic hydrocarbons mainly polymethylated alkanes, but also higher n-alkanes, form adducts with thiourea[50,51]. An adduct of n-tetradecane is formed at 0 °C, but not at 25 °C. Only from n-hexadecane upwards adducts can be prepared even at room temperature. X-ray measurements have indicated

TABLE 2.5.5

Isoalkanes adducting with thiourea

Hydrocarbon	Composition	Moles of thiourea per 1 mole of hydrocarbon
2,2-Dimethylbutane	C_6H_{14}	2.6
2,2-Dimethylhexane	C_8H_{18}	4.9
2,3-Dimethylbutane	C_6-H_{14}	2.9
2,5-Dimethylhexane	C_8H_{18}	4.9
2,7-Dimethyloctane	$C_{10}H_{22}$	5.9
2,8-Dimethylnonane	$C_{11}H_{24}$	5.9
2,9-Dimethyldecane	$C_{12}H_{26}$	6.2
2,2,3-Trimethylbutane	C_7H_{16}	3.2
2,2,4-Trimethylpentane	C_8H_{18}	4.1
2,4,6-Trimethyltetradecane and C_{18}—C_{24} homologues		7.1
2,4,7-Trimethyloctane	$C_{11}H_{24}$	6.3
2,5,9-Trimethyldecane	$C_{13}H_{28}$	7.0
2,6,10-Trimethylundecane and C_{15}—C_{18} homologues	$C_{14}H_{30}$	7.0
2,2,3,3-Tetramethylbutane	C_8H_{18}	3.1
2,2,3,5-Tetramethylhexane	$C_{10}H_{22}$	5.1
2,2,4,4-Tetramethylheptane	$C_{11}H_{24}$	4.0
2,6,9,11-Tetramethyldodecane	$C_{16}H_{34}$	7.9
2,6,10,14-Tetramethylpentadecane (pristane)	$C_{19}H_{40}$	8.8
2,6,10,14-Tetramethylhexadecane (phytane)	$C_{20}H_{42}$	9.2
2,6,11,15-Tetramethylhexadecane	$C_{20}H_{42}$	10.1
2,2,4,6,6-Pentamethylheptane	$C_{12}H_{26}$	5.5
2,6,9,12,16-Pentamethylheptadecane	$C_{22}H_{46}$	10.9
2,6,10,15,19,23-Hexamethyltetracosane (squalane)	$C_{30}H_{62}$	15.4[a]

[a] The preparation of the adduct is described on p. 261

that the length occupied by the *n*-alkane molecule in the channel is only 82 % of the maximum length of the molecule[51]; this means that the chain of the *n*-alkane is coiled to some extent in the channel, which shortens its length and the effective cross-section of the molecule is increased. Similarly as with urea very stable adducts of thiourea with *linear polymers*[5] of molecular weight above 10^5 are also formed.

Isoalkanes must have at least two side methyls in order to form an adduct with thiourea. With the exception of 2,4-dimethylalkanes that do not give adducts, other dimethylalkanes with methyl groups in positions 2,3- up to 2,10- (with terminal isopropyl groups) give them. If a further side methyl is present in any place on the chain in addition to the methyls in the unreactive positions 2,4- the hydrocarbon is again capable of adducting. The bulkier side substituents, as for example ethyl, isopropyl, etc. prevent the formation of an adduct.

Some isoalkanes the adducts of which with thiourea have been described

TABLE 2.5.6

Effect of terminal groups of isoprenoid alkanes on the stability of thiourea adducts

Hydrocarbon	Number of C atoms	Dissociation temperature °C	Equilibrium concentration C_e in wt. %
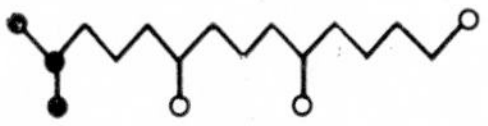 2,6,10-Trimethylpentadecane	18	100	32
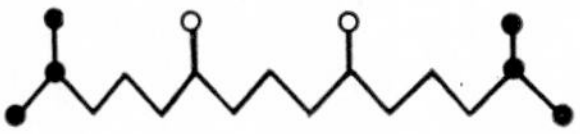 2,6,10,14-Tetramethylpentadecane	19	117	16
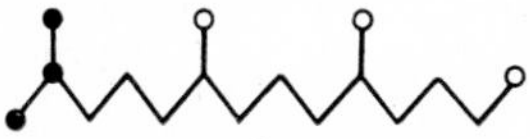 2,6,10-Trimethyltetradecane	17	101	35
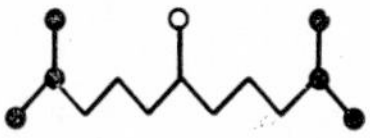 2,6,10-Trimethylundecane	14	110	9

in literature[6,18,54,55] are given in Table 2.5.5. As the longest isoalkane adducting with thiourea the literature gives 2,6,9,12,16-pentamethylheptadecane[2,55]. According to another paper[56], which is in good accordance with our own tests (see below), the C_{30} *isoprenoid* alkane *squalane* also forms the adduct. From the comparison of the length of the squalane and squalene chain in the adduct it was deduced that all six double bonds of squalene are *trans* (the isolated *cis* double bond shortens the chain by 0.088 nm, while the shortening caused by the isolated *trans* double bond is only 0.019 nm).

TABLE 2.5.7

Relative stability of thiourea adducts with alkylcyclohexanes

Hydrocarbon	Dissociation temperature °C	Equilibrium concentration C_e in wt. %
Cyclohexane	134	4
Methylcyclohexane	134	33
cis-1,2-Dimethylcyclohexane	93	68
trans-1,2-Dimethylcyclohexane		
1,3-Dimethylcyclohexane		non-adducting
cis-1,4-Dimethylcyclohexane	100	50
trans-1,4-Dimethylcyclohexane		55
Ethylcyclohexane	90	80
1,2,3-Trimethylcyclohexane		non-adducting
1-*t*-2-*c*-4-Trimethylcyclohexane	93	32
1-*c*-2-*t*-4-Trimethylcyclohexane		
1-*c*-2-*c*-4-Trimethylcyclohexane		non-adducting
1,3,5-Trimethylcyclohexane		non-adducting
Isopropylcyclohexane	100	30
1-*t*-2-*c*-4-*t*-5-Tetramethylcyclohexane		36
1-*c*-2-*t*-4-*c*-5-Tetramethylcyclohexane		40
1-*t*-2-*c*-4-*c*-5-Tetramethylcyclohexane	100	67
1-*c*-2-*c*-4-*c*-5-Tetramethylcyclohexane		74
1-*t*-2-*t*-4-*t*-5-Tetramethylcyclohexane		non-adducting

The stability of isoalkane adducts does not depend either on the chain length or on the molecular weight of the hydrocarbon. The structure of the guest has a decisive effect. The stability of the adduct is decreased by the freely moving unbranched ends of the hydrocarbon chain. If both ends are terminated with isopropyl groups that are more strongly "anchored" in the channel and cannot freely oscillate, then the resistance of the crystal lattice of the adduct to decomposition is increased. This is demonstrated by Table 2.5.6 in which dissociation temperatures of the adducts and the equilibrium concentrations of four isoprenoid alkanes are compared[53,54].

The basic *monocycloalkanes* – cyclopentane, cyclohexane and cyclooctane – form relatively stable adducts with thiourea, which dissociate only at 128 – 134 °C.

The ability of *alkyl derivatives* of cyclohexane to form adducts depends on the number and size of substituents and on their configuration. Table 2.5.7 gives a survey of some tested derivatives[53,54]. A general tendency has been observed that the increasing number of methyl groups on the ring decreases the stability of the adduct. A freely rotating ethyl group decreases the stability of the adduct more than several methyl groups. The long unbranched chain bound to the cyclohexane nucleus does not prevent adduct formation. A number of adducting *n-alkylcyclohexanes* with 3 – 14 carbon atoms in the chain have been described[55]. Some hydrocarbons of this homologous series form adducts in *stoichiometrical* mole ratio. These are hydrocarbons the ideal molecular length of which is approximately a multiple of the periodical discontinuity of the force field in the adduct's channel (0.625 nm). Homologues with 5, 6 and 7 carbons in the chain, with a length of between 1.1 and 1.3 nm, form adducts with a 6 : 1 mole ratio and homologues with 11 and 12 carbons in the chain (ideal length of the molecules about 1.9 nm) form adducts with a 9 : 1 mole ratio. The alkyl chains of these homologues are shortened in the adduct by torsion, so that the molecule fills only just two or three periodic sections in the channel (each section is formed by three molecules of thiourea).

A number of *bicyclic* hydrocarbons, both with isolated and with condensed rings, also form adducts with thiourea. The structures of some described adducting hydrocarbons[53–55,57] are given in the following survey:

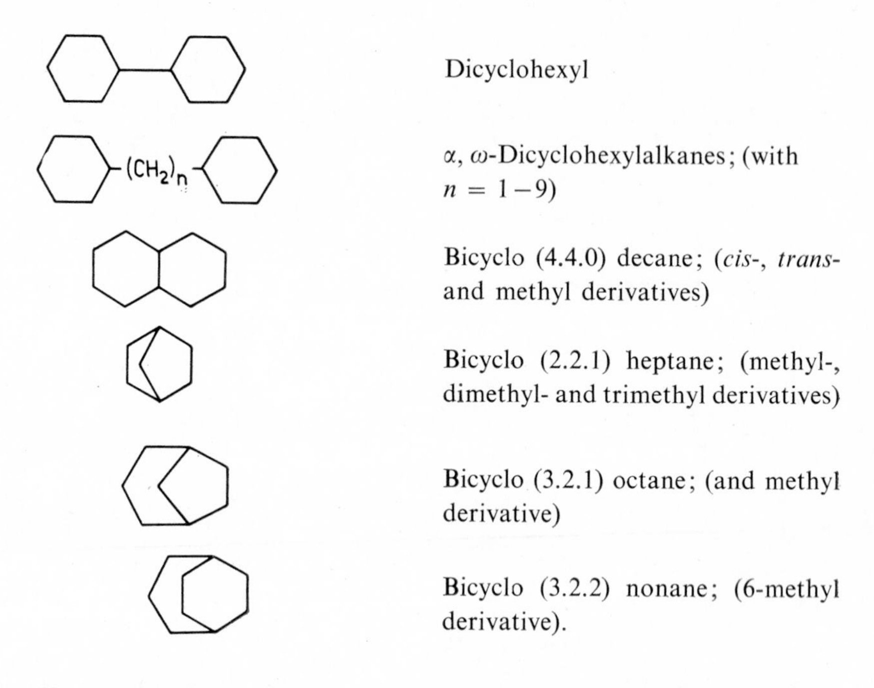

Unusually stable adducts are formed by *α,ω-dicyclohexylalkanes*. For example the dissociation temperature of 1,2-dicyclohexylethane is 172 °C. The stability of these adducts is attributed to the effect of bulky terminal groups, which similarly as the isopropyl groups in some isoprenoids anchor both ends of the alkane chain in the channel and thus prevent their oscillation. For the same reasons, mentioned in connection with *n*-alkylcyclohexanes, several members of the homologous series of α, ω-dicyclohexylalkanes form adducts in a stoichiometric mole ratio. Hydrocarbons with 1, 2 and 3 carbons in the chain, and with the ideal length of the molecule between 1.1 and 1.3 nm, have a 6 : 1 molar ratio, and homologues with 6, 7 and 8 carbons in the chain (length 1.8 – 1.9 nm) have a 9 : 1 mole ratio.

Condensed bicycloalkanes form adducts much more easily than monocycloalkanes with the same number of carbon atoms. Hydrocarbons, as for example bicyclo(3.2.1)octane, with individual equilibrium concentration $C_e = 0.5\ \%$, can therefore be isolated from various fractions by adductive crystallization in high yield.

While the substitution of bicyclic hydrocarbons with methyl groups does not prevent adduct formation, the ethyl group decreases the stability of the adduct considerably, and the vinyl and ethylidene groups prevent adducting. In contrast to this a double bond in the nucleus does not prevent adduct formation.

Among *tricycloalkanes*, the ability of which to form adducts with thiourea has been checked by qualitative tests[57], the following deserve mention:

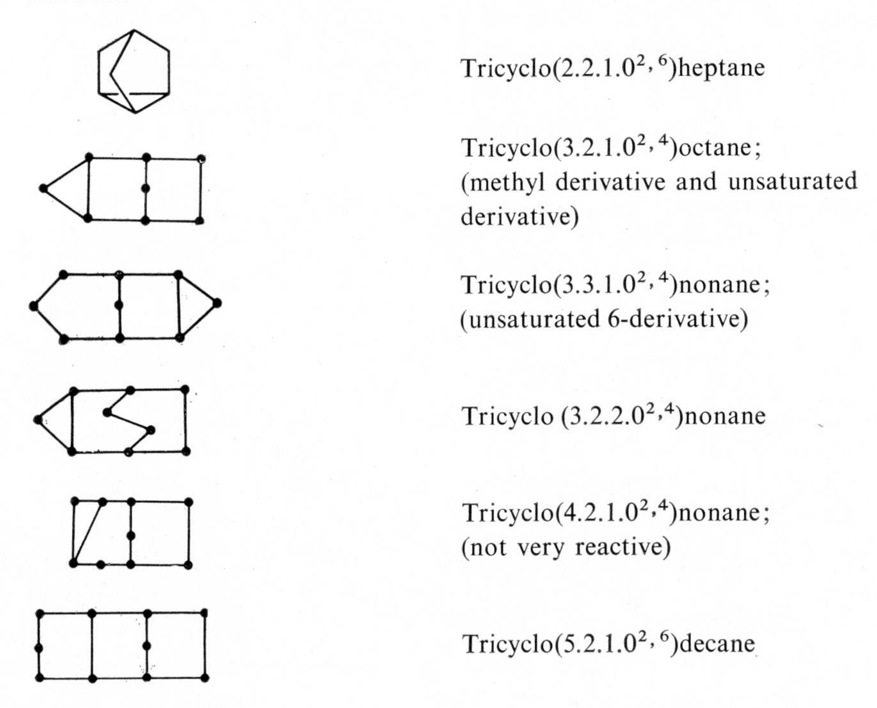

An exceptionally good ability to form adducts with thiourea was found in the authors' laboratory in tricyclo(3.3.1.$1^{3,7}$)decane (*adamantane*) and some of its derivatives[58]. Symmetrical spherical molecules of adamantane fill the channels in the crystal lattice almost perfectly, which is also reflected in the outstanding stability of the adduct[53]; dissociation temperature is 172 °C and the individual equilibrium concentration $C_e = 0.5\,\%$. The mole ratio of thiourea and adamantane is 3.4 and it is close to the mole ratio 3.1 for the cyclohexane adduct. This means that the 10 carbons of the adamantane nucleus occupy about the same space

TABLE 2.5.8

Effect of the position of alkyl on the capability of alkyladamantanes to form adducts with thiourea

Adducting hydrocarbons	Non-adducting hydrocarbons
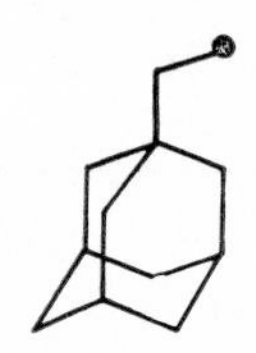 1-Ethyladamantane	 *cis*-1-Ethyl-4-methyladamantane
	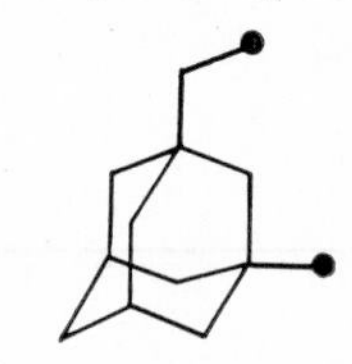 1-Ethyl-3-methyladamantane
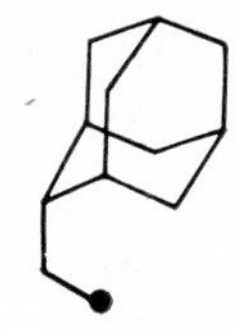 2-Ethyladamantane	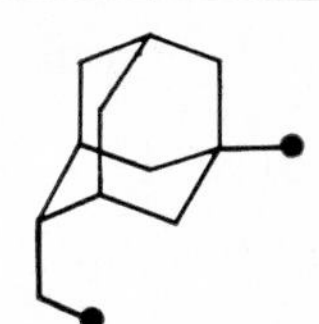 *cis*-4-Ethyl-1-methyladamantane
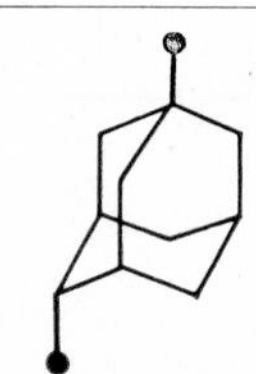 *trans*-1,4-Dimethyladamantane	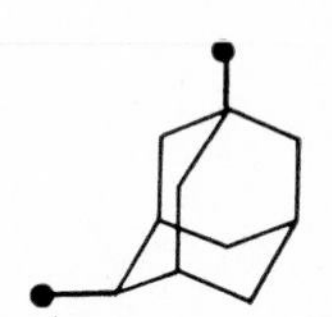 *cis*-1,4-Dimethyladamantane
	 1,3-Dimethyladamantane

in the channel as the 6 carbons of cyclohexane, so that the degree of filling of the channel space in the adamantane adduct is much higher.

The effect of the alkyls in various positions on the size of the molecule and the ability to form adducts can easily be followed on the rigid adamantane nucleus by the method of Schiessler and Flitter[52]. In Table 2.5.8 the structures of adduct-forming and non-adducting alkyladamantanes are compared. It is evident that even in disubstituted alkyladamantanes only a small difference in the orientation of the alkyls, for example the conversion of a *trans*-configuration to a *cis*-configuration, can prevent adduct formation. The dimensions of the hydrocarbon molecules, expressed as X/0.58 and Z/0.68 are practically the same, 1.05 and 0.96, for all adducting alkyladamantanes listed in the table. In contrast to this, in the case of non-adducting alkyladamantanes at least one of the dimensions attains or exceeds the value 1.1[54].

Among *polycycloalkanes* with more than three rings hydrocarbons capable of entering into adduct with thiourea were also found. Such are for example the following two tetracyclic derivatives of adamantane[58, 59]:

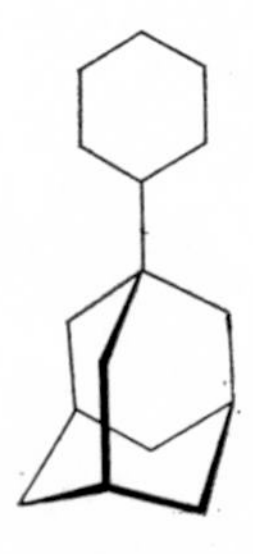

$C_{16}H_{26}$
1-Cyclohexyladamantane

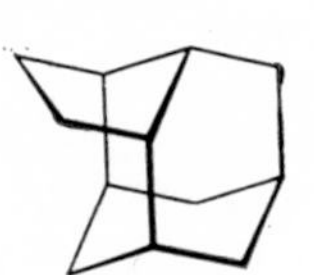

$C_{12}H_{18}$
Tetracyclo(6.3.1.$0^{2,6}$. $0^{5,10}$) dodecane; (2,4-ethanoadamantane)

In addition to the highly condensed hydrocarbons mentioned *tetracycloalkanes* with a relatively unfolded structure are also capable of adductive crystallization. Murphy, McCormick and Erlington[60] found that some *steranes* present in natural *bitumens* are also capable of forming adducts:

Cholestane; $C_{27}H_{48}$; (R = H)

Ergostane; $C_{28}H_{50}$; (R = CH_3)

Stigmastane; $C_{29}H_{52}$; (R = C_2H_5)

Adductive crystallization of these steranes takes place with high selectivity with respect to *stereoisomers*[61,62]; the adduct is formed in all cases known with isomers of *trans*-configuration of the rings A and B only, i.e. with α-steranes, while β-steranes do not enter the crystal lattice and remain in the raffinate.

Among higher polycycloalkanes pentacyclic *diamantane* and hexacyclic 1,1′-*biadamantane*[58] are known to form adducts:

$C_{14}H_{20}$
Pentacyclo(7.3.1.$1^{4,12}$. $0^{2,7}$. $0^{6,11}$)-tetradecane; (diamantane)

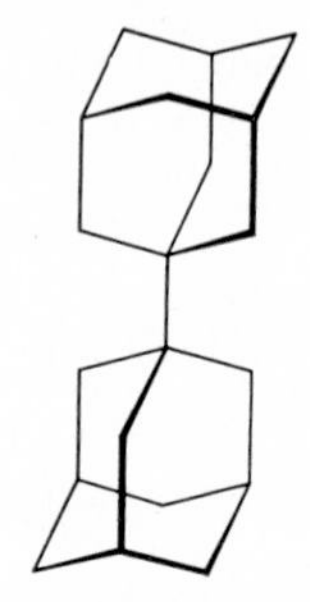

$C_{20}H_{30}$
1,1′-Biadamantane

The molecules of biadamantane have a cylindrical shape and they fill up the channels in the adduct still better than the spherical molecules of adamantane. The biadamantane adduct with a mole ratio 5.4 is the most stable of all the adducts of thiourea with hydrocarbon known so far. When heated in a sealed capillary the dry adduct of biadamantane melts (with decomposition) only at a temperature 10 K higher than the melting point of pure thiourea. The exceptionally high stability of this adduct is also documented by the DTA curve shown in Fig. 2.5.8.

In the preceding text mention was made of the fact that *benzene* and its lower homologues do not enter into the adducts with thiourea. However, if the benzene nucleus is substituted with a cycloalkyl group or a suitable branched alkyl, such hydrocarbons are again capable of forming adducts[55]. Among methylsubstituted benzenes C_8—C_{11} a stable adduct is formed only in the case of 1, 2, 4, 5-tetramethylbenzene (*durene*)[63]

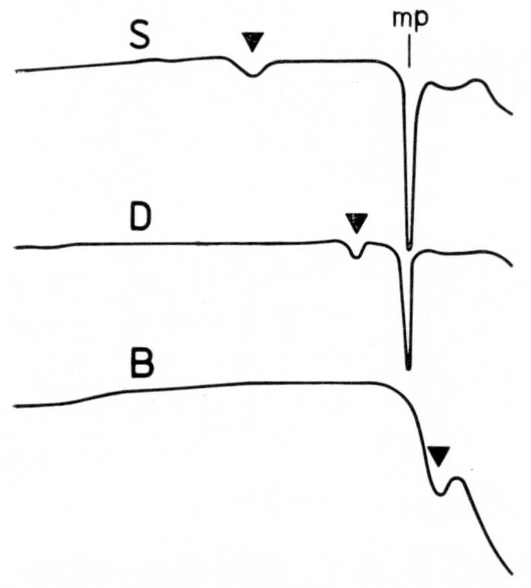

Fig. 2.5.8 Differential thermal analysis of thiourea adducts:
S — squalane, *D* — diamantane, *B* — 1,1′-biadamantane; ▼ — dissociation temperature of the adduct; *mp* — melting point of thiourea

which can be selectively isolated from a mixture of other tetramethylbenzenes in this way. *Tetralin* does not form an adduct, nor do unsubstituted polynuclear aromatic hydrocarbons naphthalene, anthracene and phenanthrene. In *binuclear* aromates the presence of a single methyl group in the nucleus suffices for adduct formation. Montgomery[64] found the ability to form crystalline adducts with thiourea in some alkylated derivatives of naphthalene, diphenyl, fluorene, anthracene and phenanthrene (even in some heterocycles). When pure compounds and model mixtures were used the ability to form adducts was detected in 2-methyl-, 2, 3- and 2,6-dimethyl-, 2,3,6-trimethyl- and 2,3,6,7-tetramethylnaphthalene;

2-methyl-, 2,3-, 2,6- and 2,7-dimethyl- and 2,3,6-trimethyl-anthracene, and in 2-methylphenanthrene. The mole ratio of thiourea and 2,3,6-trimethylnaphthalene was found to be 5.9 : 1.

A more detailed investigation of the adduct of aromatic hydrocarbons has shown that they are crystallographically different from rhomboedric adducts of saturated hydrocarbons. However, the channel-type structure of the adducts is most probably retained, and the difference consists only in the dimensions of the channels.

The critical cross-sectional dimension of the *linearly* condensed aromatic molecules (measured on models) is 0.77 nm. The methyl substituent in the position α increases the critical dimension to 0.86 nm, which evidently exceeds the diameter of the channel. In contrast to this the alkyls in the β-position may be oriented so that the original dimension of the molecule, 0.77 nm, does not increase. This may explain the higher ability for adducting in the mentioned β-methyl derivatives in comparison with their α-isomers. A similar situation also occurs in phenanthrene; the critical dimension of the molecule (0.83 nm) does not increase only on substitution in the positions 1, 2, 7 and 8. In accordance with this it was found that 2-methylphenanthrene forms adducts more easily than 3-methylphenanthrene. According to the dimensions of molecular models some suitably alkylated chrysenes and picenes also could undergo adduct formation.

So far, only very scant data is available concerning the relative stability of the adducts of aromatic hydrocarbons. Schlenk[55], for example, lists 1-methylnaphthalene and 1, 6-dimethylnaphthalene among spontaneously adducting aromates.

Applications. The techniques of the preparation of adducts with thiourea are analogous to those with urea. The hydrocarbon fraction is mixed with a concentrated or saturated thiourea solution in methanol (at 20 °C the saturated solution contains 11 wt. % of thiourea) and the adduct is isolated in the form of white needles. Well developed crystals are obtained by slow cooling of the warm mixture. The *extract* is set free by decomposing the adduct in hot water. Fractions of larger volumes are worked up with advantage by the *slurry technique*.

Before the separation proper it is advisable to test the ability of individual components of the mixture to form an adduct. For the testing of pure hydrocarbons the following procedures have proved suitable:

Preparation of thiourea adducts of pure hydrocarbons

Procedure A[6]: The tested hydrocarbon (0.5 ml) is shaken in a 10 ml ampoule with 5—6 ml of saturated thiourea solution in methanol at room temperature. Solid hydrocarbons are dissolved in a few drops of benzene. The separation of the adduct is often speeded up by the addition of a few drops of a mixture of benzene and methyl ethyl ketone (1 : 1).

Procedure B[58]: 20 mg of a hydrocarbon are dissolved in a glass ampoule in 2 ml of benzene and 10 ml of an 8 % thiourea solution in methanol are added to it. The ampoule is sealed and heated in a water bath at 60—70 °C under occassional shaking. The ampoule with the clear solution is allowed to stand overnight in a cold place (5—10 °C). The separated needles are filtered off and washed with 2 ml of cold methyl ethyl ketone.

Procedure C[57]: Some hydrocarbons which do not give an adduct with saturated thiourea solution in methanol crystallize from a saturated thiourea solution in benzene-methanol (10 : 1); for a qualitative test one to two drops of hydrocarbon are added to 0.5 ml of thiourea solution.

Procedure D: In high-boiling hydrocarbons, almost insoluble in methanol, the above described procedures fail. These hydrocarbons, however, form adducts in *n*-propanol, as shown by the examples of 1,1′-*biadamantane* (C_{20}) and *squalane* (C_{30}).

Biadamantane (20 mg) dissolved in 2 ml of hot benzene in a 100 ml flask is mixed with 40 ml of a saturated thiourea solution (at room temperature) in *n*-propanol. The mixture is refluxed for a short period, until the solution is clear. The flask is then closed and set aside in a cold place. The adduct crystallizes in long white needles (22.25 % N) with m.p. in a sealed capillary 193 °C (under decomposition); the dissociation temperature of the adduct according to DTA is 189 °C (Fig. 2.5.8).

5 drops of squalane are added into a glass ampoule containing 10 ml of a saturated (at 20 °C) thiourea solution in *n*-propanol. The ampoule is sealed and heated in a water bath at 60—70 °C under occasional shaking until a clear solution is formed. On standing in a cold place (about 5 °C) overnight the adduct separates in the form of white needles (27.05 % N). The dissociation temperature of the adduct is according to DTA 141 °C (Fig. 2.5.8).

In view of the wide choice of hydrocarbons of various structures that form adducts with thiourea the selectivity of the process is lower than with urea. On the other hand adductive crystallization with thiourea has a broader range of application. The accessibility of the crystal lattice of the thiourea adduct for hydrocarbons of various structural classes can be easily eliminated by applying adductive crystallization at later stages of the separation scheme. It is just for fractions separated previously by other methods and containing only a limited number of structural groups that the special selectivity of thiourea can be best made use of. The utility of the method is further demonstrated by various examples of applications described in literature.

The extracts isolated with thiourea from isoalkane-cycloalkane fractions of petroleum contain, according to structural type analysis, predominantly monosubstituted cyclohexanes and branched alkanes with terminal isopropyl groups and side methyls within the chain[65].

Individual isoprenoid hydrocarbons *pristane* and *phytane*[60] were isolated from *shale bitumens* with thiourea, and it is certain that this isolation technique may be applied to other isoprenoid hydrocarbons from crude oil and other raw materials. In the extract of the shale oil *squalane* was most probably also identified[66].

In the authors' laboratory adductive crystallization with thiourea was used for the isolation of *adamantane*[67,68], its homologues[69] and higher *diamondoid* hydrocarbons[59] from petroleum. Adamantane enters into the adduct so easily that it can be isolated in crystalline form even from 2 kg of naphthenic crude oil. The procedure is as follows:

Isolation of adamantane from naphthenic crude oil[67]

A fraction boiling approximately at 170—240 °C is isolated from 2 kg of crude oil by steam distillation or rectification and it is then dried and filtered through 1—2 g of silica gel. Powdered thiourea (17 g) and 22 ml of methanol are then added to 170 g of the fraction and the mixture is stirred at room temperature for 2 h. The adduct slurry is filtered off under sharp suction, then washed with a few ml of cold pentane. After half-an-hour's suction-drying the adduct is introduced into a flask containing 500 ml of water, a descending condenser is connected to the flask and about 100 ml of water are distilled over. The oily extract (10—20 ml) in the condensate co-distilled with steam, is separated from water then dried and allowed to freeze out overnight at —50 to —70 °C. The crystallized out adamantane when recrystallized from methanol and sublimated has m.p. 268—269 °C (in a sealed capillary). In the case of naphthenic crudes with a low content of gasoline the yield is usually about 0.5 g of adamantane, which often crystallizes even in the condenser during the distillation of the extract.

In the case of crude oils with a lower content of adamantane it is convenient if the obtained extract is submitted to a repeated adductive crystallization with thiourea and adamantane is isolated from the second extract.

Using adductive crystallization adamantane was determined in petroleums of various origins and its content was correlated with their geological age[70,71]. The composition of the thiourea extract from strongly naphthenic petroleum is evident from the chromatogram in Fig. 2.5.9; adamantane and both its methyl homologues are the dominant components. Among the low-boiling adducting hydrocarbons (70 – 180°) 12 bicycloalkanes C_8—C_9[72] were identified. In higher boiling fractions of the extract higher (predominantly mono substituted) alkyladamantanes[73, 74] were identified in addition to adamantane and its methyl analogues. Polyalkylated adamantanes that were identified in petroleum by other methods[75] do not pass into the extract.

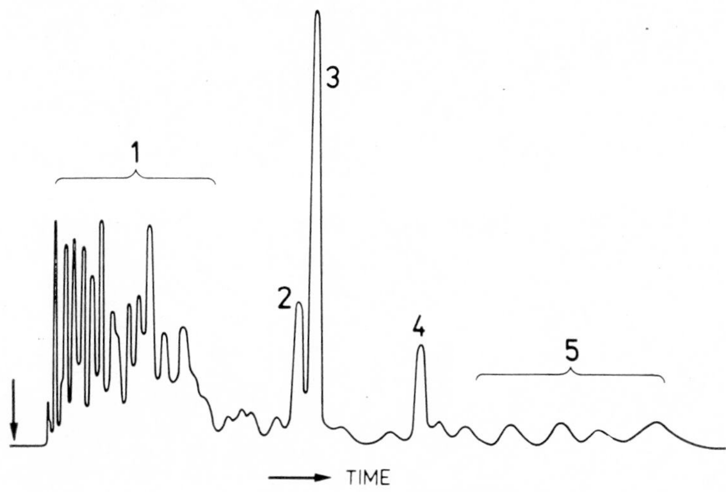

Fig. 2.5.9 Composition of the thiourea extract from naphthenic crude oil[69]
1 — monocyclic and dicyclic naphthenes; 2 — adamantane; 3 — 1-methyladamantane; 4 — 2-methyladamantane; 5 — polycycloalkanes

Recently adamantane was also isolated with thiourea from the soluble coal fraction[76]. In *bitumenaceous* coal the content of adamantane is a mere 0.16 ppm, i.e. several orders of magnitude lower than in petroleum.

Differing abilities of some isomeric alkyladamantanes to form adducts have also been made use of for the separation of synthetic mixtures. For example 1-ethyladamantane and 1,3-dimethyladamantane[77] were separated so, which are formed as main products during the isomerization of perhydroacenaphthene. Similarly, *cis*-1,4-dimethyladamantane was separated from its *trans*-isomer using thiourea[78].

Ozokerites of various origin[79] also were characterized by the method of adductive crystallization. In individual samples the presence of four fractions was detected: *A* – hydrocarbons forming adducts with urea only, *B* – hydrocarbons forming adducts both with urea and with thiourea, *C* – hydrocarbons forming adducts with thiourea only, and *D* – unreactive material (see Fig. 2.5.10). The method can be applied even to other similar raw materials and products with a predominating content of high-molecular hydrocarbons.

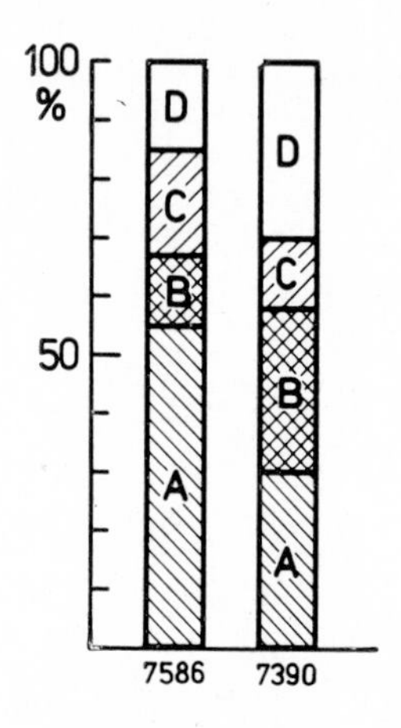

Fig. 2.5.10 Distribution of hydrocarbons adducting with urea and thiourea in samples of American (7586) and Russian (7390) ozokerite (according to W. Schlenk[79])
A — hydrocarbons forming adducts with urea only, *B* — hydrocarbons forming adducts both with urea and with thiourea, *C* — hydrocarbons forming adducts with thiourea only; *D* — unreactive hydrocarbons

The method of adductive crystallization displays surprising efficiency in the isolation of *steranes* from natural bitumens. 5α-Isomers of *cholestane, ergostane* and *stigmastane*[60,61] have been isolated from shale oils. In the extract obtained by means of thiourea from the concentrate of polycyclic naphthenes from crude oil tetracyclic *triterpane α-lanostane* ($C_{30}H_{50}$)[62] was also isolated in addition to the above-mentioned α-steranes C_{27}—C_{29}. The procedure used for the isolation is the following:

Isolation of steranes from natural bitumens by adductive crystallization with thiourea

The hydrocarbon fraction (1—20 g) with a concentrated content of polycyclic naphthenes (prepared, for example, by thermal diffusion of the dearomatized fractions boiling at 450—500 °C) is dissolved in 100 ml of chloroform and 100 ml of a saturated thiourea solution in methanol is added. The mixture is heated until it is homogeneous and it is then allowed to stand at room temperature for one day. The needles of the adduct that crystallized out are filtered off, washed twice with cold chloroform and decomposed in hot water. Thus steranes can be isolated simultaneously with isoprenoid alkanes[60] from isoalkane-cycloalkane fractions.

Linearly condensed *polynuclear aromates* methylated at the β-carbons of the nucleus[64] were isolated from the concentrate of aromatic hydrocarbons (95 wt. %) prepared from the heavy oil from catalytic cracking using thiourea. The extract was prepared in two steps: in the first the fraction was stirred with crystalline thiourea and methanol for 24 h at 24 °C. The crude extract was dissolved in a minimum amount of benzene and mixed with 33 volumes of saturated thiourea solution in methanol. The mixture was heated and the second adduct crystallized out during the slow cooling.

Among other applications the *stereospecific polymerization* of 2,3-dimethylbutadiene, 1,3-cyclohexadiene and other monomers in the channel of thiourea adduct is of interest[80]. The ability of thiourea to form adducts with *ferrocene* and some other metalocenes[80a] is of practical importance. However, it seems that the discovery of the ability of thiourea to form stable adducts with some polymers[5] remains unappreciated.

Selenourea has a similar ability to form rhomboedric adducts with hydrocarbons as thiourea. Known hydrocarbons that form adducts with selenourea[80b] are the following:

4-*t*-butyl-1-neopentylbenzene
1,4-dineopentylbenzene
trans-1,4-diisopropylcyclohexane
trans-1-*t*-butyl-4-isopropylcyclohexane
trans-1,4-di-*t*-butylcyclohexane
cis-1,4-di-*t*-butylcyclohexane
trans-1-*t*-butyl-4-neopentylcyclohexane
bicyclo(3.3.1)nonane
adamantane

All the mentioned hydrocarbons also form adducts with thiourea. In some instances, however, selenourea displays a higher selectivity in the choice of guests. For example, both isomers (*cis, trans*) of 1-*t*-butyl-4-neopentylcyclohexane form adducts with thiourea, while selenourea adducts with the *trans*-isomer, but not with the *cis*-isomer.

2.5.3 CLATHRATES OF HYDROCARBONS

In clathrates each molecule of the guest is closed from all sides by the lattice (in Latin *clatra*) formed from the molecules of the host. The cavities (cages) in the crystal lattice may be almost spherical or they can have the form of various regular polyhedric forms, or even of channels. According to the size of the cages only molecules of guests with suitable dimensions are entrapped. In contrast to adducts a part of the cavities in the clathrate lattice may remain unoccupied without this causing the lattice to collapse. The mole ratio of the guest to the host may, therefore, have various lower values than would correspond to a 100 % filling of the cavities.

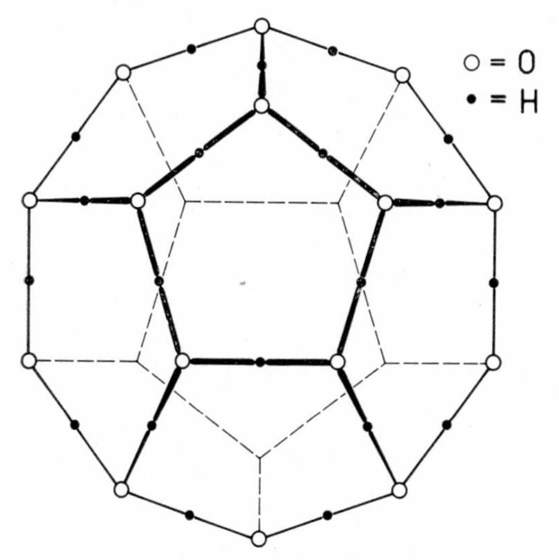

Fig. 2.5.11 A cell of crystal hydrate in the shape of a pentagonal dodecahedron within which a gas molecule can be trapped

More important types of clathrate that retain hydrocarbons will be briefly described below. Considerable differences exist among individual types of clathrates. They differ both in the character of the bonds binding the molecules of the host into the cage structure and in the shape of the cavities, stability of the complex and its capacity for the guest molecules. For this reason the boundaries between clathrates and other classes of complex compounds are not quite clear, the more so because in some clathrates the structure is not yet completely elucidated.

Among clathrates that entrap hydrocarbons belong the well known *hydrates of gases*[1, 2, 81, 82]. If water solidifies in the presence of hydrocarbon gases it forms a crystal lattice with low density in which polyhedric cavities are formed (see Fig. 2.5.11) occupied by a gas molecule. Several crystalline forms of hydrates with different content of gaseous component are known; for example the composition of the methane and propane clathrate is $5.75\ H_2O\ .\ CH_4$ and $17\ H_2O\ .\ C_3H_8$. Changing the temperature and pressure conditions the gas can be set free again from the hydrate. For example at a 3.4 MPa pressure the hydrate of ethane is formed on cooling below 14 °C, while at a temperature above 16 °C it again decomposes.

Hydrates of hydrocarbons are not of special importance for laboratory separations. However, they are given great attention from other points of view. The separation of crystalline hydrates of methane and other hydrocarbons in natural gas pipelines is a serious problem. On the other hand it opens up prospects for obtaining desalted water from the sea by using hydrates of propane and other gases (1 kg of propane binds about 7 kg of pure water in crystalline state).

The structure of clathrates formed with *β-hydroquinone*[2, 81, 83] is also being investigated in detail. The cages in the clathrate structure are of approximately spherical shape with a free space for the guest, of about 0.4 nm diameter. The elastic structure of the lattice permits a certain adjustment of the cage to the shape of the guest molecule. The molecule of the guest is surrounded by three molecules of hydroquinone. In the lattice 6 molecules of hydroquinones are always bound with hydrogen bonds to give the following structure:

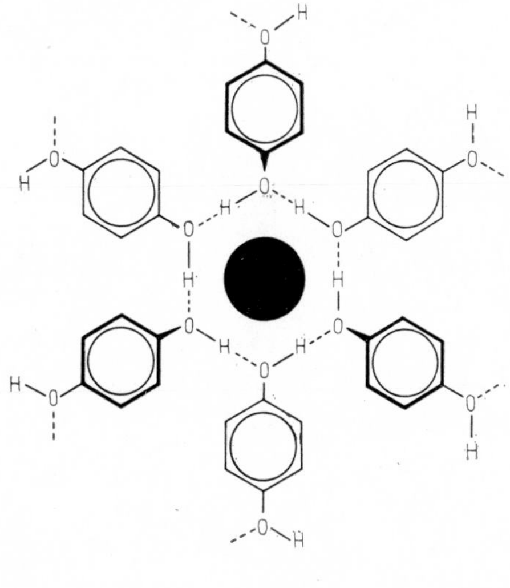

The molecules of hydroquinone are alternatingly oriented above and below the plane of the hexagon formed from the hydroxyl groups. The molecule of the guest located in the "calix" made of three molecules of hydroquinone above the hexagon plane is represented by a black circle.

Clathrates of this type are formed easily. For example the acetylene clathrate precipitates during the bubbling of a hydrocarbon through a solution of hydroquinone in ether. The clathrates of methane and propane were prepared under pressure from a saturated solution of hydroquinone in ethanol (at 30 °C). A pressure cylinder with the solution was filled under pressure at 35 °C with the hydrocarbon gas and then cooled over 8 h to 22 °C. The methane clathrate that was prepared at a pressure of 10 MPa has the extreme composition 3 $C_6H_4(OH)_2 \cdot CH_4$. The propane clathrate has at 0.9 MPa the mole ratio only 3 : 0.37. From the saturated solution of hydroquinone in n-propanol β-hydroquinone with empty cavities can be crystallized out by inoculating with a crystal of clathrate. However, on standing the structure changes spontaneously to the more stable α-modification.

Among a large number of compounds that can be guests in the clathrate structure of *tri-o-thymotide*[2, 81, 84] the hydrocarbons benzene, pentane and hexane have also been described. Smaller molecules form

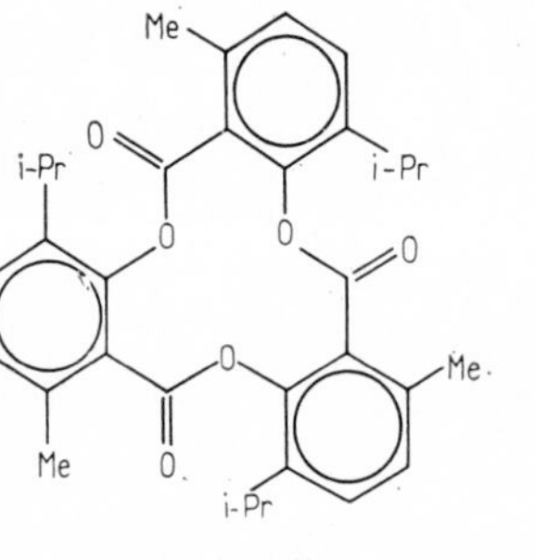

Me
O
i-Pr
i-Pr
O
O
O
O
Me
Me
O
i-Pr

Tri-*ortho*-thymotide

clathrates of the composition 2 $C_{33}H_{36}O_6 \cdot$ guest. Molecules with a longer chain are included into a different channel structure expressed by the formula $C_{33}H_{36}O_6 \cdot n$ guest, where n is a whole number, decreasing with the increasing length of the trapped molecule. Tri-o-thymotide was resolved to *enantiomers* by means of an optically active guest.

Recently a new strategy has been described for the design of clathrates which has led to the discovery of a series of *hexa-substituted benzene* hosts possessing the general formula C_6X_6, where X are bulky groups containing phenyl, naphthyl or adamantyl nuclei[84a]. Some of these "hexahosts" show remarkable selectivity for xylene isomers.

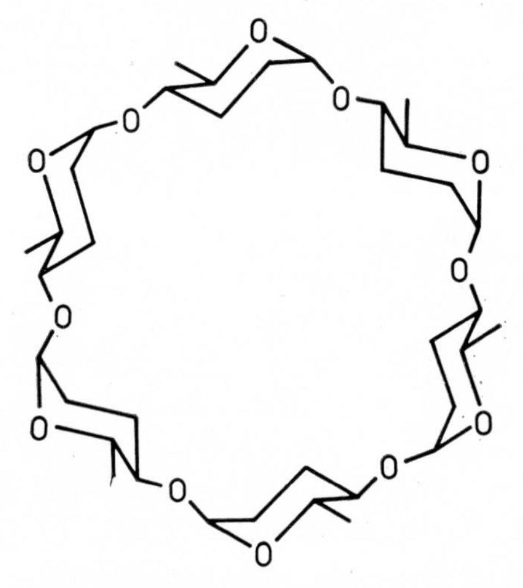

Fig. 2.5.12 Ring of α-cyclodextrin formed by six α-D-glucose molecules connected with 1,4-glucosidic bonds

The structure of *cyclodextrins*[2, 82, 85] consists of 6 to 9 *D*(+)-glucose molecules connected to a macrocycle (Fig. 2.5.12). In the crystal lattice the macrocycles are ordered so that axial channels are formed, which are either empty or foreign molecules can be guests in them.

α-Cyclodextrin consisting of six glucose cycles has the channel diameter of 0.6 nm and it forms a clathrate with benzene. In the seven-membered β-cyclodextrin the channel is broader (∅ 0.8 nm) and it is capable of retaining naphthalene. Even anthracene can be accomodated in the channel of γ-cyclodextrin (eight glucose molecules), 0.1 nm in diameter.

A remarkable feature of cyclodextrins consists in their ability to retain guest molecules even in solution, which is accompanied by development of heat of several tens of kJ/mol.

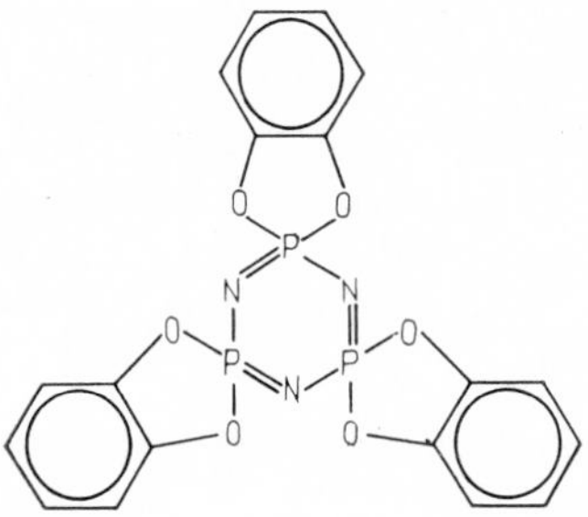

Tris (*o*-phenylenedioxy)phosphonitrile trimer

Molecules of some hydrocarbons are absorbed by *tris(o-phenylenedioxy) phosphonitrile trimer*[86] exceptionally eagerly. If this crystalline compound of m. p. 244–245 °C, sublimating under vaccuum (10 Pa) at 230 °C, enters into contact with a liquid hydrocarbon, it begins to adsorb the hydrocarbon spontaneously over its whole surface, and include it into the crystal lattice. For this property the compound has been suitably named "*crystalline molecular sponge*". In this case the clathrate is formed by direct penetration of the guest hydrocarbon through the thin crystal lattice of the host; the guest molecules fill the empty spaces in this lattice and deform it to a more stable crystal structure.

The clathrate is prepared by simple treatment of the crystalline trimer with an excess of hydrocarbon at room temperature. When the exothermic absorption accompanied by the decomposition of the crystals is over, the excess of hydrocarbon is separated and the clathrate dried in a vacuum at 10 Pa and 25 °C for 24 h. Another method of preparation of clathrate consists in the absorption of hydrocarbon vapours at atmospheric pressure and room temperature, or in the crystallization of the trimer from the corresponding hydrocarbon. The last method is limited by the very low solubility of the trimer.

The molar ratio depends on the molecular dimensions of the trapped hydrocarbon. There is no relationship between the polarity of the guest and the ease of inclusion. It is a peculiarity that the host has a higher capacity for larger molecules than for molecules of smaller size. The highest molar ratio between the guest and the host, in the 0.4–0.6 range, is achieved by cumene, *o*-xylene, *trans*-decalin, isooctane, norbornadiene and tetralin. Styrene, heptane and cyclohexane have a molar ratio about

0.4. Small molecules like acetone, carbon disulphide etc. have a molar ratio lower than 0.2 (also benzene).

The excellent separation capability of this clathrate is also remarkable. It was demonstrated by model experiments the results of which are shown in Table 2.5.9. In each case exclusively non-cyclic hydrocarbons were retained in the clathrate from the tested mixture, while cyclic hydrocarbons were completely excluded.

TABLE 2.5.9

Separation of model binary mixtures by means of clathrates of tris(*o*-phenylenedioxy)phosphonitrile trimer

Hydrocarbon mixture	Molar ratio of hydrocarbons	
	in the original mixture	in clathrate
Heptane-cyclohexane	0.74	100
Hexane-benzene	0.73	20
Hexane-cyclohexane	0.91	33

Clathrates are also known where both components -- the guest and the host – are hydrocarbons. In the temperature range 25 to –90 °C 2-*methylnaphthalene* forms with *n*-heptane and other hydrocarbons six different clathrates with a molar ratio 8 : 1, 8 : 3 and 1 : 1[87]. 1-Methylnaphthalene also has similar properties. The importance of this finding is only theoretical so far.

Another type of hydrocarbon host with wide applicability is *perhydrotriphenylene.* It is prepared by catalytic hydrogenation of dodecahydrotriphenylene under drastic conditions[88] (Pd—C, 300 °C, 10–20 MPa of hydrogen, in *n*-heptane)[244]:

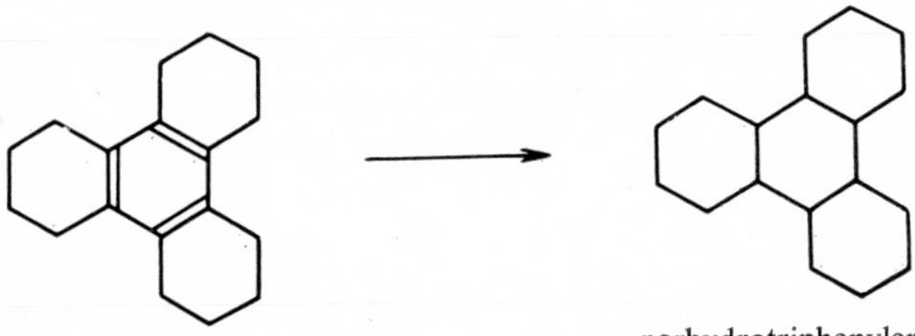

perhydrotriphenylene

The starting compound can be prepared, for example, by reaction of 1,4-dichlorobutane with benzene or tetralin. The required most stable stereoisomer *trans,anti,trans,anti,trans*-perhydrotriphenylene easily separates from the hydrogenation product even with *n*-heptane as a crystalline clathrate.

In perhydrotriphenylene clathrate there are channel cavities in which molecules of other hydrocarbons[89] or even macromolecules of linear polymers, such as polyethylene or 1,4-polybutadiene[90], can be located.

Clathrates are formed easily by crystallization of the host dissolved in the guest, or by simultaneous crystallization of both components from an inert solvent (methyl ethyl ketone) or also by melting both components. The channels formed in the hexagonal crystal lattice of perhydrotriphenylene are evident from the schematic representation of the structure of heptane clathrate in Fig. 2.5.13.

Pure perhydrotriphenylene has m.p. 128 °C. The clathrates are so stable that they have higher melting points than both components. For

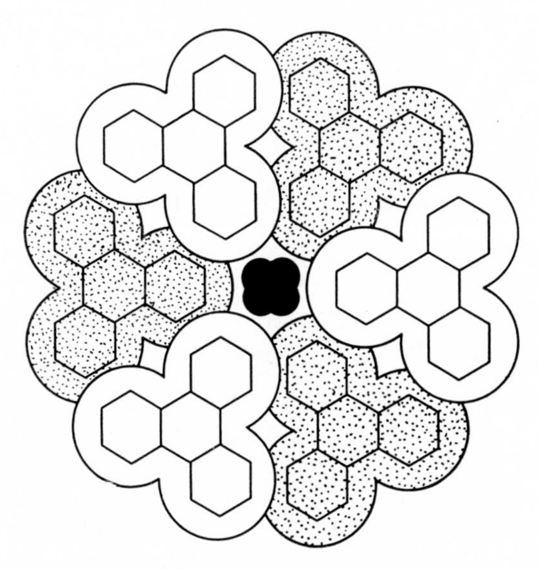

Fig. 2.5.13 Scheme of the location of the molecules of perhydrotriphenylene in the clathrate lattice
The location of the *n*-alkane molecule in the channel is represented by a black area

example the melting points of the clathrates of *n*-alkanes C_{12}—C_{36} are between 140 – 164 °C.

The highly symmetrical structure (group D_3) of the mentioned all-*trans*-perhydrotriphenylene is *chiral*, so that the hydrocarbon can be resolved to two optically active *enantiomers*. Some special applications are based on this fact. Similarly as in the previously mentioned urea adducts the molecules of the monomers retained in the perhydrotriphenlene clathrate can be *polymerized stereospecifically*[91]. Interesting asymmetric polymerizations have been carried out in the channels of the optically active host. For example, optically active isotactic 1,4-*trans*-poly-(met-hylbutadiene) was obtained from 1,3-pentadiene.

A large group of clathrates and one which is most important for the separation of hydrocarbons is the group that contains *atoms of metals* in the lattice of the host. These clathrates are used mainly for the separa-

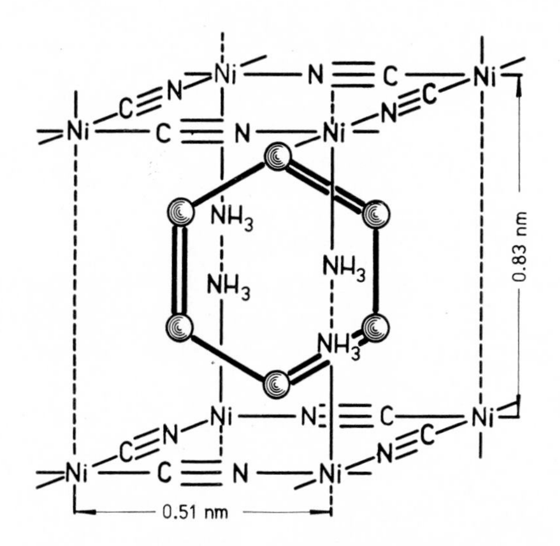

Fig. 2.5.14 Cage with a molecule of benzene in the clathrate $Ni(CN)_2NH_3 \cdot C_6H_6$

tion of aromatic hydrocarbons and they have the advantage that their shape-selectivity can be finely regulated by the composition of the host.

As early as 1897 Hofmann and Küspert[92] described a complex of the composition $Ni(CN)_2 \cdot NH_3 \cdot C_6H_6$ which is formed on addition of benzene to a solution of nickel cyanide in ammonia. Fifty years later X-ray, magnetochemical and other investigations of this complex showed unambiguously that it possesses clathrate structure[93]. The bonds between the nickel atoms and the CN and NH_3 groups form cages in the crystal, in which the molecules of benzene are captured without any bond binding them to the structure of the host (Fig. 2.5.14). The cages can even be filled with other compounds the *molar volume* of which is not higher than 91–93 ml; in addition to benzene (88.6 at 15 °C) these compounds also include aniline (90.5), phenol (88.8), pyridine (80.0), thiophene (78.5) and others. Compounds with a larger molar volume, as for example toluene (105.6), naphthalene (111.0) etc., no longer form clathrates.

A procedure for the preparation of well developed crystals of benzene clathrate (for the study of X-ray diffraction etc.) has been published by Jacobs[94].

Preparation of benzene clathrate with amocomplex of nickel dicyanide

$Ni(CN)_2 . 4\,H_2O$ (9.0 g) is added to 300 ml of concentrated ammonium hydroxide and the complex is stirred for 1—2 h and filtered through a fritted glass filter. The deep blue solution of the amocomplex of nickel dicyanide is carefully overlayered with benzene. Within a few hours the crystals of the clathrate begin to grow at the interface of both phases and after having attained a certain size they fall to the bottom of the vessel. The crystals of the clathrate $Ni(CN)_2 . NH_3 . C_6H_6$ are insoluble in water. After filtering them off under suction they are dried in air.

Biphenyl and some of its derivatives also crystallize from the amocomplex of nickel dicyanide[95], but the clathrate retains only one half of the molar amount of biphenyl in comparison with benzene. The molecules of biphenyl which evidently always fill two neighbouring cages bind the individual layers of the complex to each other and so strengthen the crystal lattice. This is also indicated by the higher temperature of the decomposition of the biphenyl clathrate (180 °C) in comparison with benzene clathrate (150 °C).

On evacuation of a sample of Hofmann aniline (or pyridine) clathrates the guest molecule replaces the coordinated ammonia to give $Ni(C_6H_5NH_2)_2Ni(CN)_4$ as the final product. The new complexes themselves can act as host lattices forming clathrates containing guest molecules such as benzene or aniline[95a].

Nowadays a long series of combinations of organic molecules and inorganic complexes is already known that form inclusion compounds similar to *Hofmann's clathrates*. Even though the structures of all these compounds are not yet completely elucidated, they are usually classified among clathrates. From binary cyanides of the type $MNi(CN)_4$, where M = Cu, Cd, Zn, Mn, Fe and Co, clathrates[96,97] of the Hofmann type with various combinations of metals have been prepared.

The selectivity of clathrates of the type: $Ni(SCN)_2$ (org. base)$_4$. guest that originate from the so-called *Werner complexes* is known in detail. They have been prepared in a great number of variants. Instead of nickel the complex can contain Mn, Zn, Co, Fe or Cu. The SCN^- may be replaced by Cl^-, CNO^- or some other anion. According to the organic base used these complexes are classified into two large groups. The first one, investigated mainly in the laboratories of Union Oil Co. (California), contains substituted *pyridines*[98] as a base. A typical representative is $Ni(SCN)_2$(4-methylpyridine)$_4$.

The second group was developed in the Belgian Labofina S. A. and it contains substituted *benzylamines*[99] as organic base. In addition to this further complexes were described containing other nitrogen bases, such as quinoline, ethylenediamine[100], 4,4′-dipyridyl[101], etc.

The preparation of clathrates of Werner's complexes is relatively simple. The following two procedures are mainly used: A – Dissolution of the Werner complex in a suitable warm solvent (Methylcellosolve, propylene glycol) and addition of hydrocarbon. The clathrate separates on cooling the mixture. B – Stirring of the suspension of the solid complex with the liquid hydrocarbon or with a solution of the hydrocarbon in unreactive hydrocarbon solvent. The trapped hydrocarbon can be separated from the filtered off and washed clathrate by steam distillation or by decomposition of the clathrate with hydrochloric acid and extraction of the liberated hydrocarbon phase. These preparative procedures are illustrated by concrete examples:

Preparation of the $Ni(SCN)_2$(4-methylpyridine)$_4$ complex[98]

KCNS (178 g) is added to a solution of 216 g of $NiCl_2 . 6 H_2O$ in 1700 ml of water and 340 g of 4-methylpyridine are then added under stirring to the resulting green solution over 5 min. A blue precipitate is formed. The mixture is stirred for another 15 min and then filtered. The blue precipitate on the filter is dried in air for 24 h. Yield, 480 g of complex (96 % of the theory).

Preparation of xylene clathrate

Methylcellosolve (38.0 ml) and 4-methylpyridine (4.0 ml) are added into a 150 ml flask containing 20.0 g of the above mentioned complex and the mixture is heated at about 105 °C until dissolved. A mixture of xylenes (16.5 ml) containing 20 % of *p*-isomer is added to the hot solution which is then cooled under stirring to room temperature. During the cooling blue crystals of clathrate separate. The suction-dried clathrate is washed with Methylcellosolve and decomposed with hydrochloric acid. The hydrocarbon phase set free (extract) contains about 70 % of *p*-xylene.

Preparation of xylene clathrate with $Ni(SCN)_2(\alpha$-butylbenzylamine$)_4$[99]

$NiCl_2 . 6 H_2O$ (9.51 g) and KSCN (7.78) are dissolved in 40 ml of water and the solution is cooled with ice under stirring and simultaneous gradual addition of 29.33 g of α-butylbenzylamine diluted with 100 ml of a mixture of xylenes (34 mol % of *o*-isomer). After 20 min stirring the blue precipitate is filtered off, washed with two 100 ml portions of cold *n*-heptane and dried at room temperature for 1 h. The clathrate is decomposed with 6 N HCl and the liberated hydrocarbon phase is extracted. The hydrocarbon mixture trapped in the clathrate (11 % per weight of clathrate) contains 84 mol % of *o*-xylene.

Preparation of $Ni(SCN)_2$(1-phenylethylamine$)_4$ complex[99]

$Ni(SCN)_2$ (0.2 mol) is dissolved in 100 ml of distilled water and 100 ml of *n*-heptane are added to the solution. 1-Phenylethylamine (11.8 ml) dissolved in 100 ml of *n*-heptane is then added to the mixture under stirring. After 15 min a blue precipitate separates which is filtered off and washed with 100 ml of *n*-heptane. The yield of the complex is quantitative. *Preparation of clathrate*: The solid complex is added to the hydrocarbon mixture and the suspension is stirred at room temperature. The clathrate is formed within a few minutes.

These inorganic complexes can be applied in the separation of hydrocarbon mixtures in various ways. The possibility of separating isomeric aromatic hydrocarbons is especially attractive. No wonder therefore that some of these separation processes were applied years ago on an industrial scale[102].

Differing abilities of hydrocarbons to enter into clathrates can be made use of, in the simplest case, for purification. For example benzene can be purified from the contaminating hydrocarbons to 99.99 % purity by a single crystallization *via* the Hofmann adduct.

The selectivity of Werner's complexes is closely connected with the structure of the organic nitrogen base. In a series of pyridine derivatives the following complexes have high selectivity:

$Ni(SCN)_2$(3-ethyl-4-methylpyridine$)_4$ – for *m*-xylene
$Ni(formate)_2$(4-ethylpyridine$)_4$ – for *o*-xylene
$Ni(SCN)_2$(4-methylpyridine$)_4$ – for *p*-xylene

The last complex also separates naphthalene from biphenyl, 1-methylnaphthalene from 2-methylnaphthalene, and anthracene from phenanthrene. The enrichment of the extracts in the *p*-isomer during the separation of xylenes and ethyltoluenes with 4-methylpyridine complex is evident from Table 2.5.10.

TABLE 2.5.10

Separation of isomeric dialkylbenzenes by complexing with $Ni(SCN)_2$(4-methylpyridine)$_4$[98]

	Isomers in vol. %		
	para	*meta*	*ortho*
Xylenes			
in the original mixture	26.9	50.7	22.2
in the extract	84.2	12.3	3.5
Ethyltoluenes			
in the original mixture	33.4	47.3	10.0
in the extract	58.4	25.1	16.5

In Table 2.5.11 the results of the separations of individual isomers of dialkylbenzenes by means of benzylamine complex are shown. According to various combinations of the substituents R and Y in benzylamine

H
|
Y–C_6H_4–C—NH_2
|
R

R = H or alkyl;
Y = various substituents in *o*-, *m*- or *p*-position

the complexes have sharply differentiated selectivity.

The selectivity of Werner's complexes was also investigated by chromatographic techniques. On columns containing powdered complexes as the *stationary phase* in *gas chromatography* the affinity of complexes for hydrocarbons of a certain shape is clearly reflected in the order of retention of individual isomers[103]. Monomethylnaphthalenes could be separated by the technique of *liquid chromatography* on a column containing $Ni(SCN)_2$(4-methylpyridine)$_4$ as the stationary phase; as the mobile phase a solution of 4-methylpyridine and NH_4SCN in methanol was used[104].

TABLE 2.5.11

Separation of isomeric dialkylbenzenes by formation of complexes of the type $Ni(SCN)_2(Y{-}C_6H_4 . CH(R) . NH_2)_4$[99]

Substitution of benzylamide		Composition of extract			Selectivity	Capacity[a] wt. %
R	Y	*ortho*	*meta*	*para*		
		xylenes[b]				
i-C_4H_9	*p*-Cl	85	6	9	*ortho*	11.3
C_6H_{13}	*p*-F	6	84	10	*meta*	8.0
CH_3	*m*-Br	9	9	82	*para*	7.5
		ethyltoluenes[b]				
i-C_4H_9	*p*-Cl	97	3	0	*ortho*	10.4
C_6H_{13}	H	8	84	8	*meta*	10.4
CH_3	*m*-Br	6	7	87	*para*	13.2
		diethylbenzenes[b]				
CH_3	H	93	4	3	*ortho*	8.6
C_6H_{13}	H	18	68	14	*meta*	10.7
CH_3	*m*-Br	3	4	93	*para*	10.8

[a] Hydrocarbon content in clathrate
[b] The ratio of isomers in the starting mixture was approximately 1 : 1 : 1

A distinct separation ability for isomers of aromatic hydrocarbons was also found in the complex with *copper-I trifluoromethanesulphonate*[105]. The form of the retention of aromatic hydrocarbons in the structure of the complex is not yet clear, but the much higher selectivity of the complex with respect to the shape of the hydrocarbon partner than with respect to its *π-basicity* indicates that a clathrate structure is very probable. With aromatic hydrocarbons the complex gives microcrystalline clathrates of the composition $(CuSO_3CF_3)_2$. guest. If a suspension of clathrate in inert alkane hydrocarbons is acted upon by another aromatic hydrocarbon that has better conditions for being a guest, an exchange of both hydrocarbons takes place.

TABLE 2.5.12

Selectivity of cuprous trifluoromethanesulphonate for some aromatic hydrocarbons[104]

Hydrocarbon	Separation factor α^{a}	Hydrocarbon	Separation factor α^{a}
	1	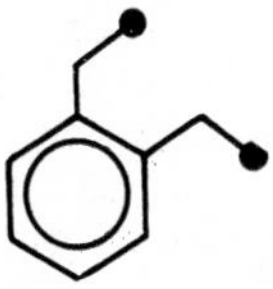	1.9
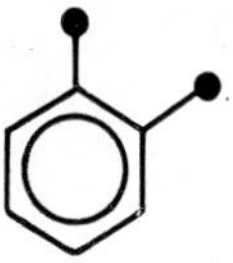	1.24	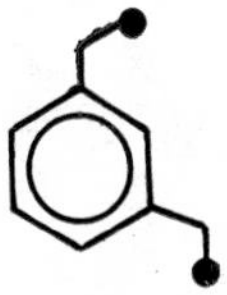	0.3
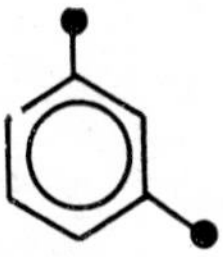	0.46		6.5
	16.0	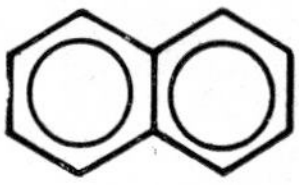	25.0
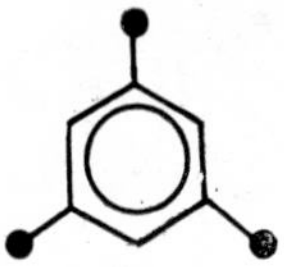	0.03	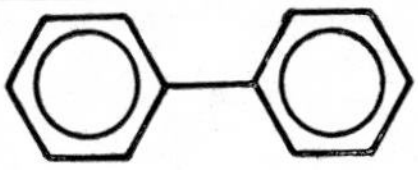	43.0
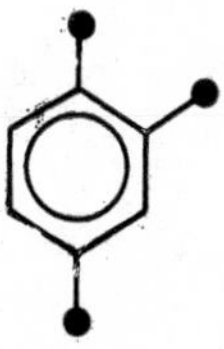	1.2		

[a] Ratio of equilibrium constants (benzene = 1)

Preparation of the copper-I trifluoromethanesulphonate complex

Cuprous oxide (16.7g) is added to 35 g of the anhydride of trifluoromethanesulphonic acid in 50 ml of octane and the mixture is refluxed overnight. The resulting slurry is filtered off and dried. The yield of the dark-red powdery complex of $CuSO_3CF_3$ is 44.4 g (90 %); in a sealed capillary it decomposes at 300 to 305 °C.

Direct preparation of the benzene clathrate

Anhydride of trifluoromethanesulphonic acid (127 g) and cuprous oxide (48.2 g) are added to 250 ml of benzene and the mixture is refluxed for one day, then cooled and filtered off in a dry box. A white, microcrystalline clathrate (155 g) $(CuSO_3CF_3)_2 . C_6H_6$ is obtained decomposing in a sealed capillary at 125 °C.

The selectivity of the complex is evident from Table 2.5.12 in which the separation factors of some aromates, referred to benzene ($\alpha = 1$), are given. The preference of *p*-substituted isomers is clear. For example, when a suspension of the benzene clathrate is stirred with *p*-xylene the exchange takes place within a few minutes.

2.5.4 MOLECULAR COMPOUNDS OF AROMATIC HYDROCARBONS

The ability of aromatic hydrocarbons to form stoichiometric molecular compounds with various substances has been and still is given great attention. In addition to theoretical questions the possibilities of utilization of these compounds for industrial separations of aromatic raw materials are especially investigated. Practical applications are mainly directed to the separation of xylenes[1]. An easier separation of C_8-isomers by crystallization can be achieved by means of molecular compounds of $SbCl_3$ with *p*-xylene and *m*-xylene. Similarly the yield of *p*-xylene can also be increased by crystallization of the molecular compound of *p*-xylene with CCl_4 (melting at 74 °C). CCl_4 forms equimolar molecular compounds of the same type with pseudocumene and durene.

For laboratory separations of hydrocarbons molecular compounds with electron *donor-acceptor* bond are more important even from a general analytical point of view. The first such π-*molecular* compounds among aromatic hydrocarbons and picric acid was already described 120 years ago. Later a number of other substances were discovered that, as electron acceptors, form molecular compounds with aromatic hydrocarbons; they are surveyed by Andrews[106] in a detailed review. Some of the π-molecular compounds will be mentioned in detail below.

The *dianhydride of pyromellitic acid* (PMDA)[107] is one of the strongest known electron acceptors (π-acids). It is a solid substance only slightly soluble in the majority of organic liquids. In contact with

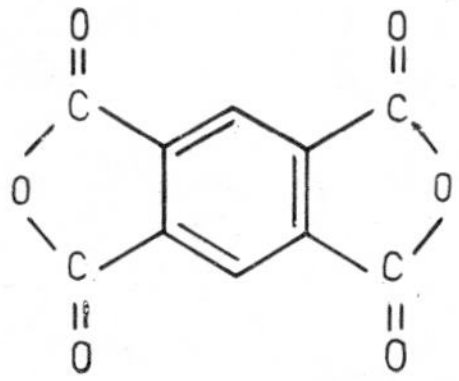

Dianhydride of pyromellitic acid (PMDA)

a suitable aromatic hydrocarbon (electron donor) the solid pyromellitic dianhydride begins to react in a remarkable manner, i.e. it grows in volume and weight, it gets warm and changes colour from white through yellow to red. In such a manner PMDA forms solid molecular compounds in a 1 : 1 molar ratio with benzene and all methylated benzenes. In contrast to this it does not react with ethylbenzene, cumene, *tert.*-butylbenzene, *o*- and *p*-ethyltoluene and diisopropylbenzenes at room temperature. The order of stability of PMDA molecular compounds with C_8 aromates is: *o*-xylene $>$ *p*-xylene $>$ *m*-xylene $\gg$ ethylbenzene.

The comparison of the stabilities of molecular compounds has shown that their stability increases with increasing number of methyls on the benzene nucleus up to a maximum in hexamethylbenzene; bulky alkyls have a destabilizing effect, as evident from the cases of cumene, *t*-butylbenzene etc.

Solid π-molecular compounds with PMDA are also formed with polynuclear aromates such as naphthalene, anthracene, pyrene and perylene[108]. The ability to form molecular compounds increases in the order benzene – naphthalene – anthracene. Pyrene and perylene (similarly as *o*-xylene) form molecular compounds with 2 moles of hydrocarbon and one mole of PMDA in addition to the normal 1 : 1 compounds[109].

Solid molecular compounds of PMDA with aromatic hydrocarbons are prepared either by suspension technique at room temperature or by crystallization from cold or hot solutions. The procedure is evident from the following examples:

Preparation of molecular compounds of aromatic hydrocarbons with pyromellitic acid dianhydride

A — isolation of *p*-xylene from ethylbenzene[107]. 10 g of PMDA are mixed with 20 ml of a mixture of *p*-xylene and ethylbenzene (50 : 50) at 25 °C and the mixture is allowed to stand overnight and then filtered. The yellow molecular compound weighs 18.5 g. The filtrate contains 33 % of *p*-xylene, which shows that 4.38 g *p*-xylene entered the complex. This corresponds to a 87 % yield of the 1 : 1 molecular compound. The adsorbed hydrocarbon can be distilled off from the molecular compound at 2.5 kPa pressure and 60—135 °C (end of the distillation at 200 °C).

B — well developed crystals of π-molecular compounds can be obtained by slow cooling of hot equimolar mixtures of both compounds in a suitable solvent. For PMDA 100 % acetic acid, acetic anhydride or dry methyl ethyl ketone are suitable solvents.

The benzene-PMDA molecular compound is prepared[108] by carefully overlayering a hot solution of PMDA in acetic acid with boiling benzene; the crystals of the molecular compounds grow on the phase boundary.

The pyrene-PMDA molecular compound with a 2 : 1 ratio crystallizes from a highly concentrated solution: 6.1 g of pyrene and 1.4 g of PMDA in 60 ml of almost boiling acetic anhydride[109].

C — molecular compounds of PMDA with some hydrocarbons of naphthalene series were prepared by crystallization from cold acetone solutions[110]: An equimolar amount of the aromatic hydrocarbon is added to a solution of 2.18 g of PMDA in 36 ml of acetone. After 2—3 h standing at room temperature the solutions are allowed to stand for another 12—14 h at 5 °C. The separated, brightly coloured crystalline complex is filtered off, washed with a small amount of cold acetone and ether, and dried at room temperature. The yields of the molecular compounds of methylated naphthalenes decrease in the following order: 2,3-dimethyl-(88 %), 1,8-dimethyl-(80 %), 2,6-dimethyl-(70 %), unsubstituted naphthalene (50 %), 2-methyl-(18 %), 1,6-dimethylnaphthalene (0 %).

Another unusually strong electron acceptor is *tetracyanoethylene* (TCNE)[111,112]. It is a solid substance melting at 197 – 200 °C (in sealed capillary), having an exceptionally high thermal stability.

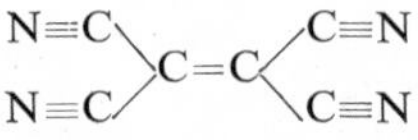

$$(N{\equiv}C)_2C{=}C(C{\equiv}N)_2$$

Tetracyanoethylene

In the planar structure shown the four electronegative CN groups decrease the electron density around the ethylene bond and thus cause a high electroaffinity of the compound.

As a strong electron acceptor TCNE forms intensively coloured molecular compounds with a number of aromatic donors. These *charge transfer* compounds are already formed in solution and in many cases they can be isolated as coloured crystalline substances. The majority of aromatic hydrocarbons give molecular compounds in a 1 : 1 molar ratio, but compounds formed in other ratios (1 : 2, 1 : 3 and 2 : 3) are also known. They are prepared by simple crystallization of an equimolar mixture of TCNE with the donor from a concentrated solution. In the crystals (mostly monoclinic) the molecules of TCNE and hydrocarbon are located in alternating parallel planes and they form a *sandwich structure* with the donor and acceptor centres above them.

Even though the crystalline molecular compounds of TCNE can be isolated[108], the properties of TCNE are utilized mainly for analytical purposes. The intensive coloration of molecular compounds enables, for example, detection of submicrogram amounts of hydrocarbons in *paper* and *thin-layer chromatography*[113,114]. Another wide use of coloured molecular compounds with the charge transfer, formed in solution, is in quantitative *spectrophotometric* analysis or aromatic hydrocarbons and related substances[115]. A solution of aromatic hydrocarbons in dichloromethane can be titrated directly with a solution of TCNE in dichloromethane, and the end-point is determined photometrically[116].

Electrical properties of molecular compounds of TCNE with polynuclear aromatic hydrocarbons are also of great interest. In the crystals of these donor-acceptor compounds the electrons are more mobile than in the components alone, and the crystals have *semiconductor* properties.

Many aromatic hydrocarbons form coloured well crystallizing π-molecular compounds with various *aromatic polynitro derivatives*. Their molar ratio is mostly 1 : 1 and only in some exceptional cases two moles of nitro derivative combine with 1 mole of hydrocarbon. The cohesion of both molecules depends both on the nature of the aromatic hydrocarbon and of the complexing agent. Some molecular compounds are stable crystalline substances with high melting points, while others are decomposed easily or they exist in solution only.

Formerly, owing to their sharp melting points, the stable crystalline molecular compounds were used frequently for the identification and characterization of aromatic hydrocarbons and their derivatives. The different inclinations of aromatic hydrocarbons for the formation of π-molecular compounds can also be utilized, however, for effective separation of hydrocarbon mixtures by adductive crystallization or by some other technique. Similarly as in the preceding case the intensive coloration of the molecular compounds is also used for various analytical applications.

The most often employed nitrated complexing agents are *picric acid, styphnic acid,* 1,3,5-trinitrobenzene and 2,4,7-trinitro-9-fluorene.

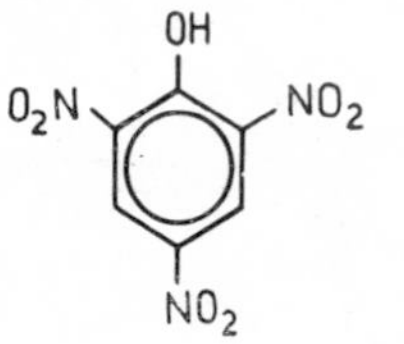

Picric acid

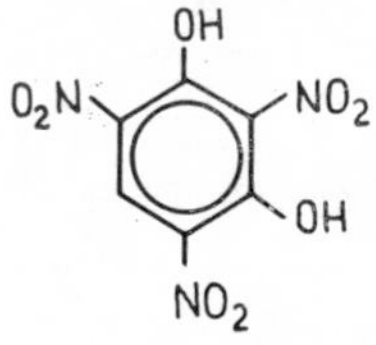

Styphnic acid (2,4,6-trinitroresorcinol)

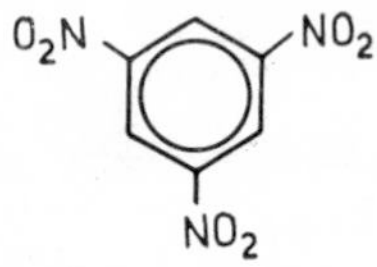

1,3,5-Trinitrobenzene

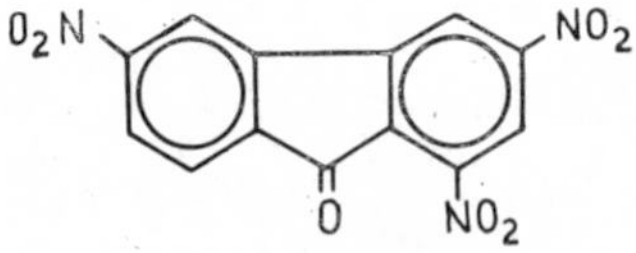

2,4,7-Trinitro-9-fluorenone

For the preparation of the molecular compounds the hydrocarbon is simply mixed with the complexing agent in an equimolar ratio. The procedure can be modified in various ways, as we shall demonstrate by several examples of the preparation of picrates.

Preparation of molecular compounds of aromatic hydrocarbons with picric acid

A — Equimolar amounts of hydrocarbon and picric acid are dissolved in a minimum amount of a suitable hot solvent (ethanol, chloroform, ether, benzene, ethyl acetate, glacial acetic acid etc.) and the solution is allowed to stand overnight in a cool place. The separated crystals of picrate are filtered off, washed with a small amount of ether (ethanol) and dried. If the picrate is stable, it can be recrystallized. Less stable picrates are better recrystallized from non-polar solvents.

The procedure can be modified so that the hydrocarbon and the picric acid are dissolved in the same solvent separately, and then both hot solutions are mixed.

B — When the solubility of the hydrocarbon and of picric acid in the same solvent is very different, it is advantageous[117] to use the mixture of two solvents (for example ethanol-benzene) in such a ratio that both components have an approximately equal molar solubility.

C — Picrates of liquid hydrocarbons can also be prepared by saturating the hydrocarbon with picric acid under heating and then allowing the solution to cool slowly.

D — Solid hydrocarbons can be melted carefully with an equimolar amount of picric acid in a test tube. The well mixed melt is then allowed to cool and the picrate is recrystallized from a suitable solvent.

Unstable picrates cannot be recrystallized because they decompose in the absence of an excess of hydrocarbon. In air these picrates decompose spontaneously within 1 – 5 days, in dependence on the volatility of the hydrocarbon component.

The hydrocarbon component can be set free from stable picrates by heating in dilute ammonia. Another practical method consists in the filtration of a benzene solution of the picrate through a column of alkaline alumina. Picric acid remains adsorbed at the beginning of the column, while the hydrocarbon is eluted with benzene quantitatively.

Benzene gives only a low-melting unstable picrate which cannot be recrystallized. In the group of methyl benzenes the stability of the picrate increases with the number of methyl substituents, up to pentamethylbenzene and hexamethylbenzene, the picrates of which can be isolated and have high melting points, from 131 °C to 170 °C.

In the series of *n*-alkylbenzenes the stability of picrates decreases with the prolongation of the alkane chain[118]. From ethylbenzene up the picrates are still less stable than that of benzene. Here the *alternating* effect of the chain with an even and an odd number of carbon atoms becomes evident. The picrates of branched alkylbenzenes (isopropyl-, isobutyl-, *tert.*-butylbenzene) also have a lower stability than the picrate of benzene itself.

Dicyclic hydrocarbons with one aromatic nucleus and one saturated ring (benzocycloalkanes) – tetralin, indane and benzocyclobutane – give unstable picrates only (similar to alkylbenzenes) which decompose in air[119].

The stability of picrates generally increases with the size of the aromatic system. Naphthalene and other condensed binuclear aromates already form very stable picrates (the melting point of the naphthalene picrate is 150 °C). Coloured π-molecular compound between naphthalene and picric acid is formed very easily even in a solid state[120]. The reaction can be carried out in a glass capillary which is filled from one side with powdered picric acid and from the other with powdered hydrocarbon, so that the small columns of both components are in contact. At room temperature the reaction starts within 7 – 10 min. The kinetics of the reaction can be followed easily on the basis of the movement of the coloured zone through the picric acid layer.

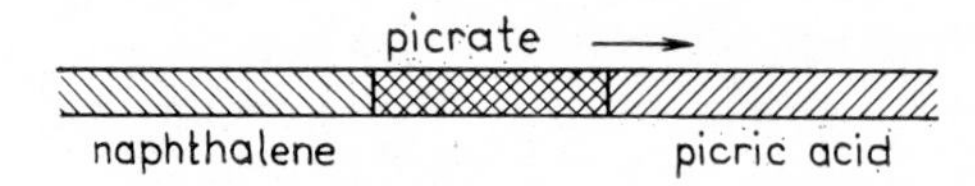

In contrast to naphthalene and its homologues, non-condensed binuclear aromates, biphenyl and biphenylmethane, do not form picrates. Therefore the mixtures of both types of binuclear aromates can be separated effectively by means of π-molecular compounds.

A large number of trinuclear and polynuclear aromatic hydrocarbons give stable π-molecular compounds with picric acid, styphnic acid and other nitro derivatives in a predominantly 1 : 1 molar ratio. The stability of the picrates of polynuclear aromates is reflected in high melting points that are higher as a rule than those of both pure components. Detailed data were published on some such systems – for example the formation of solid solutions and the conversions of picrates in the system picric acid-anthracene/phenanthrene[121] was investigated by means of *calorimetry* and *X-ray diffraction.*

It has been demonstrated by some types of tetracyclic hydrocarbons that the stability of picrate depends, *inter alia,* on sterical factors[122]. Among twelve methyl-1,2-benzanthracenes ten isomers form stable picrates melting higher than the parent hydrocarbons. Two isomers behave exceptionally; 9-methyl-1,2-benzanthracene and 1′-methyl-1,2-benzanthracene. In the first case the melting point of the picrate (115 °C) is lower than of the parent hydrocarbon (136 °C), and in the second case the picrate has an anomalous composition,

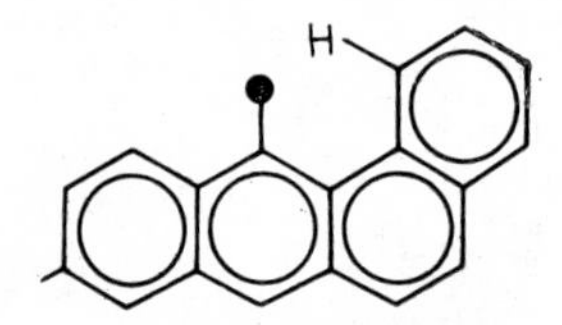

9-methyl-1,2-benzanthracene

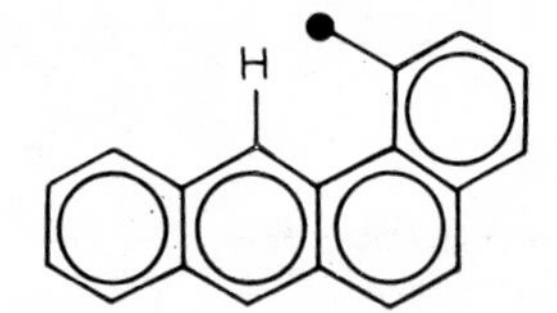

1′-methyl-1,2-benzanthracene

i.e. 2 moles of picric acid per 1 mole of hydrocarbon. The explanation can be found in the fact that for sterical interference of the methyl with the hydrogen atom in the positions 9 and 1′ neither of these two hydrocarbons has a *planar* structure as the remaining ten isomers have. It is evident that non-planar molecules of the aromatic hydrocarbon make the deposition of both components into parallel planes in the crystal lattice of the π-molecular compound more difficult, and that they prevent a maximum approach of the donor and the acceptor centres. The result is either a lower stability of the picrate or a higher molar ratio of both components.

An analogous case has been found in the group of monomethylbenzo(c)-phenanthrenes. Among the six possible isomers five form stable picrates melting higher than pure hydrocarbons. Only with 1-methylbenzo(c)phenanthrene which is non planar in consequence of the interaction of the methyl with the opposite hydrogen, the picrate is not formed.

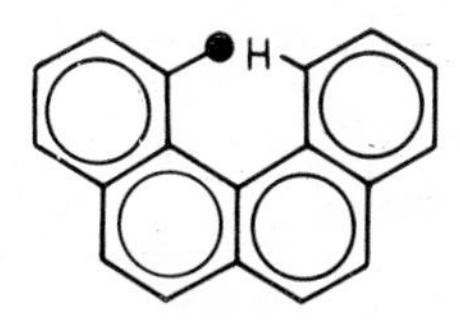

1-methylbenzo(c)phenanthrene

Among other non-planar aromatic hydrocarbons *cis*-stilbene and 1-phenylnaphthalene do not give picrates. In contrast to this, 2-phenylnaphthalene, that can be completely planar, does.

In Table 2.5.13 the melting points are compared of molecular compounds of basic dinuclear and trinuclear aromatic hydrocarbons with some nitrated reagents. More detailed data can be found in literature (Beilstein Vol. VI); the properties of picrates of a larger number of aromates are also surveyed by Baril and Hauber[123] and Gavat and Irimescu[124] (trimethylnaphthalenes) while the molecular compounds of 2,4,7-trinitrofluorenone are described in the paper by Orchin and Woolfolk[125].

Methylpicric, dimethylpicric and ethylpicric acids also form molecular compounds with aromates[126]. However, their melting points are as a rule lower than those of picrates. In contrast to this styphnates and molecular compounds of trinitrofluorenone mostly have higher melting points.

In the group of the nitrated reagents used, the relative ability to form π-molecular compounds with aromates increases in the following order: styphnic acid $<$ picric acid $<$ 1,3,5-trinitrobenzene $<$ 2,4,7-trinitrofluorenone. The asymmetric acceptors 3- and 4-nitrophthalic anhydrides[125a]

TABLE 2.5.13

Melting points of molecular compounds of some aromatic hydrocarbons with nitro compounds

Hydrocarbon	Melting point °C	Melting point of the molecular compound in °C[a]			
		Picric acid	Methyl-picric acid	Styphnic acid	Trinitro-fluorenone
Naphthalene	80	150	114	168	152
Acenaphthene	95	162	122	154	175
Anthracene	217	138	142	180	194
Phenanthrene	100	133	120	142	197
Fluorene	114	84	108	134	179

[a)] round values

form charge-transfer complexes with a smaller number of polynuclear aromatic hydrocarbons than the more common symmetrical acceptors. The complexing selectivity of the 3-nitrophthalic anhydride is better than that of 4-nitro isomer.

Trinitrofluorenone – a reagent which most easily gives molecular compounds[116,125,127] – is prepared by nitration of fluorenone. The reagent is less soluble in organic compounds than picric acid, and therefore it is very suitable for the preparation of molecular compounds with poorly soluble aromatic hydrocarbons of high molecular weight.

The effects of the structure of aromatic hydrocarbon on the stability of picrate, mentioned above, are roughly similar for other nitroderivatives. Specific differences in the reactivity of individual reagents can be used in various ways for the solution of special separation problems. Generally π-molecular compounds can be used for the isolation of polynuclear aromatic hydrocarbons from various fractions or for the separation of aromatic isomers that differ in their complexing ability.

The hydrocarbons with a distinct inclination for molecular compound formation can be *extracted* from hydrocarbon fractions with a solution of an aromatic polynitro derivative. For example, in a procedure worked out for the isolation of naphthalene from the fraction boiling at 100–

200 °C a 12 % picric acid solution in 90 % methanol, or a 45 % 3,5-dinitrosalicylic acid solution in 80 % aqueous triethylene glycol are used[128].

A survey of older papers concerning the isolation and the determination of aromates in petroleum fractions by means of molecular compounds has been published by Pasquinelli[129]. The efficiency of the method has also been demonstrated in the *API Research Project 6* for the isolation and identification of C_{13}—C_{17} alkylnaphthalenes and alkylbiphenyls from a dinuclear aromatic petroleum fraction[127]. In this paper alkylnaphthalenes were separated from unreactive alkylbiphenyls in the form of crystalline π-molecular compounds. For this purpose 2,4,7-trinitrofluorenone was selected as reagent because it has the highest ability to make complexes and because an almost complete separation of both classes of hydrocarbons can be achieved with it. For the separation of isomers within the class of alkylnaphthalenes, however, the "milder" reagent, 1,3,5-trinitrobenzene, was used.

The effect of nitrated complexing reagents was also tested for the separation of polynuclear aromates by *thin-layer chromatography*[130]. An outstanding improvement was achieved on alumina treated with 2,4,7-trinitrofluorenone. Later the procedure was modified for micropreparative scale so that the layer was treated with the nitrated reagent on the lower part of the layer only[131]. The hydrocarbons separated in the form of molecular compounds on the lower part of the layer, split from the reagent in the upper part of the thin layer and can be detected under the UV light and isolated in pure form. The combination of trinitrofluorenone charge-transfer complexation with column liquid chromatography was developed by Jewell[132] for the isolation of petroleum aromatics and by Giger and Blumer[133] for the analysis of very complex mixtures of polycyclic aromatic hydrocarbons from soils and marine sediments. Chemically bonded charge-transfer compounds as stationary phases for LC have been described by Porath[134] and Nondek and Málek[135].

Some applications of nitrated reagents are purely analytical. Pasquinelli[129] has elaborated a method of rapid determination of aromates in petroleum fractions based on the finding that a solution of picric acid in nitrobenzene (20 g of picric acid in 100 ml of nitrobenzene) when shaken with the fraction gives an intensive and stable coloration dependent on the content and the nature of the aromates present. In the test 20 ml of petroleum fractions are shaken in a test tube with 1 ml of reagent for 10 s; the colour is measured after 10 min standing. The method was also modified for quantitative measurement.

Microscopic amounts of polynuclear aromates may also be identified on a *Kofler block* using a method[135] in which the melting point of the pure hydrocarbon is measured in one heating cycle together with the melting point of the molecular compound with 2,4,7-trinitrofluorenone and the melting point of eutectic mixtures. The method is suitable, for example, for the identification of crystalline aromates retained from the *effluent* of chromatographic columns.

REFERENCES (2.5)

1. R. A. Findlay, Adductive Crystallization, in New Chemical Engineering Separation Techniques. H. M. Schoen (Ed.), Vol. 1, Interscience, New York—London, 1962.
2. M. Hagan, Clathrate Inclusion Compounds, Reinhold Publ. Corp., New York, 1962; E. C. Makin, *Separ. Sci.* **9** (1974) 541.
3. D. F. R. Gilson and C. A. McDowell, *Nature* **183** (1959) 1183.
4. H. U. Lenné, H. C. Mez and W. Schlenk, *Liebigs Ann. Chem.* **732** (1970) 70.
5. F. E. Bailey and H. G. France, *J. Polymer. Sci.* **49** (1961) 397.
6. W. J. Zimmerschied, R. A. Dinerstein, A. W. Weitkamp and R. W. Marschner, *Ind. Eng. Chem.* **42** (1950) 1300.
7. O. Redlich, C. M. Gable, A. K. Dunlop and R. W. Millar, *J. Amer. Chem. Soc.* **72** (1950) 4153.
8. W. Schlenk, *Ann.* **565** (1949) 204.
9. R. C. McLauglin, Separation of Paraffins by Urea and Thiourea, in Chemistry of Petroleum Hydrocarbons, B. T. Brooks et al. (Eds.), Vol. 1, Reinhold Publ. Corp., New York, 1954.
10. H. B. Knight, L. P. Witnauer, J. E. Coleman, W. R. Noble and D. Swern, *Anal. Chem.* **24** (1952) 1331.
11. H. G. McAdie, *Can. J. Chem.* **40** (1962) 2195; **41** (1963) 2144.
12. A. V. Topchiev, L. M. Rozenberg, N. A. Nechitailo and E. M. Terent'eva, *Dokl. Akad. Nauk SSSR* **98** (1954) 223; **109** (1956) 1144.
13. D. Swern, *Ind. Eng. Chem.* **47** (1955) 216.
14. B. Casu, *Nature* **191** (1961) 802.
15. M. Kisielow, *Roczniki Chemii* **29** (1955) 888.
16. R. Kaiser, *Chem. Tech.* **10** (1965) 388.
17. C. Karr and J. R. Comberiati, *J. Chromatogr.* **18** (1965) 394.
18. E. Terres and S. N. Sur, *Brennstoff-Chem.* **38** (1957) 330.

19. J. Radell, J. W. Connolly and L. D. Yuhas, *J. Org. Chem.* **26** (1961) 2022, *Can. J. Chem.* **43** (1965) 304.
20. W. Schlenk, *Angew. Chem.* **62** (1950) 299.
21. V. E. Kovjazin and S. Hála, *Coll. Czech. Chem. Commun.* **38** (1973) 2938.
22. C. Kajdas and R. Tümmler, *Erdöl u. Kohle* **21** (1968) 213; **22** (1969) 459.
23. S. Landa and S. Hála, *Coll. Czech. Chem. Commun.* **24** (1959) 203.
24. G. Heinze, *Erdöl u. Kohle* **14** (1961) 179.
25. G. Hessler and G. Meinhardt, *Fette u. Seifen* **55** (1953) 441; 855.
26. K. Stránský, M. Streibl and F. Šorm, *Coll. Czech. Chem. Commun.* **33** (1968) 416.
27. K. Stránský, M. Streibl and V. Kubelka, *Coll. Czech. Chem. Commun.* **35** (1970) 882.
28. G. Kovachev, D. Delova and K. Stránský, *Coll. Czech. Chem. Commun.* **37** (1972) 4106.
29. G. Kovachev, K. Ubik, T. Mincheva and R. Nikolov, *Coll. Czech. Chem. Commun.* **40** (1975) 3728.
30. H. Schlief, *Chem. Tech. (Berlin)* **6** (1954) 456.
31. G. G. Rumberger, Symposium on Composition of Petroleum Oils, New Orleans, 1957; published by ASTM No. 224, 1958, p. 283.
32. W. Kisielow, M. Grochowska and M. Rutkowska, *Nafta (Katowice)*, **26** (1970) 372.
33. E. F. Shevchenko in "Novye metody issledovanii sostava neftei"; VNIGNI, Moscow, 1972.
34. W. Kisielow and C. Kajdas, *Fette-Seifen-Anstrichmittel* **73** (1971) 4.
35. V. V. Veselov, *Khim. Tekhnol. Topliv i Masel* **1** (1960) 47.
36. GOST 15095-69, Gosudarstvennyi standard Soyuza SSR, Moscow.
37. A. K. Seleznev, L. V. Pavlov, N. D. Desyatova, T. M. Lyuko and V. I. Stepanova, *Neftepererabotka i neftekhimiya* **6** (1973) 40.
38. I. Orság and J. Báthory, *Acta Chim. Hung.* **40** (1964) 367.
39. J. R. Marquart, G. B. Dellow and E. R. Freitas, *Anal. Chem.* **40** (1968) 1633.
40. Phillips Petroleum Company, Method 7214-AZ (1972).
41. T. A. Bolshova and G. L. Starobinets, *Khim. i Tekhnol. Topliv i Masel* **5** (1961) 17.
42. V. M. Bhatnagar and A. Liberti, *J. Chromatogr.* **18** (1965) 177.
43. K. Mařík and E. Smolková, *Chromatographia* **6** (1973) 420; *J. Chromatogr.* **91** (1974) 303.
44. M. Schlenk, *Liebigs Ann. Chem.* (1973) 1145, 1156, 1179, 1195.
45. M. Schlenk, *Chem. Ber.* **101** (1968) 2445.
46. M. Schlenk, *Experientia* **8** (1952) 337.
47. H. U. Lenné, H. C. Mez and G. Schlenk, *Chem. Ber.* **101** (1968) 2437.
48. D. M. White, *J. Amer. Chem. Soc.* **82** (1960) 5679.
49. V. S. Ivanov, T. A. Sukhikh, Yu. V. Medvedev, A. Kh. Breger, V. B. Osipov and V. A. Goldin, *Vysokomol. Soed.* **6** (1964) 782.
50. R. L. McLaughlin and W. S. McClenahan, *J. Amer. Chem. Soc.* **74** (1952) 5804.
51. H. U. Lenné, H. C. Mez and W. Schlenk, *Liebigs Ann. Chem.* **732** (1970) 70.
52. R. W. Schiessler and D. J. Flitter, *J. Amer. Chem. Soc.* **74** (1952) 1720.

53. G. G. Kakabekov, N. A. Nechitailo, E. I. Bagrii and P. I. Sanin, *Neftekhimiya* **13** (1973) 889.
54. G. G. Kakabekov, E. I. Bagrii and P. I. Sanin, *Neftekhimiya* **12** (1972) 290; *Dokl. Akad. Nauk SSSR* **199** (1971) 342.
55. W. Schlenk, *Liebigs Ann. Chem.* **573** (1951) 142.
56. N. Nicolaides and F. Lanes, *J. Amer. Chem. Soc.* **76** (1954) 2596.
57. A. L. Liberman, D. N. Furman and E. M. Milvickaya, *Neftekhimiya* **13** (1973) 145.
58. S. Landa and S. Hála, *Coll. Czech. Chem. Commun.* **24** (1959) 93.
59. S. Hála, S. Landa and V. Hanuš, *Angew. Chem.* **78** (1966) 1060.
60. M. T. J. Murphy, A. McCormick and G. Erlington, *Science* **157** (1967) 1040.
61. E. Gelpi, P. C. Wszolek, E. Yang and A. L. Burlingame, *Anal. Chem.* **43** (1971) 864
62. S. D. Pustilnikova, N. A. Abryumina and A. A. Petrov, *Neftekhimiya* **15** (1975) 183.
63. J. W. Teter and W. P. Hettinger, *J. Amer. Chem. Soc.* **77** (1955) 6695
64. D. P. Montgomery, *J. Chem. Eng. Data* **8** (1963) 432.
65. A. J. Kuklinskii and R. A. Pushina, *Neftekhimiya* **13** (1973) 467.
66. G. C. Speers and E. V. Whitehead, Crude Petroleum in Organic Geochemistry, G. Erlington and M. T. J. Murphy (Eds.), Springer-Verlag, Berlin, 1969, p. 660.
67. S. Landa and S. Hála, *Erdöl u. Kohle* **11** (1958) 698.
68. S. Hála, M. Kuraš and S. Landa, *Neftekhimiya* **6** (1966) 3.
69. S. Hála and S. Landa, *Erdöl u. Kohle* **19** (1966) 727.
70. E. I. Bagrii, E. I. Amosova and P. I. Sanin, *Neftekhimiya* **6** (1966) 665.
71. J. M. Slobodin, V. E. Koviazin, P. I. Motovilov and V. F. Vasileva, *Neftekhimiya* **9** (1969) 921.
72. E. I. Bagrii, P. I. Sanin, N. S. Vorobeva and A. A. Petrov, *Neftekhimiya* **7** (1967) 515.
73. A. A. Petrov, Khimiya naftenov, Nauka, Moscow 1971.
74. L. D. Melikadze, E. A. Usharauli, A. A. Dzamukashvili and M. A. Machabeli, *Bulletin of the Academy of Sciences of the Georgian SSR*, **68** (1972) 81.
75. P. I. Sanin, E. I. Bagrii, N. N. Cicugina, A. A. Suchkova, I. A. Musaev and E. Kh. Kurashova, *Neftekhimiya* **14** (1974) 333.
76. Imuta Kazutoshi and Ouchi Koji, *Fuel* **52** (1973) 301.
77. S. Hála, K. Cejnar and S. Landa, *Scientific Papers of the Institute of Chemical Technology, Prague*, D **22** (1971) 105.
78. J. Vais, J. Burkhardt and S. Landa, *Z. Chem.* **9** (1969) 268.
79. W. Schlenk, *Analyst* **77** (1952) 867.
80. J. F. Brown and D. M. White, *J. Amer. Chem. Soc.* **82** (1960) 5671.
80a. A. N. Nesmeyanov, G. B. Shulpin and M. I. Rybinskaya, *Dokl. Akad. Nauk SSSR* **221** (1975) 624.
80b. H. van Bekkum, J. D. Remijnse and B. M. Wepster, *Chem. Commun.* **2** (1969) 67.
81. L. Mandelcorn, *Chem. Rev.* **59** (1959) 827; "Non-Stoichiometric Compounds", L. Mandelcorn (Ed.), Academic Press, New York-London, 1964.
82. W. Schlenk, *Chemie in unserer Zeit* **3** (1969) 120.
83. G. Peryonel and G. Barbieri, *J. Inorg. Nucl. Chem.* **8** (1958) 582.
84. D. Lawton and H. M. Powell, *J. Chem. Soc.* (1958) 2339.

84a. D. D. MacNicel and D. R. Wilson, *J. Chem. Soc. Chem. Commun.* **13** (1976) 494. *Chemistry and Industry* **2** (1977) 84.
85. F. Cramer, *Angew. Chem.* **68** (1956) 115.
86. H. R. Allcock and L. A. Siegel, *J. Amer. Chem. Soc.* **86** (1964) 5140.
87. J. Milgrom, *J. Phys. Chem.* **63** (1959) 1843.
88. M. Farina and G. Audisio, *Tetrahedron* **26** (1970) 1827.
89. G. Allegra, M. Farina, A. Immirzi, A. Colombo, U. Rossi, R. Broggi and G. Natta, *J. Chem. Soc.* B **10** (1967) 1020; 1028.
90. M. Farina, G. Allegra and G. Natta, *J. Amer. Chem. Soc.* **86** (1964) 516.
91. M. Farina, *Macromolecules* **3** (1970) 475; **4** (1971) 265.
92. K. A. Hofmann and F. Küspert, *Z. anorg. Chem.* **15** (1897) 204.
93. U. M. Bhatnagar, *J. Chem. Educ.* **40** (1963) 647.
94. G. D. Jacobs, *J. Chem. Educ.* **47** (1970) 394.
95. J. Leicester and J. K. Bradley, *Chem. Ind.* (1955) 1449.
95a. S. Akyuz et al., *J. Mol. Struct.* **38** (1977) 43; **17** (1973) 105.
96. R. Baur and G. Schwarzenbach, *Helv. Chim. Acta* **43** (1960) 842.
97. T. Nakano, T. Miyoshi, T. Iwamoto and Y. Sasaki, *Bull. Chem. Soc. Japan* **40** (1967) 1297.
98. W. D. Schaeffer, W. S. Dorsey, D. H. Skinner and C. G. Christian, *J. Amer. Chem. Soc.* **79** (1957) 5871.
99. P. Radzitzky and J. Hanotier, *Erdöl u. Kohle* **15** (1962) 892; *Ind. Eng. Chem. Proc. Des. Develop.* **1** (1962) 10.
100. T. Miyoshi, T. Iwamoto and Y. Sasaki, *Inorg. Nucl. Chem. Lett.* **6** (1970) 21.
101. G. M. Mamedaliev, N. N. Kapitsov and A. I. Gasanov, *Neftekhimiya* **13** (1973) 601.
102. P. M. Sherwood, *Brennstoff-Chem.* **43** (1962) 263.
103. A. C. Bhattacharyya and A. Bhattacharjee, *Anal. Chem.* **41** (1969) 2055; *J. Chromatogr.* **41** (1969) 446.
104. J. Lipkowski, K. Lesniak, A. Bylina, D. Sybilska and K. Duszczyk, *J. Chromatogr.* **91** (1974) 297.
105. M. B. Dines, *Separ. Sci.* **8** (1973) 661.
106. L. J. Andrews, *Chem. Rev.* **54** (1954) 713.
107. L. L. Ferstandig, W. G. Toland and C. D. Heaton, *J. Amer. Chem. Soc.* **83** (1961) 1151.
108. J. C. A. Boeyens and F. H. Herbstein, *J. Phys. Chem.* **69** (1965) 2153; 2160.
109. I. Ilmet and L. Kopp, *J. Phys. Chem.* **70** (1966) 3371.
110. G. P. Naletova, D. F. Varfolomeev, L. V. Osintseva and V. V. Prokofev, *Neftekhimiya* **13** (1973) 199.
111. D. N. Dhar, *Chem. Rev.* **67** (1967) 611.
112. V. Kubáň and J. Janák, *Chem. listy* **63** (1969) 639.
113. J. Janák, *J. Chromatogr.* **15** (1964) 15; **16** (1964) 494.
114. N. Kucharczyk, J. Fohl and J. Vymětal, *J. Chromatogr.* **11** (1963) 55.
115. G. H. Schenk, P. W. Vance, J. Pietrandrea and C. Mojzis, *Anal. Chem.* **37** (1965) 373.
116. G. H. Schenk and M. Ozolins, *Anal. Chem.* **33** (1961) 1562.

117. E. K. Andersen, *Acta Chem. Scand.* **8** (1954) 157.
118. H. D. Anderson and D. L. Hammick, *J. Chem. Soc.* (1950) 1089.
119. J. B. F. Lloyd and P. A. Ongley, *Chem. Ind.* (London) **28** (1964) 1267.
120. R. P. Rastogi, P. S. Bassi and S. L. Chadha, *J. Phys. Chem.* **66** (1962) 2707; **67** (1963) 2569.
121. Yui Norio, Kurokawa Yoichi, Otsuki Hiroyoki, *Bull. Chem. Soc. Japan* **47** (1974) 247.
122. M. Orchin, *J. Org. Chem.* **16** (1951) 1165.
123. O. L. Baril and E. S. Hauber, *J. Amer. Chem. Soc.* **53** (1931) 1087.
124. I. Gavat and I. Irimescu, *Ber.* B, **75** (1942) 820.
125. M. Orchin and E. O. Woolfolk, *J. Amer. Chem. Soc.* **68** (1946) 1727; **69** (1947) 1225.
125a. F. Casellato, C. Vecchi and A. Girelli, *Chemistry and Industry* **2** (1977) 83.
126. R. P. Mariella, M. J. Gruber and J. W. Elder, J. Org. Chem. **26** (1961) 3217.
127. Foch Fu-Hsie Yew and B. J. Mair, *Anal. Chem.* **38** (1966) 231.
128. F. Veatch, R. W. Foreman and J. A. Gecsy, U.S. 2, 941.017 (1960).
129. A. Pasquinelli, *Anal. Chem.* **26** (1954) 329.
130. A. Berg and J. Lam, *J. Chromatogr.* **16** (1964) 157.
131. H. Kessler and E. Müller, *J. Chromatogr.* **24** (1966) 469.
132. D. M. Jewell, *Anal. Chem.* **47** (1975) 2048.
133. W. Giger and M. Blumer, *Anal. Chem.* **46** (1974) 1663.
134. J. Porath, *J. Chromatogr.* **159** (1978) 13; **133** (1977) 180.
135. L. Nondek and J. Málek, *J. Chromatogr.* **155** (1978) 187.
136. D. E. Laskowski, D. G. Grabar and W. C. McCrone, *Anal. Chem.* **25** (1953) 1400

2.6 Molecular sieves

Adsorption of hydrocarbon molecules on the surface of solid porous substances is affected mainly by the nature and the area of the surface, diameter of the pores and their relative frequence with respect to their size. Adsorbents with a narrow distribution of small pores possess some specific properties which can be utilized with advantage in the separation of complex mixtures. The most important is the *molecular-sieve* effect which operates with different intensity if the diameters of the most numerous pores agree with the molecular dimensions of the adsorbates.

Among the known adsorbents synthetic *zeolites*, called *molecular sieves*, possess the highest shape selectivity. Zeolites are crystalline aluminosilicates the three-dimensional crystal lattice of which, composed of tetrahedral SiO_4 and AlO_4 groups (Fig. 2.6.1), forms a regular system of cavities connected mutually with openings of uniform dimensions.

The fundamental structural unit is a *cubo-octahedral* cell containing

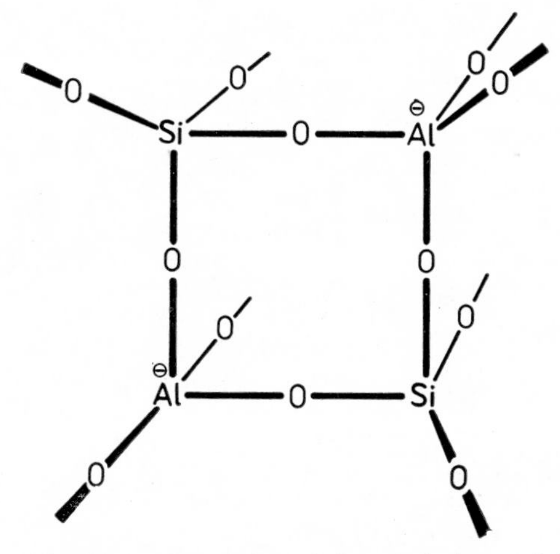

Fig. 2.6.1 Connection of the tetrahedrons of SiO_4 and AlO_4 in the crystal lattice of molecular sieves

totally 24 ions of Si and Al bound by 36 atoms of oxygen (Fig. 2.6.2). Two neighbouring cubo-octahedral cells in the crystal lattice of synthetic zeolite can be bound together either by means of four oxygen bridges (type A) or by means of six oxygen bridges (type X). In the first case each cubo-octahedron is connected with six other cubo-octahedrons

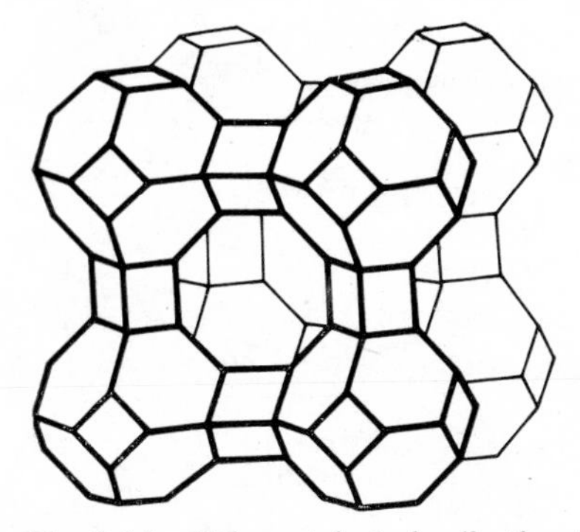

Fig. 2.6.2 Cubo-octahedral cell—the structural unit of molecular sieves

in a cubic structure. In the crystallographic structure of the type X each cubo-octahedron is bound with four further cubo-octahedrons in a tetrahedral arrangement.

The cavities in the crystal lattice are filled with water molecules after the zeolite has crystallized. Dehydration of zeolite at elevated temperature and a decreased pressure brings about the elimination of water from the cavities, so that a porous crystalline structure with a large internal surface is formed ($700-800\ m^2/g$), containing two types of cavities: the smaller cavities inside the cubo-octahedrons have diameters of 0.66 nm and they are not accessible for hydrocarbon molecules; the second type of cavities is formed between the cubo-octahedrons and these are larger (in the type A their diameter is 1.14 nm) and hydrocarbon molecules of suitable size can penetrate them. In the proximity of the openings into the larger cavities those cations are localized that compensate the negative charge of the anionic lattice. The outstanding sieving effect arising during the adsorption of substances in the larger cavities is given by the fact that the access of foreign molecules into the inner structure of the sieve is controlled by openings of uniform size. In the case of type A sieve the openings connecting the cavities have the form of rings of eight oxygen anions; in the case of the sieve of type X these openings are larger because they are formed by twelve-member oxygen rings. In addition to the diameter of the openings, that is given by the basic crystallographic structure of zeolite the penetrability of the openings is also affected by the number and type of cations (by their ionic radius) present in a unit cell. This enables a change in the *effective diameter* of the openings by ion exchange and thus a modification of the adsorption selectivity of the sieve.

Among the large number of types of molecular sieves described, the commercially produced types A and X are of greatest importance for the separation of hydrocarbons. The basic sodium sieves have the following approximate composition:

type 4A – $Na_2O \,.\, Al_2O_3 \,.\, 2\, SiO_2$
type 13X – $Na_2O \,.\, Al_2O_3 \,.\, 3\, SiO_2$

When the sodium cation is replaced by potassium the sieve of type 4A is converted to a drying sieve

type 3A – $K_2O \,.\, Al_2O_3 \,.\, 2\, SiO_2$

The replacement of the sodium cation by calcium is the basis of the conversion of the sieves of the type 4A and 13X to the following:

type 5A – $CaO . Al_2O_3 . 2\ SiO_2$ and
type 10X – $CaO . Al_2O_3 . 3\ SiO_2$

These formulae do not express the composition of the sieves exactly, because the ratio of $SiO_2 : Al_2O_3$ varies within certain limits (1.9 – 2.0 in type A and 2.3 – 3.3 in type X) and the exchange of the cations is also not always complete (it is usually above 75 %). In addition to the mentioned types, various forms of sieves of type Y and L are also produced for catalytic and other purposes.

The producers give various commercial names to synthetic zeolites; for example in the U.S.A.: MOLECULAR SIEVES, MICROTRAPS, ZEOLON; in the U.S.S.R.: ZEOLITH, in G. B.: ZEOSORB, ZEOLOX; in France: SILIPORTE etc. – A more detailed list of the names used is given in ref.[1]

The effective diameters of the openings, that are a resultant of the crystallographic diameter and the shielding effect of the cations, have the following values for the main types:

type of sieve	– 3A	4A	5A	10X	13X
effective diameter of the openings in nm	– 0.38	0.42	0.5	0.8	0.9

Adsorbates with a critical diameter of the molecule (see p. 246) lower or comparable to the effective diameter of the openings penetrate the inner cavities easily and are adsorbed rapidly. At drastic conditions, when the oscillations of the crystal lattice are larger, the openings can also permit the passage – but at a much lower rate – of molecules with a slightly larger critical diameter. Compounds for which the openings are unpenetrable under all conditions are excluded from adsorption.

Simultaneously with the sieving effect the charges of the cations located inside the lattice cavities also play a role in adsorption. The cations (as centres with a strong positive charge) attract polar and polarizable molecules by electrostatic forces and facilitate their penetration into the inner structure of the sieve. Among the molecules of adsorbates of otherwise equal dimensions *polar substances* are therefore preferred for adsorption, among hydrocarbons those that are more unsaturated; water is one of the most strongly adsorbed substances.

The selectivity of common types of molecular sieves is surveyed in

Table 2.6.1 on a "scale" of compounds arranged according to increasing critical diameter of the molecule.

TABLE 2.6.1

Adsorption properties of molecular sieves of type A and X

Compound	Critical diameter of the molecule nm	Molecular sieve				
Helium	0.20					
Hydrogen	0.24					
Water	0.26	3A				
Oxygen	0.28		4A			
Nitrogen	0.30	0.38		5A		
Methane	0.40					
Ethene	0.42					
Ethane	0.44		0.42			
n-Butane	0.49				10X	
n-Alkanes	0.49					13X
Cyclopropane	0.50			0.5		
Isobutane	0.56					
Isoalkanes						
Cyclohexane						
Benzene	0.65					
Naphthalene					0.8	
Triethylbenzene						
Decahydrochrysene						0.9

During their production the molecular sieves are formed as a white microcrystalline powder of particle size from one tenth to several tens of μm. For technological purposes the sieves are adjusted by compressing them in the presence of a binder (clays) to pearls of various sizes. This deteriorates their selectivity a little, because a secondary porous structure is formed in the pearls in dependence on the type and the amount of the granulation additive used, in which non-selective adsorption may take place.

Before the first use of a molecular sieve, it should be *activated*, i.e. the crystal water should be eliminated from the lattice. This is carried out by heating the sieve at 350–450 °C (molecular sieves can stand a heating up to 550 °C without adverse consequences). Activation can be speeded up by decreasing pressure or by blowing a stream of dry inert gas through the heated sieve. After the adsorption cycle the sieves can be *regenerated* and reused. The procedure is similar to that during activation – the adsorbate is eliminated at elevated temperature and decreased pressure or in a stream of inert gas. The adsorbed substance may also be first displaced from the sieve by a more strongly adsorbed, but more volatile, substance, which is then eliminated from the sieve in the same manner as water during activation.

Today molecular sieves are widely used both for laboratory and analytical purposes[1a], as well as in many industrial processes, and the development of new types of sieves increases continually[2, 3]. In view of the hydrophilic character of the inner surface and the high adsorption capacity for water (24 wt. % per mass of sieve) they represent the most important industrial dessiccants, especially in the area of "superdrying". Their high affinity for polar substances enables their use in the purification of gases and liquids from undesirable components (desulphuration etc.). The molecular sieve effect is also made use of in the separation of hydrocarbon mixtures, mainly in the petroleum industry. Processes have been introduced for the isolation of *n*-alkanes (MOLEX, ENSORB, ISOSIV, T. S. F., PAREX) and olefins (OLEX, OLEFIN-SIV) from various fractions and for the isolation of some aromates (PAREX-UOP, EBEX). Not less important are the applications of molecular sieves as catalysts, ion exchangers, and carriers of special chemicals.

The following sections are devoted to the applications of molecular sieves in laboratory separations of hydrocarbon mixtures only and to the analytical aspects following from them.

2.6.1 SEPARATION OF *n*-ALKANES ON THE MOLECULAR SIEVE 5 A

The total volume of the cavities in the crystal lattice of the molecular sieve 5A is 0.32 cm^3/g; of this the volume of larger cavities adsorbing *n*-alkanes makes 0.27 cm^3/g. The limit adsorption volume for *n*-butane is a little lower, i.e. 0.23 cm^3/g.

Each of the larger cavities inside the crystal lattice is connected with six equal cavities by means of six "windows" composed of eight-member

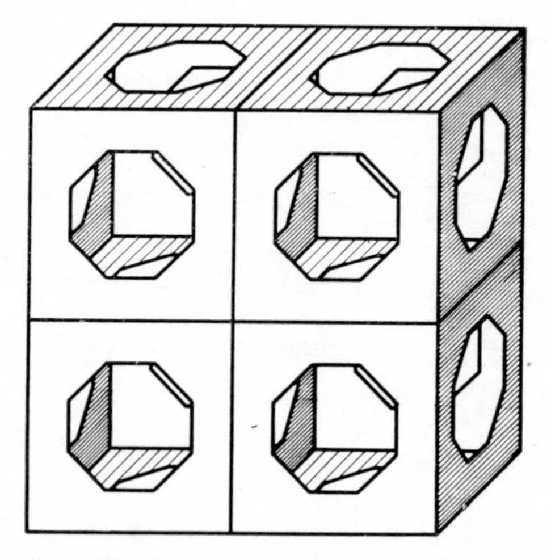

Fig. 2.6.3 Schematic representation of the cavities between cubo-octahedrons in the molecular sieve 5A
Each corner of the cube corresponds to the centre of a cubo-octahedral cell (which are not however represented)

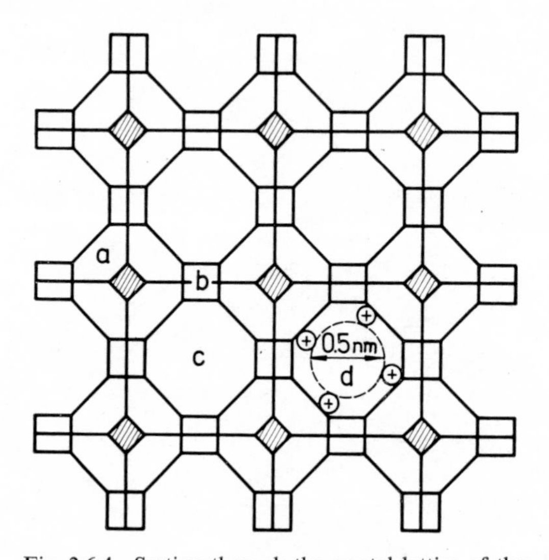

Fig. 2.6.4 Section through the crystal lattice of the molecular sieve 5A
a — cubo-octahedral cell composed of 12 Al, 12 Si and 48 O, b — four oxygen bridges connecting neighbouring cubo-octahedrons, c — pores of eight-membered oxygen rings connecting large cavities in the cell, d — representation of the shielding effect of cations

oxygen rings. The spatial arrangement of the cavities and the connecting openings can be represented in a simplified manner as neighbouring cubes connected by openings in their walls (Fig. 2.6.3). A cross-section of a schematic crystal lattice of the sieve 5A is shown in Fig. 2.6.4. The openings into the cavities, with an effective diameter of 0.5 nm, permit adsorption of *n*-alkanes (and *n*-olefins), but exclude branched alkanes, aromates and cycloalkanes – with the exception of cyclopropane.

The equilibrium adsorption capacity of the sieve 5A for *n*-alkanes C_6—C_{16} is (at 25 °C) within the 10–13 g of hydrocarbon per 100 g of sieve range[4]. In an average crystal of zeolite of 2 μm size the adsorbed *n*-alkane molecule must diffuse through about 1 000 cavities before it can reach the centre of the crystal. The length of the alkane chain has a great influence on the rate of adsorption. *n*-Alkanes with more than 14 carbon atoms are adsorbed very slowly at room temperature. For example, the rate constant of zeolitic diffusion at 25 °C is more than 100 times lower for *n*-hexadecane than for *n*-hexane[4].

The adsorption of *n*-alkanes is slowed down if gases are adsorbed in the cavities. If an activated sieve is left in contact with air, nitrogen or some other gas, it is saturated within 4–6 hours. The adsorbed amounts of some gases are given in Table 2.6.2. During the adsorption of *n*-alkanes from liquid phase the gases from the inner cavities of the sieve are displaced and they form gaseous "plugs" decreasing the rate of penetration of *n*-alkanes into the sieve. This can be prevented by activating the sieve in a vacuum. The adsorption of *n*-alkanes from the gaseous phase takes place much more rapidly but the capacity of the sieve decreases at ele-

TABLE 2.6.2

Amounts of some gases adsorbed by the molecular sieve 5A at 18 °C

	H_2	O_2	N_2	CH_4	CO_2
ml of gas in 1 g of sieve	0.64	3.15	7.25	10.00	57.70
mg of gas in 1 g of sieve	0.089	4.5	9.1	7.1	113.0

vated temperatures; at 200 °C the adsorption capacity of the sieve 5A for *n*-nonane is 5.6 wt. %, and at 300 °C it is about 3.5 wt. %.

Laboratory separations of *n*-alkanes on the sieve 5A can be carried out – as in industrial processes – both in the gas phase and in the liquid phase. Adsorption from the gas phase is mostly used as a part of some analytical techniques of *gas chromatography,* mentioned in Chapter 4.1. In the liquid phase *n*-alkanes are separated either by filtration of the fraction through a column packed with a powdered sieve, or – more usually – by refluxing a solution of the hydrocarbon fraction in an inert solvent in the presence of an excess of sieve. In both instances *n*-alkanes remain adsorbed on the sieve and other types of hydrocarbons pass into the filtrate.

The adsorption on the sieve takes place with maximum selectivity only within a certain range of conditions. If correct conditions are not observed, both the capacity of the sieve and the selectivity of the separation process may deteriorate. The main disturbing effects are discussed below.

The rate of adsorption of *n*-alkanes and the adsorbed amount (capacity of the sieve) clearly deteriorate in the presence of *polar substances* that are adsorbed preferentially on the surface of the sieve where they obstruct the openings. When testing the disturbing effect of compounds of various nature Wiel[5] found that strongly polar nitrogen and oxygen-containing substances, for example pyridine, organic acids, ketones, etc., considerably deteriorate the adsorption of *n*-alkanes even at a 1 % concentration. In contrast to this even high concentrations of polycyclic naphthenes, monocyclic aromates, and weakly polar sulphur-containing substances do not disturb adsorption. Therefore the main condition for a successful separation of higher *n*-alkanes with a molecular sieve is the elimination of strongly polar substances from the separated fraction; for this purpose column chromatography on silica gel is most suitable.

The second important circumstance is the effect of the conditions on the sieve *selectivity.* It was observed that the selectivity of the sieve 5A for *n*-alkanes is not absolute, because under strong conditions even some isoalkanes penetrate the inner structure of the sieve. This is demonstrated by the results of Wiel[5], shown in Table 2.6.3. At a higher temperature (177 °C) and a sufficiently long time the sieve 5A also adsorbs monomethylalkanes. The sieve's capacity for monomethylalkanes is the

TABLE 2.6.3

Rate of adsorption of model hydrocarbons on the molecular sieve 5A at various temperatures (according to van der Wiel[5])

Hydrocarbon	Time of adsorption in hours	Adsorbed amount in wt. %[a]	
		at 99.3 °C[b]	at 177 °C[c]
n-Heptacosane	1	99.5	99.4
2-Methylpentacosane	1	3.5	22.5
2-Methylpentacosane	24	28.2	99.4
4-Methylpentacosane	24	4.2	97.3
1-Cyclohexyloctadecane	24	1.3	5.7

[a] calculated per hydrocarbon used at a sieve to hydrocarbon ratio 10 : 1
[b] under reflux in isooctane solution
[c] under reflux in decalin solution

same as for *n*-alkanes, which means that isoalkanes are not only adsorbed on the outer surface, but they also penetrate the inner structure of the sieve. The oscillations of the crystal lattice at elevated temperatures evidently increase the openings to such an extent that they can let through alkanes with a side-methyl. In contrast to *n*-alkanes which can be desorbed to 90 – 95 % with boiling *n*-heptane from the inner cavities of the sieve (after 24 h), methylalkanes are desorbed under the same conditions only negligibly, and they are completely liberated only when the sieve is decomposed in an acid. Aliphatic chains terminated by cyclohexane rings that are deformed with difficulty do not enter the sieve even at elevated temperatures or after prolonged periods of adsorption.

The main factors affecting adsorption can be summarized as follows: the selectivity of adsorption increases at a low temperature, a low excess of sieve with respect to adsorbate, and at a short adsorption time. In order to achieve a satisfactory rate of adsorption of high-boiling *n*-alkanes an increased temperature is indispensable, and it is advantageous to use a finely powdered sieve suspended in an inert hydrocarbon. For medium and heavy distillates a temperature between 80 – 100 °C represents a reasonable compromise, because – as is evident from Table 2.6.3 – at the boiling temperature of isooctane (99.3 °C) the high boiling *n*-alkanes are adsorbed practically quantitatively within 1 h, while the adsorption of methylalkanes is at least 20 times slower. At 80 °C adsorption takes place somewhat more slowly, but in this case practically pure *n*-alkanes are separated.

The method of adsorption of *n*-alkanes on the sieve 5A in the liquid phase is used in various modifications for the isolation of *n*-alkanes and for quantitative elimination of *n*-alkanes from hydrocarbon fractions. The difference of weight or volume of the fraction before and after adsorption or the increase in weight of the sieve after adsorption, correspond – under controlled conditions – to the content of *n*-alkanes in the fraction. Therefore the majority of separation processes can be used as a simple method for the analytical determination of *n*-alkanes as a group.

The two original methods[6,7] for the determination of *n*-alkanes in petroleum distillates by means of sieves are rapid and accurate, it is true, but they are limited to fractions with a narrow boiling range only or to low-boiling fractions. Later O'Connor and Norris[8] described a method utilizable for wide boiling fractions: gasoline, jet fuel, kerosene and light gas oil. For analysis only the saturated part of the fraction obtained by the FIA method is used (see Section 3.5.1). The sample (0.5 – 1 g) is allowed to soak into a short column packed with 15 g of powdered sieve 5A (activated at 450 °C in vacuum for 6 h) and the whole column is weighed. Then unadsorbed hydrocarbons are eluted with isopentane and the remains of isopentane is evaporated from the column under reduced pressure. The increase in the weight of the column corresponds to the content of *n*-alkanes in the sample. Standard deviation is ± 0.8 %. The adsorbed *n*-alkanes are isolated from the sieve in a continuous extractor with *n*-pentane; the isolation of alkanes C_{12}—C_{23} requires 100 h.

For serial determinations of *n*-alkanes in gasolines two modifications of the column technique have been elaborated, in which the adsorption and the desorption cycle and the regeneration of the sieve take place directly in the column provided with a heating jacket.

Pavlova et al.[9] employ an adsorption column packed with the sieve 5A of 0.25 – 0.5 mm particle size and a sample to sieve ratio of 1 : 5. Adsorption takes place at 20 – 25 °C under reduced pressure, 4 – 6 kPa (for higher boiling fractions at 30 – 40 °C and 0.3 kPa). Non-adsorbed hydrocarbons pass through the column and condense in the cooled receiver. The adsorbed alkanes are desorbed by heating the column at 350 – 360 °C at the same reduced pressure, and they are collected in a separate receiver. The sieve is regenerated at the desorption temperature for 30 to 40 minutes and then allowed to cool in a vacuum.

Kvitkovskii and Gruschetskaya[10] use a similar apparatus consisting of a column packed with the sieve 5A to 10 cm of its height (8 – 9 g), a heating jacket, and a cooled receiver. The sieve is regenerated in the column at 375 °C in a weak stream of hydrogen for 60 – 90 min. The use of hydrogen as a rinsing gas secures a maximum weight stability of the sieve during the determination (see Table 2.6.2). When the regeneration is over the column is taken out of the jacket, it is allowed to cool under hydrogen and it is then closed and weighed. A sample of 0.2 to 1.5 ml volume is injected directly onto the sieve and the column with the sample is weighed again. It is then replaced in the heating jacket, connected with the source of hydrogen and heated in a hydrogen stream at an operating temperature that is 10 – 20 K higher than the end of distillation of the sample (usually 210 – 220 °C) . The operation temperature is maintained for 5 minutes during which adsorption of *n*-alkanes is terminated as well as the flushing of the vapours of unadsorbed hydrocarbons into the receiver. The column is allowed to cool under hydrogen and weighed again. The error of the determination does not exceed 0.5 wt. %.

Both methods are relatively simple and do not require expensive apparatus. For the determination of *n*-alkanes in gasolines and other light petroleum distillates (without isolation) gas chromatography is usually more convenient (Chapter 4.1). A column technique for direct determination of *n*-alkanes in petroleum distillates 200 – 470 °C was devised by Sista and Skrivastava[11]. The separation of *n*-alkanes from heavy petroleum distillates (up to C_{40}—C_{50}) by adsorption in the liquid phase has been described by Brunnock[12]. In a 100 ml flask 1 g of fraction (or paraffin wax) is dissolved in 60 ml of benzene and 20 g of the pearls of activated sieve 5A are added. The contents of the flask is refluxed for at least 48 h until adsorption is complete (according to a GLC control). The sieve is filtered off using a sintered glass filter and both the flask and the sieve are washed with hot benzene. The non-adsorbed hydrocarbon fraction is obtained from the combined filtrates by evaporation of benzene from a water bath under reduced pressure. The pearls of the sieve are then transferred into a Soxhlet extractor and extracted again with benzene for at least 48 h in order to eliminate non-linear hydrocarbons adsorbed on the outer surface of the sieve. The sieve is then dried in air and decomposed in 40 ml of a 1 : 1 mixture of water and hydrofluoric acid. The mixture is evaporated on a water bath to dryness and the *n*-alkanes set free are extracted with tetrachloromethane.

This method requires a lot of time, but in view of the relatively low temperature of adsorption (80 °C) and the thorough elimination of hydrocarbons from the outer surface of the sieve the isolated *n*-alkanes are exceptionally pure.

In the isolation of *n*-alkanes from paraffin waxes O'Connor et al.[13] employed a more rapid procedure in which alkanes are adsorbed on the sieve in a medium of boiling isooctane. The method is widely used both for the isolation of *n*-alkanes from various fractions (cf. ref.[14]), and for the determination of the content of *n*-alkanes as a group. The analytical procedure is as follows:

Determination of n-alkanes in paraffin waxes by adsorption on the molecular sieve 5A[13]

A — Paraffin wax (2.0—2.1 g) is weighed into a flask with an accuracy of up to 0.5 mg and 100 ml of isooctane are added to it. After dissolution of the sample 40 g of molecular sieve 5 A in the form of pearls are added (activated at 250 °C in a vacuum for 6 h) and the contents are refluxed intensively for 4 h. The sieve is filtered off while hot (under suction) and the flask and the sieve are washed twice with 50 ml portions of hot isooctanet The filtrates are combined and transferred into a 400 ml flask, previously weighed with a 0.5 mg accuracy. The filtration flask is rinsed with two 25 ml portions of isooctane. which is added to the filtrate. Isooctane from the filtrate is evaporated on a water bath in a stream of dry air. The evaporation is continued until two weighings at 10 minute intervals do not differ by more than 2.0 mg. The weight difference of the original sample and the hydrocarbons isolated from the filtrate is equal to the weight of adsorbed *n*-alkanes; standard deviation is 1.1 %.

Isolation of n-alkanes: The sieve is transferred into a ground flask, *n*-pentane is added to cover it completely and the mixture is allowed to stand in the stoppered flask for at least 15 days. After sep' ration of the sieve by filtration extraction can continue with fresh *n*-pentane.

B — On the basis of the testing of optimum conditions of adsorption van der Wiel[5] recommended the following conditions for this method: 1 g of paraffin wax is dissolved in 100 ml of isooctane and 10 g of powdered sieve 5 A are added (0.5—6 μm, activated at 350 °C for 24 h). The content of the flask is stirred at the isooctane boiling point for 1 h.

Isolation of η-alkanes: a) — The sieve is stirred in a flask with an excess of boiling *n*-heptane for 24 h (yield 92—95 % of adsorbate). b — The sieve is decomposed using hot dilute hydrochloric acid and it is extracted repeatedly with *n*-heptane (yield 97 % of adsorbate).

Determination of the content of *n*-alkanes in microcrystalline paraffin waxes fails if the aromates are not eliminated completely. The specific structures of the aromates present – probably monoaromates highly substituted with alkyl chains – inhibit adsorption of *n*-alkanes very effectively even at concentrations of about 5 % (unsubstituted aromates do not prevent adsorption even at concentrations of up to 15 %).

Washall et al.[15] developed a technique enabling the determination of *n*-alkanes in paraffin waxes and microcrystalline waxes (C_{30}—C_{55}) without previous elimination of aromates. The adsorption of *n*-alkanes on the sieve 5A takes place from a decalin solution at 180 °C (without refluxing) for 16 h. The results are comparable with those of the isooctane method of dearomatized samples. In view of the high temperature of adsorption it should be reckoned that the adsorbed *n*-alkanes will be contaminated with a certain amount of non-normal hydrocarbons. *n*-Alkanes in gas oils with a high content of aromatic compounds are also determined[16] by a similar technique of adsorption at 200 °C.

The adsorption of *n*-alkanes with a sieve from a liquid phase is accompanied by a volume contraction of the liquid. The method of Larson and Becker[17] for the determination of *n*-alkanes (in vol. %) in gasolines is based on this principle. Later on Bacúrova[18] extended this method to fractions boiling in the 150 – 350 °C range. The procedure is the following:

Determination of n-alkanes in kerosene and diesel fuel by adsorption on the molecular sieve 5A[18]

The determination takes place in parallel in two volumetric flasks of 50 ml content with a long neck (provided with a mark) and a ground glass joint for a reflux condenser.

A sample is measured (0.5—5 ml) into the flask and exactly 20 ml of isooctane are added to it from a burette to dissolve the sample. Activated sieve 5A (0.8—1 mm granulation, 20 g) is added into the flask that is then half filled.

Into a second flask 20 ml of isooctane are measured first, followed by so many ml of isooctane as corresponds to the sample volume in the first flask. Again exactly 20 g of molecular sieve are added into the blank.

Both flasks are provided with reflux condensers and put into a heated aluminium block where their content is kept refluxing for 4 h. When adsorption is over, the flasks are taken out of the block and allowed to cool to room temperature (without taking off the reflux condensers). When the temperatures are equalized the volume in both flasks is made up to the mark from a burette with isooctane. The volume % of *n*-alkanes in the sample is calculated using equation

$$\text{vol \% of } n\text{-alkanes} = \frac{a - b}{c} \cdot 100$$

where a is the sum of ml of sample + volume of isooctane consumed in the first volumetric flask; b is the sum of ml of isooctane consumed in the blank; c is the volume of sample in ml. The accuracy of the determination varies within the ± 2 vol. % range.

Kisielow et al.[19] compared the mentioned weight[13] and volume[18] method and came to the conclusion that for adsorption of *n*-alkanes from refined wide boiling-range oil fractions the optimum ratio of sieve to the sample is dependent on the molecular weight of alkanes; for light fractions it should be 10–15: 1, and for the heaviest fractions the ratio should be increased to 22–25 : 1. A 2–3 h adsorption at reflux temperature of isooctane is sufficient. For fractions distilling above 300 °C both methods give equal results. The weight method is more advantageous for poorly soluble high-molecular hydrocarbons. With fractions that begin to boil below 300 °C only the volume method gives correct results. The error of the weight method in the case of lighter fractions is due to the impossibility of completely separating isooctane by distillation from the relatively volatile hydrocarbons in the filtrate.

A rapid and versatile microtechnique for the isolation of *n*-alkanes from heavy fractions and their quantitative determination has been described by Mortimer and Luke[20]. Similarly to the determination of *n*-alkanes with urea (Section 2.5.1) among the methods using the sieve 5A, the combined procedure consisting of the isolation of *n*-alkanes with the sieve and subsequent determination of the distribution and the content of *n*-alkanes by gas chromatography has the greatest advantages. *n*-Alkanes are adsorbed from a few mg of sample in the vapour phase in a microadsorber containing the sieve. The determination takes 2–4 h and it is suitable for distillates boiling within the 170–500 °C range. The procedure is as follows:

Determination of n-alkanes in petroleum products by adsorption on the molecular sieve 5A and by gas chromatography[20]

The microadsorption unit consists of a Pyrex tube 15 cm long and 3 mm I.D. that is narrowed at one end to 1 mm I.D. The tube is packed with a 7 cm long filling of the 40—60 mesh sieve 5A (400 mg or a forty-fold amount of the assumed weight of *n*-alkanes) and 2 cm layer of 60—80 mesh celite. The tube is fastened tightly into a brass holder in which a septum for injection is fixed and a nitrogen inlet at the side. Then the microadsorption unit is inserted into a metallic block provided with electric heating.

n-Alkane boiling outside the boiling range of the fraction is then introduced accurately to the sample of the analysed fraction as internal standard. The solid sample is dissolved in isooctane or cyclohexane. The stream of nitrogen passing through the layer of celite and the bed of molecular sieve in the adsorption tube is adjusted to 10 ml/min and a sample volume containing about 10 mg of *n*-alkanes is injected through the septum into the celite layer at a sufficient depth. The heating of the block is then started which is adjusted so that 300 °C temperature should be attained from the room temperature within 20 min, and maintained there. Under these conditions non-linear hydrocarbons are eluted with nitrogen, while *n*-alkanes remain adsorbed on the sieve. When the vapours of non-adsorbed hydrocarbons stop coming out of the narrowed end of the tube the adsorption unit is withdrawn from the block and allowed to cool in streaming nitrogen. The sieve is poured into a test tube with a screw closure, containing 1 ml of water. The tube is then cooled in solid CO_2 in order to remove the liberated heat rapidly. When water in the test tube is frozen 1 ml of 40 % hydrofluoric acid is put on the ice surface. The test tube is closed and allowed to warm up slowly until the ice melts, and the content is shaken (under cooling if necessary). The test tube is then shaken mechanically for another 30 min to decompose the sieve completely. Isooctane (0.3 ml) is then added to the mixture, followed by a peck of KOH in order to neutralize excess acid. After 15 min of mechanical shaking the test tube is centrifuged and the isooctane solution of *n*-alkanes separates as a clear layer above the sediment of the decomposed sieve. The isooctane solution is analysed by programmed GLC and the content of *n*-alkanes is calculated against the area of the internal standard.

Adsorption of n-alkenes on the sieve 5A. O'Connor and Norris[8] studied adsorption of model hexenes on the sieve 5A and they found that all three types of *n*-alkenes are adsorbed on it: *cis*-isomers, *trans*-isomers, and *terminal n*-olefins. Diolefins are adsorbed in the same manner, but iso- and cyclo-olefins are excluded from adsorption on the sieve 5A.

These results are valid for adsorption from concentrated solutions. Fenselau and Calvin[21] have found that various isomers of *n*-octadecene are adsorbed on the sieve 5A from very dilute solutions (0.4 %) in various manners. In addition to this, adsorption is dependent on the type of solvent used; from a refluxing benzene solution only *trans*-9-octadecene is adsorbed, while the *cis*-isomer and the *terminal n*-octadecene are excluded. In contrast to this, from a cyclohexane solution all three isomers are adsorbed to different extents (1-octadecene least) and the adsorption of *n*-octadecane also takes place much faster. This so far little investigated effect opens up further prospects for a fine regulation of the adsorption selectivity of molecular sieves.

In contrast to adsorption in the gas phase, polymerizations do not take place during adsorption of olefins from the liquid phase on the molecular sieve. However, terminal and *cis*-isomers adsorbed on the sieve isomerize to a considerable extent and *trans*-isomers with a double bond shifted to the centre of the chain are formed predominantly. The olefins which are in contact with the sieve under the same conditions (i.e. with their outer surface) but which are not adsorbed, do not polymerize.

Recently the separation of methane and lower olefins on heterogeneous molecular-sieve membranes[22] is also being investigated. They are prepared from a polymer binder foil filled to a high content (55 – 75 %) with the molecular sieves 4A or 5A.

The sodium cations of the molecular sieve can be exchanged by Ag^+ ions using the conventional ion exchange technique. The affinity of the sieve to olefins is thus considerably increased. Silver zeolites can therefore be applied for example for the purification of saturated hydrocarbons from olefinic contaminants[23].

2.6.2 SEPARATION OF AROMATES ON MOLECULAR SIEVES 10X AND 13X AND OTHER APPLICATIONS

So far, there are not many examples described of the use of molecular sieves 10X and 13X in laboratory separations of hydrocarbons. Shortly after the sieves 10X and 13X were put on the market, Mair and Shamaiengar[24] developed a method for the fractionation of aromatic hydrocarbons on these adsorbents. They tested the sieve selectivity on model aromatic hydrocarbons C_{12}—C_{20}. All further mentioned tested hydrocarbons are adsorbed in the liquid phase by the sieve 13X. On the sieve 10X only *n*-decylbenzene is adsorbed completely; 6-decyltetralin and 2-butyl-5-hexylindane are adsorbed only partly (50 – 60 %) and 1,3,5-triethylbenzene and dodecahydrochrysene are excluded almost completely.

The fractionation technique was then applied to concentrates of mononuclear, dinuclear and trinuclear aromates C_{18}—C_{25} from heavy gas oil and light oil distillate. The fraction was introduced at the bottom of a column packed with the molecular sieve and the non-adsorbed fraction was eluted with isooctane. The hydrocarbons adsorbed on the sieve were desorbed from the column with overheated steam (at 175 – 220 °C) or with ethanol at 75 °C. Fractionations on the sieve 13X and 10X were combined according to the following scheme:

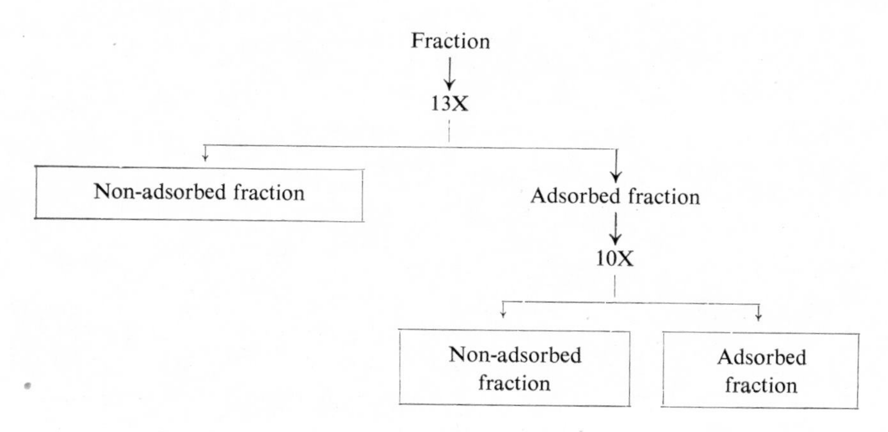

Three subfractions were separated from each fraction. In all separated fractions the non-adsorbed part on the sieve 13X has a higher viscosity (up to 61 times in the case of dinuclear aromates) than the adsorbed part on the sieve 10X. This confirms that the hydrocarbons separated in the two parts differ substantially in their molecular shape.

The main limitation of this technique consists in the fact that isooctane used as an eluent is also adsorbed both by the sieve 13X and the sieve 10X. Therefore it can be used for the separation of aromatic hydrocarbons only, that are much more strongly adsorbed than isooctane. For the separations of saturated hydrocarbons on the sieve 10X e.g. 1,3,5-trimethylcyclohexane could come into consideration as an eluent.

In a perfected form the technique of adsorption on a column of the sieve 10X was used later in *API Research Project 6* for the separation of mononuclear aromatic hydrocarbons C_{24}—C_{27}[25]. The procedure was the following: 15–20 g of the separated oil were introduced at the bottom of a column packed with 125 g of the sieve 10X (42–60 mesh). The oil was then slowly forced upward through the column by introducing CCl_4 at the bottom. In this arrangement the oil came in contact with fresh adsorbent only. The rate of introduction of CCl_4 was adjusted so that the front of the oil should move in the column at a rate of approximately 1 cm per hour. (The whole experiment lasted about 100 h.) When the experiment was over the sieve from the upper half and the lower half of the column was extracted separately with boiling toluene in a Soxhlet extractor. Hydrocarbons are adsorbed in the sieve cavities very strongly, and for their isolation to a 94–98 % extent a continuous extraction is indispensable, lasting 120 – 150 h. A concentrate of alkylbenzenes and monocyclanobenzenes was isolated from the oil fraction adsorbed in the bottom half of the column; in the non-adsorbed part di-, tri-, tetra- and higher cyclanobenzenes were concentrated. In these experiments it was also found that isoalkane-cycloalkane fractions C_{24}—C_{27} cannot be separated on the sieve 10X, because both types of molecules are adsorbed in the sieve to the same extent[25].

Kajdas and Ligezowa[26] applied this column technique to the separation of monoaromates isolated by elution chromatography from narrow fractions of a refined oil distillate (320 – 500 °C). On a column with 100 g of sieve 13X (at a loading of the column of 0.18 g of aromatic fraction per 1 g of sieve) aromatic hydrocarbons were separated to an adsorbed part (14 – 21 wt. %), containing monocyclanobenzenes, and a non-adsorbed part (eluted with CCl_4), containing polycyclanobenzenes. With increasing boiling point of the fraction the part of the aromates that are not adsorbed on the sieve 13X increases.

Another example of the separation of aromates on molecular sieves is the selective adsorption of 2,7-dimethylnaphthalene from its *eutectic* mixture with 2,6-dimethylnaphthalene. The selectivity of various types of molecular sieves is evident from Table 2.6.4 that summarizes the results of batch adsorption experiments[27] in which 10 g of a dimethylnaphthalene concentrate, 2.5 g of isooctane and 5.0 g of sieve were heated at 100 °C for 2 h. The non-adsorbed fraction was filtered off, the sieve washed with isooctane at room temperature and the adsorbed part isolated with refluxing benzene. The results show that 2,7-dimethylnaphthalene is adsorbed with greatest selectivity by the sodium form of sieve Y. On the contrary, on the sieve of type L 2,6-isomer is adsorbed preferentially. In separations of this type the differences in adsorptivities due to differing basicity of aromates present in the mixture also play an important role in addition to the sieve effect of zeolites.

The development of new types of molecular sieves is constantly continuing and it may be expected that the series of available selective adsorbents suitable for the separations of hydrocarbons will be substantially extended. For example the possibility emerges that molecular sieves could be used for the separation of isoalkane-cycloalkane fractions. Current types of molecular sieves cannot be used for this purpose because both types of hydrocarbons are excluded by the sieve 5A and completely

TABLE 2.6.4

Selectivity of various types of molecular sieves in the separation of 2,6-dimethylnaphthalene and 2,7-dimethylnaphthalene[27]

Type of molecular sieve	Capacity of the sieve g of hydrocarbon per 100 g of sieve	Separation factor[a] α
Sodium form of type Y	16.6	8.0
Ammonium form of type Y	13.2	5.8
Type 13X	9.2	2.8
Type 10X	6.6	2.4
Potassium form of type L	11.4	0.53[b]

[a] $\alpha = \frac{\text{\% 2,7-DMN adsorbed/\% 2,7-DMN unadsorbed}}{\text{\% 2,6-DMN adsorbed/\% 2,6-DMN unadsorbed}}$

[b] 2,6-dimethylnaphthalene is adsorbed preferentially

adsorbed by the sieve 10X. It is evident, however, that new types of sieves with an effective diameter of the inlet openings between 0.5 and 0.8 nm can contribute to the solution of this difficult separation problem.

Mair and Barnewall[25] indicated in their study the potentiality of an experimentally prepared sieve, labelled 7A; this sieve has suitable properties for differentiated adsorption of saturated hydrocarbons according to the size and the shape of the molecules. Curran et al.[28] described the usefulness of the sieve 7A in the separation of the isoalkane-cycloalkane fraction from *shale-oil* (Green River). They chose 1,3,5-triisopropylbenzene as solvent for the fraction, because mesitylene and 1,3,5-triethylbenzene were found unsuitable. A solution of 27 mg of the fraction in 2.5 ml of triisopropylbenzene was stirred with the molecular sieve 7A (*Mordenite* with pore diameters 0.66 nm; activated at 300 °C in a vacuum for 48 h) under nitrogen for 2 h. The non-adsorbed part was withdrawn with a pipette and the sieve was washed 5 times with triisopropylbenzene at 80 – 90 °C for 30 min. The combined solutions were eluted with *n*-hexane from a column containing silica gel impregnated with 10 % $AgNO_3$ (2 g of $AgNO_3$ are dissolved in a minimum volume of water, 50 ml of methanol and 20 g of silica gel are added, and the liquids are evaporated; the remaining powder is activated at 100 °C for 1 h). Triethylbenzene remained adsorbed in the column and the hexane eluate afforded, after evaporation, 10 mg of saturated cyclanic fraction containing predominantly triterpanes; branched alkanes were completely absent. Water was added to the washed sieve just to cover it, followed by an equal volume of hexane, and decomposition was brought about by addition of 40 % hydrofluoric acid. The excess of the acid was decomposed by addition of a saturated solution of boric acid and the adsorbed part of hydrocarbons was extracted with hexane. After evaporation of hexane the residue was eluted from the column containing $AgNO_3$ in the above described manner. From the eluate 2.5 mg of branched alkanes (*phytane*, *pristane*, etc.) were then isolated.

2.6.3 CARBONACEOUS MICROPOROUS ADSORBENT AND POROUS GLASSES

Activated *charcoal* with small pores is formed by thermal decomposition of some synthetic polymers. For this purpose the polymers and copolymers of vinylidene chloride (plastics of the Saran type) are especially suitable because they begin to decompose thermally already at temperatures of 150 – 180 °C setting free hydrogen chloride according to the equation

$$(C_2H_2Cl_2)_n \rightarrow 2n\,C + 2n\,HCl.$$

Highly pure carbon with microporous structure in the form of irregular narrow cracks and channels is the unvolatile product of the reaction. It can be further modified by special procedures. Compounds with small molecules (for example *n*-pentane) are adsorbed on such activated charcoal rapidly and completely, while larger molecules (neopentane) are adsorbed much more slowly or they are excluded from adsorption completely (α-pinene)[29]. Owing to their unique adsorption properties these saran charcoals are called "*carbon molecular sieves*". For example the adsorbent developed by Kaiser[30] for chromatographic columns is produced under the name Carbosieve-B; its surface is about 1000 m^2/g and the pore diameter within the 1 – 1.2 nm range. Another type of microporous adsorbent has been prepared from coal and called "Molekularsiebkoks"[31]. A microporous activated charcoal was also prepared experimentally by carbonization of phenolformaldehyde resins[32]; according to the degree of activation it has a molecular sieve effect for compounds with a critical diameter of 0.4 – 0.8 nm.

Saran charcoals and other types of microporous activated carbon have -- it is true – special adsorptive properties, but generally they do not display such an exclusive shape selectivity for adsorbates as zeolites do. Using model adsorbates with various critical diameters of molecules it was found[33] that under the conditions of gas chromatography some activated charcoals do not show any sieve effect in the range of the given pore size; for example hydrocarbons C_5 and higher, with a critical diameter exceeding the declared pore size, are not excluded, but – on the contrary – adsorbed so strongly that they cannot be thermally desorbed without decomposition.

The properties of saran charcoals are advantageous from many viewpoints and they have numerous practical applications as adsorbents in gas, column and thin-layer chromatography[30]. In gas chromatography Carbosieve-B was found suitable, for example, for the separation of permanent gases and C_1—C_4 hydrocarbons. It has the unique property that it has a highly non-polar surface which causes water to be eluted before methane. This is made use of in the analysis of traces of water and other applications. The chromatographic properties of active charcoals are more thoroughly described in Chapter 4.1.

Porous glasses. When the structure of special multicomponent (for example borosilicate) glasses is destroyed by mineral acids under controlled conditions glasses are formed with a narrow distribution of pores of the required dimensions. Further thermal and chemical treatment can modify the properties of porous glasses suitably[34]. Some types of porous glasses are convenient as carriers and adsorbents in gas chromatography[35–38]; the widely used Corning Type 7930 Vycor Glass has a 173 m^2/g surface area and 2.5 nm average pore diameter. For molecular sieve chromatography of high-molecular compounds in liquid phase Controlled-Pore Glass and Bio-Glass are available. For example the type CPG-10 is commercially available in a number of types with graded average pore size from 7.5 nm to 20 nm. The types with smaller pore diameters have larger surface areas; a glass with pores about 7.5 nm has an internal surface area of 240 – 340 m^2/g.

In the case of porous glasses prepared in the laboratory and having a narrow distribution of pores of small size it was shown that they adsorb selectively compounds with small molecules[39]. The capacity and the selectivity of porous glasses does not by far attain the molecular sieve properties of synthetic zeolites. On the other hand, porous glasses are inert, chemically stable even in acid medium and they are well utilizable as packings of columns, especially for chromatographic separations of polymers and various biological materials.

REFERENCES (2.6)

1. H. Knoll, *Chem. Tech.* **26** (1974) 391.
1a. N. G. McTaggart and L. A. Luke, *Z. Anal. Chem.* **290** (1978) 1.
2. "Molecular Sieve Zeolites-II", *Advances in Chemistry Series* **102** (1971).
3. "Molecular Sieves", *Advances in Chemistry Series* **121** (1973).
4. P. V. Roberts and R. York, *Ind. Eng. Chem. Proc. Des. Develop.* **6** (1967) 516.
5. A. van der Wiel, *Erdöl u. Kohle* **18** (1965) 632.
6. R. D. Schwartz and D. J. Brasseaux, *Anal. Chem.* **29** (1957) 1022.
7. K. H. Nelson, M. D. Grimes and B. J. Heinrich, *Anal. Chem.* **29** (1957) 1026.
8. J. G. O'Connor and M. S. Norris, *Anal. Chem.* **32** (1960) 701.
9. S. N. Pavlova, Z. V. Driatskaya and M. A. Mkhchiyan, *Khim. i Tekhnol. Topliv i Masel* No. **3** (1962) 58.
10. L. N. Kvitkovskii and E. V. Grushetskaya, *Khim. i Tekhnol. Topliv i Masel* No. **3** (1962) 61.
11. V. R. Sista and G. C. Skrivastava, *Anal. Chem.* **48** (1976) 1582.
12. J. V. Brunnock, *Anal. Chem.* **38** (1966) 1648.
13. J. G. O'Connor, F. H. Burow and M. S. Norris, *Anal. Chem.* **34** (1962) 82.
14. P. Jarolímek, V. Wollrab, D. Streibl and F. Šorm, *Coll. Czech. Chem. Commun.* **30** (1965) 880.
15. T. A. Gashall, S. Blittman and R. S. Mascieri, *J. Chromatogr. Sci.* **8** (1970) 663.
16. N. Y. Chen and S. J. Lucki, *Anal. Chem.* **42** (1970) 508.
17. L. P. Larson and H. C. Becker, *Anal. Chem.* **32** (1960) 1215.
18. Bacúrová, *Ropa a uhlie* **7** (1965) 171.
19. W. Kisielow, M. Grochowska and D. Rutkowska, *Nafta* (*Katowice*) **27** (1971) 79.
20. J. V. Mortimer and L. A. Luke, *Anal. Chim. Acta* **38** (1967) 119.
21. C. Fenselau and M. Calvin, *Nature* **212** (1966) 889.
22. F. Wolf, W. Hentschel and E. Krell, *Z. Chem.* **16** (1976) 107.
23. H. W. Quinn in Progress in Separation and Purification, E. S. Perry and C. J. van Oss (Eds.), Vol. 4, Wiley, Interscience, New York, 1971.
24. B. J. Mair and W. Shamaiengar, *Anal. Chem.* **30** (1958) 276.
25. B. J. Mair and J. M. Barnewall, *J. Chem. Eng. Data* **9** (1964) 282; 12 (1967) 126.
26. C. Kajdas and S. Ligezowa, *Nafta* (*Katowice*) **29** (1973) 371.

27. J. A. Hedge in "Molecular Sieve Zeolites-II", *Advances in Chemistry Series* **102** (1971) 238.
28. R. Curran, G. Eglinton, I. Maclean, A. G. Douglas and G. Dungworth, *Tetrahedron Letters* No. **14** (1968) 1669.
29. J. R. Dacey and D. B. Thomas, *Trans. Faraday Soc.* **50** (1945) 740.
30. R. Kaiser, *Chromatographia* **3** (1970) 38.
31. H. Jüntgen, K. Knoblauch, H. Müntzner and W. Peters, *Chem., Ing., Tech.* **45** (1973) 533.
32. T. G. Plachenov, V. P. Musakina, L. B. Sevryugov and V. M. Falchuk, *Zh. prikl. khim.* **42** (1969) 2020.
33. V. Patzelová, O. Kadlec and P. Seidl, *J. Chromatogr.* **91** (1974) 313.
34. D. M. W. Anderson, I. C. M. Dea and A. Hendrie, *Talanta* **18** (1971) 365.
35. S. P. Zhdanov, A. V. Kiselev and Ya. I. Yashin, *Zh. fiz. khim.* **37** (1963) 1432; *Neftekhimiya* **3** (1963) 417.
36. H. L. Macdonell, J. M. Noonan and J. P. Williams, *Anal. Chem.* **35** (1963) 1253.
37. R. W. Ohline and R. Jojola, *Anal. Chem.* **36** (1964) 1681.
38. O. Lysy and P. R. Newton, *Anal. Chem.* **36** (1964) 2514.
39. S. P. Zhdanov and E. V. Koromaldi, *Dokl. Akad. Nauk SSSR* **138** (1961) 870.

2.7 Thermal diffusion

If a mixture of liquid hydrocarbons is placed into a narrow slit between two parallel vertical walls of which one is heated and the other cooled, a temperature gradient is formed across the slit that causes thermal diffusion of the molecules of the mixture in the direction perpendicular to the walls. As the size and the shape of the molecules have a great influence on the thermal diffusion of substances in the liquid phase, some types of hydrocarbons are concentrated close to the hot wall, and others at the cold one. The differences in the density of the liquid mixture caused by the temperature gradient bring about, however, an additional movement of the molecules, i.e. an upward convection along the hot wall and downward convection along the cold wall (see Fig. 2.7.1). By this the molecules that are concentrated at the hot wall are transported to the upper part of the slit, and the molecules that more easily diffuse to the cold wall move to the lower part of the slit. The result of this simultaneous effect of thermal diffusion and vertical convection is an effective separation of the molecules of different structural types.

The first apparatus based on the *thermogravitational* principle was developed by Clusius and Dickel[1] for the separation of gas mixtures;

it consisted of a hot wire axially disposed in a cooled vertical tube. The simple construction and the high separation ability of the apparatus, which was demonstrated by the isolation of pure isotopes of chlorine, aroused a great interest in this technique and stimulated the construction of similar apparatuses for the separation of liquid mixtures[2].

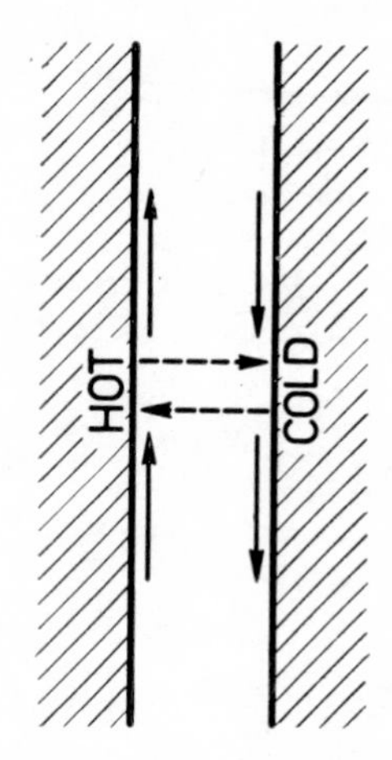

Fig. 2.7.1 Flow directions of currents in the thermodiffusion column slit
Broken arrows—ordinary and thermal diffusion current; full arrows—convection current

For laboratory separations of liquid hydrocarbon fractions columns made of coaxial tubes or simple plate apparatuses are mainly used. A scheme of a thermogravitational tubular column is represented in Fig. 2.7.2. Cooling water flows through the inner tube, while the outer tube is heated. The narrow slit between the inner surface of the outer tube and the outer surface of the inner tube represents the operational space proper, that is filled with the liquid mixture. The operational space is closed with a ring-shaped seal inserted between the two tubes. After a certain time a longitudinal concentration gradient is attained in the slit, and the separated fractions are withdrawn through the ports (not represented) located at various heights of the slit.

Plate columns are composed of two metallic plates of rectangular shape between which a seal is inserted that closes the operational space and keeps the selected distance between the plates constant. One plate is heated and the other cooled. Both plates with the inserted seal are pressed together with screws. The plate apparatuses usually have small dimensions and they are used predominantly as testing units during the checking of theoretically inferred relationships, during the investigation of the effect of working conditions on the separation etc.

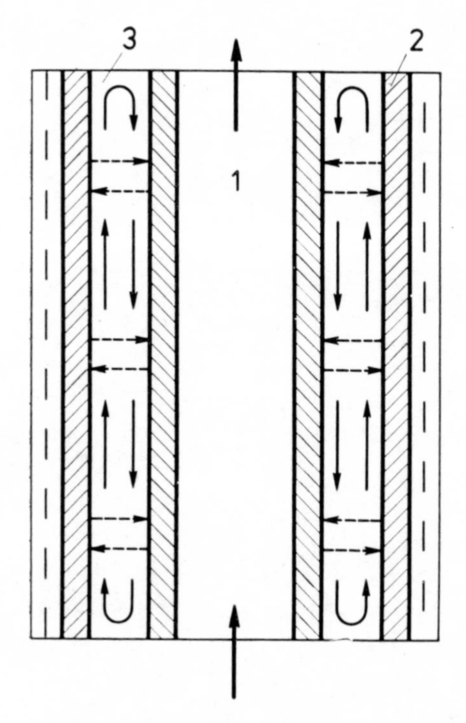

Fig. 2.7.2 Scheme of thermogravitational column for liquids, made of concentric tubes 1 — Inner tube cooled with streaming water, 2 — Outer tube with electric heating, 3 — Operation space; full arrows—convectional current, broken arrows—currents of regular and thermal diffusion

The mathematical description of the separation of liquid mixtures in a thermogravitational column is extremely complex. The process includes interactions between molecules, their regular and thermal diffusion connected with heat transfer, and all this under complex hydrodynamic conditions. For thermal diffusion in liquid phase a satisfactory theoretical explanation has been attained in the development of phenomenological theories[3] which are in qualitative agreement with experimental data. Theoretically derived relationships explain the effect of the column dimensions and operational conditions on the separation achieved, and they are a good basis for an optimalization of the apparatus.

If the separation of a binary mixture at a steady state is expressed as

$$q = \frac{\left(\frac{C}{1 - C}\right)_{\text{top}}}{\left(\frac{C}{1 - B}\right)_{\text{bottom}}} \tag{2.7.1}$$

where C is the concentration fraction of the component that is concentrated at the column top, then the general relation for a batch thermogravitational vertical column can be expressed by the equation:

$$\ln q = k \frac{\alpha D \mu L}{\beta T \omega^4} \tag{2.7.2}$$

where k is a constant, α the thermal diffusion coefficient, D regular diffusion coefficient, μ is viscosity, L the column length, β the coefficient of thermal dependence of density, T the mean temperature of the sample and ω is the width of the slit.

The direct dependence of ln q on column length was confirmed experimentally for columns of various dimensions. Of the geometrical factors the width of the slit has the greatest effect on separation. According to a theoretically derived relationship (2.7.2) the separation deteriorates with the fourth power of the slit width. A better fit with the experiment can be achieved if the exponent has a lower value, between 3 and 4; according to Powers[4] the right corrected value of the exponent is 3.3.

The important selectable factor of thermal diffusion and the driving force of the process, the temperature gradient ΔT between the hot and the cold wall, does not appear in the equation (2.7.2). Indeed the experiments confirm that in batch columns with a narrow slit the final degree of separation (at a steady state) is independent or only weakly dependent (in columns with a broader slit) on ΔT. However the temperature difference of the walls has a considerable effect on the rate of approaching equilibrium. For the so-called relaxation time (t_r) of the apparatus, i. e. the time necessary for attaining a certain degree of separation before equilibrium, the following relationship has been derived;

$$t_r = k' \frac{D\mu^2 L^2}{\beta^2 \omega^6 (\Delta T)^2} \tag{2.7.3}$$

from which it is evident that the rate of approaching equilibrium increases with the square of ΔT. The increase in temperature difference is limited by the physical properties of the separated mixture; the temperature of the hot wall should not exceed that of the boiling point of the lowest-boiling component and the temperature of the cold wall should be above the solidification point of the components. In the separation of hydrocarbon fractions temperature differences of 30 – 150 K are used. From equation (2.7.3) it also follows that the rate of approaching equilibrium decreases with the square of the column length. The steep concentration difference between both ends of the column at the initial phase of thermal diffusion is formed at a rate independent of the column length. Among other variables the approach to equilibrium is accelerated mainly by a decrease in sample viscosity, while among constructional parameters the width of the slit has a specially strong effect; in equation (2.7.3) this factor occurs in its sixth power.

2.7.1 COLUMNS FOR LIQUID THERMAL DIFFUSION

For practical separations of hydrocarbon fractions tubular thermogravitational columns are mainly used. They are simple and easy to construct in the laboratory. The tubes are either metallic or glass; they must have a perfect surface and accurate dimensions so that the slit formed on coaxial mounting of the tubes should be uniform all along the column. The heating of the tube can be electric or by a thermostated jacket. Flowing water is used for cooling. More often the inner tube is cooled and the outer heated, but a reverse arrangement is also possible. A uniform temperature along the column is maintained with difficulty, because the upper part of the vertical column has a tendency to overheating. Therefore the electric heating of longer columns is composed of several sections regulated separately, or the resistance winding is wound so that its output continually decreases in an upward direction.

Thermodiffusion columns are constructed either for a batchwise operation (Fig. 2.7.3a, b) or as continuous flow columns (Fig. 2.7.3c). During a batchwise operation the hydrocarbon mixture is pressed into a column through the bottom port until the whole annular space is filled and the meniscus of the liquid appears in the venting tube. The cooling water is then introduced, the heating of the column switched on, and when the required temperature gradient is formed equilibrium is allowed to form in the column. Depending on the construction of the column this takes from a few hours up to several days. The process of attaining equilibrium

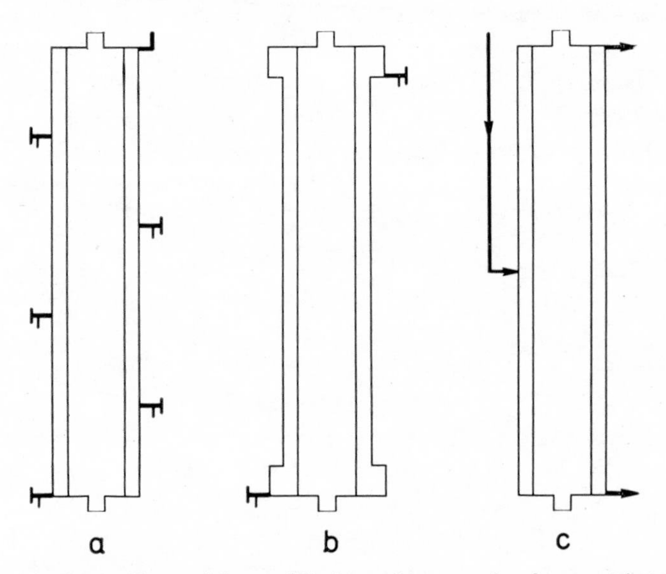

Fig. 2.7.3 Types of thermodiffusion columns made of concentric tubes
a — Batch column with five withdrawal ports, b — Batch column with a top and a bottom reservoir, c — Continuous flow-through column

can be followed so that at selected time intervals a few drops of sample are withdrawn from the bottom port, and the composition is determined on the basis of its refractive index or GLC analysis. When equilibrium is attained, i.e. when the concentration difference at both ends of the column does not change further, the heating of the column is switched off and the sample is let out. In the column shown in Fig. 2.7.3a the separated sample is withdrawn in five fractions (of 20 % of the total volume each) beginning with the upper port downwards. In columns with reservoirs (Fig. 2.7.3b) first the upper fractions are let out and then the bottom fractions.

In flow-through columns the sample is introduced continuously into the middle part (Fig. 2.7.3c) or some other part of the column and the separated fractions are withdrawn from the upper and the lower end of the column.

The column capacity is determined by the size of the operational space, i.e. by the slit width, column length and its diameter. The slit width cannot be increased too much, because – as follows from equation (2.7.2) – the separation deteriorates approximately with the fourth power of the slit width. For batch columns the optimum width of the slit is 0.2 – 0.4 mm; the flow-through columns have a wider slit, 0.5 – 1.5 mm, for practical reasons.

The column length has an important effect on the column efficiency. In the selection of the column dimensions it should be born in mind that not only the separation efficiency increases with increasing column length, but also the time necessary for attaining equilibrium.

Generally thermodiffusion columns can have very different dimensions, depending on whether they are intended for analytical measurements, for laboratory separations, or as production equipment. For example the equipment for the production of lubricating oils with a high viscosity index, developed at Standard Oil Company (Ohio)[5], includes bands of thermodiffusion columns of concentric tubes over 10 m long with a 1.1 mm slit. Columns for laboratory separations of hydrocarbon fractions are built much smaller – with a length of between 15 – 250 cm and an operational volume of between 10 and 100 ml. For the separation of very small samples thermodiffusion *semimicroapparatuses* have been described. For example Thompson[6] constructed a through-flow column with an internal electrically heated tube made of steel capillary used for the manufacture of syringe needles (1.47 mm O. D.) and an external cooled glass tube of 2 mm I. D. The operational space of the column is 0.95 ml only and its separation efficiency is excellent.

Another interesting arrangement for thermal diffusion of small samples is the so-called *paper thermodiffusion* developed by Vámos and Major[7], It consists of a hot and a cold plate, between which a strip of filter paper (30 × 5 cm) wrapped in an aluminum foil is compressed, soaked with 0.5 to 2 g of sample. After the end of the experiment the filter paper strip is cut to the required number of pieces from which the separated components are extracted with a suitable solvent.

Continuous apparatus. Promising technological potentialities of thermal diffusion have stimulated the development of continuous columns. Considerable progress has been achieved both in the construction of flow-through units and in theoretical development. The results obtained are also important for experimental separations of hydrocarbons, because the installation of a laboratory thermodiffusion column as an automatically operating continuous flow unit is easy and permits the working up of relatively large volumes of fractions.

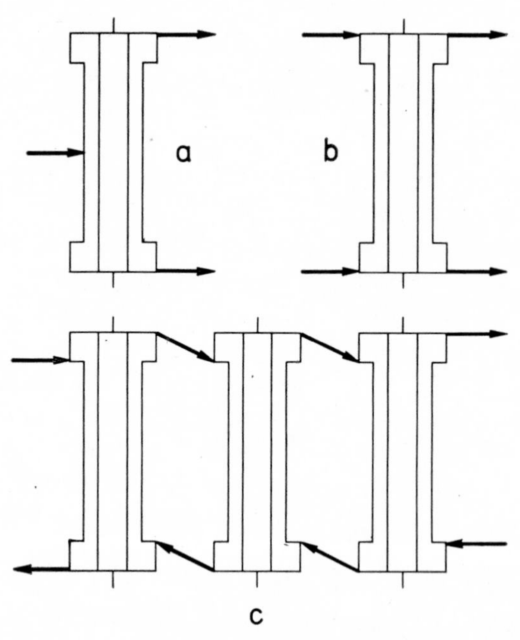

Fig. 2.7.4 Systems of the flow of continuous liquid thermodiffusion columns
a — Centre-feed system, b — Mixed-end feed system, c — Transverse-flow system

Theoretical and experimental contributions of many investigators to the development of continuous columns are summarized in review articles[2,3]. The theory of separation on a flow-through horizontal plate-type apparatus was developed by Butler and Turner[8] who checked it on a small apparatus of 5×15 cm dimensions with a 0.3 mm slit. An excellent study concerning a flow-through vertical plate-type apparatus with a central feed was put forward by Wilke and Powers[9]. Important progress in the theoretical description of continuous types of columns made of concentric tubes was recently achieved in the work of Yeh and co-workers[10-14].

In columns with a continuous flow the separation deteriorates with increasing flow-rate. At high rates the separation is almost independent of the column length. From theoretically deduced relationships it follows that the flow rate of a fraction through the column is optimum when the separation, expressed as ln q (see eq. 2.7.2), has just one half of the value for zero velocity (i.e. in batchwise separation).

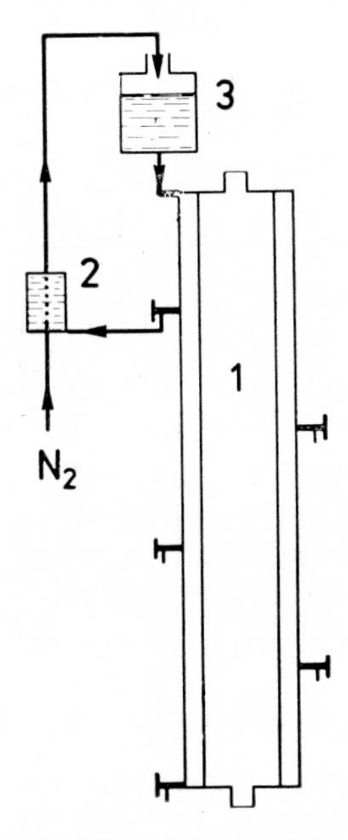

Fig. 2.7.5 Scheme of circulation thermal diffusion
1 — Thermodiffusion column, 2 — Bubble pump, N_2 — Nitrogen inlet, 3 — Reservoir

The continuous flow columns can be connected with advantage to *batteries* and *cascades*[3]. Various systems of feed stock introduction and product removal[15,16] were studied in detail. Among the three systems represented in Fig. 2.7.4 the system of transverse-flow is most suitable. Of the two remaining systems that of centre-feed is more effective at a relatively high separation level, while the system of mixed-end feed is more effective at a lower level.

The *circulatory* arrangement of the thermogravitational column represents a hybrid system connecting the elements of a batchwise operation and of continuous flow. This simple method of increasing the column capacity, originally described by Melpolder et al.[17] is shown is Fig. 2.7.5. The separated mixture is circulated by means of a bubble pump between the reservoir (volume about 200 ml) and the upper two ports of the thermodiffusion column. In the course of incessant circulation lasting several days up to a ten-fold larger amount of fraction can be worked up than the operational volume of the column. The circulatory arrangement shown in Fig. 2.7.5 is suitable for the separation of components present at low concentration and passing into the bottom part of the column. In the author's laboratory, for example, circulatory thermodiffusion was employed for the isolation of adamantane from paraffinic crude oil[18].

Improvement of the separatory effect. Much effort has been devoted to experiments aiming at the increase of the separation efficiency of thermodiffusion columns. Constructional improvements aimed at both an acceleration of the convective flow and also at the regulation of the flow in the column so that undesirable back-mixing of the countercurrent convectional streams should be decreased.

In order to accelerate circulation, moving walls and other devices were tested. Columns were proposed provided with a rotating tube, plate apparatuses with an infinite moving band passing through a slit and carrying both liquid fractions[22, 22], and also originally constructed vertical apparatuses with a moving wall formed by the down-flowing film of an immiscible liquid[24].

The possibilities of suppressing the back-mixing on the contact surface of convectional streams have been studied theoretically and experimentally. The effect of a semipermeable membrane inserted in the slit[25-27], inclination of the apparatus[10, 28], and other arrangements have also been tested. A practically important improvement of the separatory effect can be achieved with the following three types of apparatuses.

Rotary columns. Originally a batch-type column made of concentric tubes with a rotating inner tube was described[29]. The theory of separation on continuous rotary thermodiffusion columns was developed later[12, 13]. The arrangement in which the outer tube rotates and the inner does not is better from the point of view of the stability of flow than the construction with a rotating inner tube. According to general equations derived by Yeh and Tsai[12] it is possible to compare the separation on various types of continuous columns consisting of concentric tubes. An improvement of the separation on a rotary column, in comparison with a column with stationary walls, is less pronounced at higher flow rates.

The disadvantage of the rotary columns consists in their more complex construction and the increased requirements of the accuracy of their parts. The rotating tube must be accurately centered and balanced in order to avoid vibration of the apparatus. For current laboratory separations rotary columns are not generally used.

Packed columns. Sullivan et al.[28] tested batch-type thermodiffusion columns made of concentric tubes with an 0.76 to 3.17 mm annular space filled with glass wool. They found that the separations on a packed column are generally better than on a column of the same dimensions with an open annulus. The separation improves with increasing density of the packing, but the equilibrium attainment is simultaneously slowed. A theory for packed columns was elaborated both for batch-type[30] and for continuous flow apparatuses[14,31]. The already mentioned paper thermodiffusion[7] also belongs in the category of packed columns. In contrast to the columns with an open annular space the packed columns are constructionally simpler because at an equal separation they may have a broader slit. However, they are not used often in laboratory experiments owing to difficulties connected with a quantitative withdrawal of fractions from individual sections of the column during batchwise operation.

Wire columns. A simple and very efficient way to limit the contact surface of countercurrent convection streams in a concentric tube column was developed by Washall and Melpolder[32]. The back-mixing is restricted by winding a few threads of wire of a diameter equal to the slit width onto the inner tube. The operational space is thus limited with a steep helix that directs both countercurrent streams: the ascending stream is compelled to proceed along the bottom side of the wire and the descending stream descends along the upper side of the wire (see Fig. 2.7.6). The convective streams guided by the spiral have a much smaller contact area for back-mixing, which is also reflected in an improved separation of the components.

The separation efficiency of the column is dependent on the angle at which the wire is wound on the inner tube. The best angle of inclination of the wire is different for columns of different dimensions; for example, on a column 91 cm long a maximum separation is achieved if the angle of the wire ascension is 40°, while maximum separation of a column 183 cm long was found[32] at an angle of 64°.

For continuous wire columns a separation equation was deduced which enables the calculation of the best angle of inclination of the wire neces-

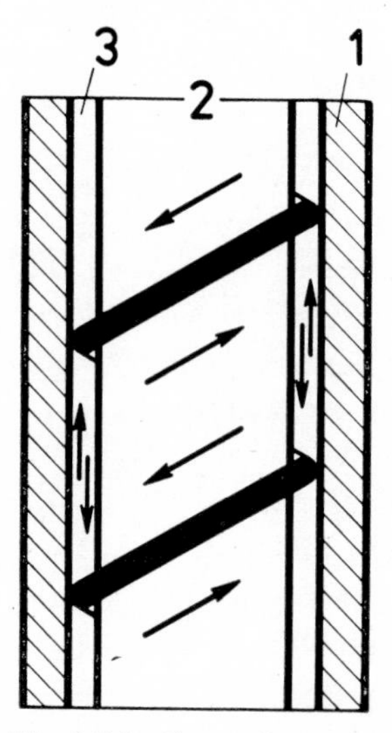

Fig. 2.7.6 Convection currents in wire thermodiffusion column
1 — Outer (hot) tube; 2 — Inner (cold) tube with a helically wound wire; 3 — Slit

sary for maximum separation[11]. Experimental results obtained with binary testing mixtures are in a qualitatively good agreement with theoretical assumptions. In continuous columns the best angle of inclination of the wire is also dependent on the flow rate, in addition to the geometrical factors of the column. Some calculated values of separation of a binary mixture on a column with an open slit and on a wire column at various flow rates are given in Table 2.7.1. From the Table it is evident that the improvement of the separation on a wire column becomes more pronounced with decreasing flow rate.

For laboratory separations of hydrocarbon mixtures the wire columns are most advantageous. Usually they have a higher separation efficiency

TABLE 2.7.1

Separation of benzene-heptane mixture (1 : 1) on a continuous thermodiffusion column at various flow rates (Yeh and Ward[11])

Flow rate (g/min)	Difference in concentrations of the top and the bottom product in %		The best angle of the inclination of the wire (degrees)
	open column	wire column	
0.125	6.6	16.6	61
0.25	6.3	11.7	55
0.5	5.9	8.2	47

than the columns with an open slit of the same dimensions, simultaneously preserving the advantages of a simple construction, without moving parts. This is of importance mainly for the reliability of a prolonged operation of the column without supervision. Recently Rabinovich et al.[32a] demonstrated that on a short (63 cm) perfectly heated column the wire in the slit (angle 15 – 30°) accelerates the equilibrium attainment only. The wire spiral practically does not diminish the operational space and does not interfere with the withdrawal of fractions and washing of the slit either. Another advantage consists in the fact that the wire spiral can be inserted into any thermodiffusion column made of coaxial tubes, which can further improve their efficiency. The spiral is wound so that the inner tube is fixed in a horizontal position and the selected slope of the spiral is indicated (the angle of inclination is usually between 40 and 65 °C). The selected wire of the same diameter as the width of the slit is glued with a drop of an epoxy resin to the end of the tube and it is then firmly wound according to the indications in as many loops as necessary. At the other end the wire is again glued. Then the outer tube is heated and the inner tube with the wound-on spiral is lightly lubricated with a well soluble oil and inserted carefully in a vertical position into the outer tube.

The dimensions and the characteristic parameters of several laboratory thermodiffusion columns made of concentric metallic tubes, that have proved efficient in the separations of hydrocarbon fractions, are given in Table 2.7.2. Columns *A* and *C* are commercially produced types.

According to column *A* described by Jones and Milberger[33] columns *D* were constructed that are used in the group of Zimina[34] and that of Petrov[35] in the Soviet Union. Washall and Melpolder[32] tested the effect of a wire spiral on column *C*; the column was also used in *API Research Project 6* for the separation of petroleum fractions[36] and later Brooks and Smith[37] separated hydrocarbon groups from low-temperature tar on the same type of column.

In the authors' laboratory a wire column *E* with a tube length of only 75 cm proved useful for many years [18,38,39]. Its simple construction is evident from Fig. 2.7.7. A detail of the inner tube with the wound-on wire spiral (angle of inclination is about 60°) is given in Fig. 2.7.8. In

Fig. 2.7.7 Parts of a thermodiffusion column made of concentric tubes: inner tube, outer tube with withdrawal ports, upper and bottom head with a teflon annular seal

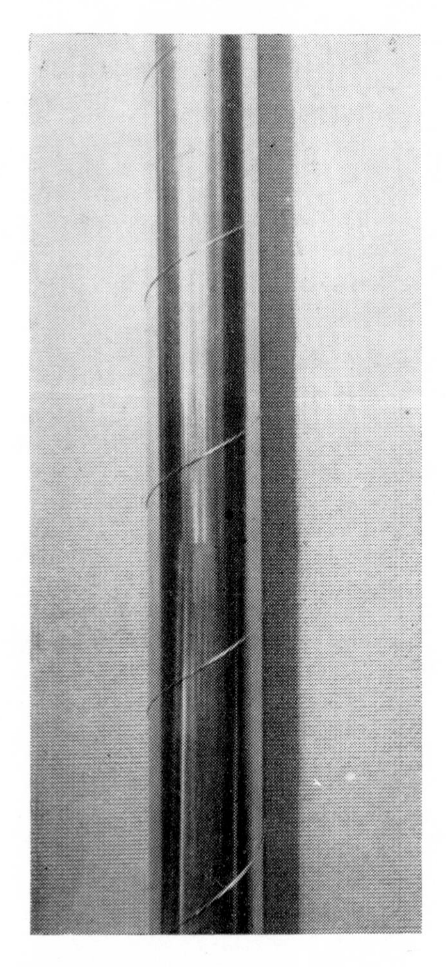

Fig. 2.7.8 Inner tube of a thermodiffusion column with a wound wire of 0.4 mm diameter, directing the convectional currents in the slit

view of the directed convection and short tubes the column E combines a good separation efficiency with a rapid equilibrium attainment. Common separations can be completed on the column overnight. This is very advantageous for repeated and multiple thermodiffusion fractionations which increase the separation effect. Columns of similar dimensions have also been constructed in several laboratories in the Soviet Union where they are used for the separations of petroleum fractions[40, 41].

TABLE 2.7.2

Parameters of some laboratory thermodiffusion columns made of metallic concentric tubes, used for the separation of hydrocarbons

Column parameters	A[a]	B	C[a]	D	E	F
Column height in cm	152	91	183	150	75	56
O. D. of the inner tube in cm	1.75	2.84	1.57	1.94	3.96	3.34
Width of the slit in mm	0.3	0.3 (wire)	0.3 (wire)	0.3—0.4	0.4 (wire)	0.8 (wire)
Volume of the operational space in ml	22	26	30	31	38	48
Number of withdrawal ports	10		20—30	10	5	2 (continual)
Time of separation in h	48[b]	ca 100	64[b]	48[b]	12	—
Temperature difference in K	30	66	48	94	90	60
References	33	32	32, 36, 37	34, 35	18, 38, 39	11

[a] Commercial product (M. Fink Co., Ohio)

[b] Time of a typical experiment; equilibrium attainment requires 120—160 h or more

Column *F* is a continuous flow-through device on which Yeh and Ward[11] tested the validity of theoretically derived separation equations.

2.7.2 SEPARATION OF HYDROCARBONS

The movement of liquid hydrocarbons in the slit of the thermogravitational column is affected by various molecular properties and interactions in a complex not yet fully known manner. There is no doubt that in the liquid phase the structure (molecular shape) has a much greater effect on thermal diffusion, i.e. the movement of the molecules in the direction of temperature gradient, than its mass. Those molecules that have a smaller area of the cross section, a smaller molecular volume or a smaller surface area make better use of the kinetic energy acquired at the hot wall for movement towards the cold wall. Only in the absence of differences in shape molecules with a higher mass concentrate near the cold wall.

The circulatory flow in the column is affected by other factors. Here the main role is played by: difference of densities of the separated components, temperature coefficients of the densities, viscosities and their temperature coefficients. The common effect of physical properties of the molecules and other influences have not yet been expressed mathematically accurately enough to enable a prediction of thermodiffusional behaviour on the basis of the properties of pure components alone.

So far the best applicable quantitative criterion of the separation of hydrocarbons is the temperature coefficient of viscosity[2]. During the separation in a thermogravitational column hydrocarbons with the highest *viscosity index* are concentrated in the upper part of the column almost without exception, while the viscosity index decreases in the downward direction. It is known that the viscosity index depends strongly on the structure of the substance, mainly on the ratio of the length of the molecule to its cross section. The highest viscosity index, i.e. least change of viscosity with temperature, is exhibited by long flexible molecules. Therefore *n*-alkanes concentrate most at the column top during the separation of a complex mixture of hydrocarbons. On the contrary rigid and compact cyclic or polycyclic molecules, the viscosity index of which is low to negative, concentrate at the column bottom. This is theoretically substantiated by the work of Dougherty and Drickamer[42] who deduced that the coefficient of thermal diffusion is a function of the activation energy for a viscous flow and the molar volume.

Binary mixtures. In order to elucidate the mechanism of separation by thermodiffusion many model mixtures were tested. From the binary mixtures of hydrocarbons of close molecular weights but of differing structures *n*-alkanes separated well from isoalkanes, alkanes from cyclic hydrocarbons (saturated and aromatic), and cyclic hydrocarbons with various numbers of rings. Generally the separation takes place by the aliphatic hydrocarbons concentrating in the upper part of the column and the cyclic hydrocarbons in its lower part. When hydrocarbons with a differing number of cycles are separated those with a larger number of cycles in the molecule concentrate at the column bottom. Alkanes separate from alkenes, and cycloalkanes from aromates relatively poorly.

A few examples of well and poorly separated pairs of hydrocarbons

TABLE 2.7.3

Separation of binary hydrocarbon mixtures by thermodiffusion (according to refs[2,7,18,43])

Top	Bottom
	Good separation[a]
n-Heptane	2,2,3-Trimethylbutane
1-Octene[b]	*o*-Xylene
n-Nonane	Isopropylbenzene
1-Undecene[b]	1-Methylnaphthalene
n-Hexadecane	1-Methylnaphthalene
n-Hexadecane	Decahydronaphthalene
n-Amylbenzene	1-Methylnaphthalene
Isopropylbenzene	Indane
Decahydronaphthalene	1-Ethyladamantane
	Poor separation[c]
n-Hexadecane	*n*-Hexadecene
n-Hexadecane	Cyclohexane
n-Octadecane[d]	Benzene
n-Hexadecane	Dodecylbenzene
2,2,3-Trimethylbutane	Cyclohexane
Benzene	Cyclohexane
Toluene[e]	Methylcyclohexane
Decahydronaphthalene	1-Methylnaphthalene
Tetrahydronaphthalene	1-Methylnaphthalene

[a] Purity of the top-product higher than 80 %
[b] Unpublished results from the authors' laboratory
[c] Unseparable or poorly separable pairs with the concentration of the top-products lower than 60 %
[d] Unseparable pair if *n*-octadecane prevails in the mixture[2,13]
[e] This pair displays the so-called "forgotten effect"[43] (cf. ref.[44,45])

are presented in Table 2.7.3. The separation of aliphatic and cyclic hydrocarbons takes place even at a considerable difference in molecular weights; *n*-hexadecane separates from 1-methylnaphthalene in the upward direction even though its molecular weight is higher by 84 units. Only with such pairs as *n*-hexadecane – cyclohexane, and *n*-octadecane – benzene, in which the differences in their molecular weights are 142 or 176 mass units respectively, the high molecular weight of the alkane begins to compensate for the structural difference; these pairs are inseparable or poorly separable, with an anomalous course of the process[2, 43].

The fundamental disadvantage of batch-type thermogravitational columns consists in the fact that even at a good separation only one part (20 – 60 %) of the separated mixture is obtained in the form of pure components. The equilibrium concentration gradient set in the slit is continual, so that a certain part of the unseparated mixture always remains in the central part of the column. This is shown by the composition of ten fractions obtained by thermodiffusion of a binary mixture (50 – 50 %) of *n*-hexadecane with 1-methylnaphthalene on column *E* (Table 2.7.4). From 38 ml of the introduced mixture 30 vol. % of 1-methylnaphthalene are obtained in this case in a purity higher than 97 %, and 30 vol. % of *n*-hexadecane of a purity higher than 95 %; 40 vol. % remain as an intermediate fraction. However, when the column is operated in a continuous manner, the problem of the intermediate fractions does not occur.

TABLE 2.7.4

Composition of thermodiffusion fractions after separation[a] of a 50 % mixture of *n*-hexadecane—1-methylnaphthalene

Fraction number	Content of n-hexadecane in vol. %	Content of 1-methylnaphthalene in vol. %
1	98.6	1.4
2	97.5	2.5
3	95.3	4.7
4	87.0	13.0
5	67.0	33.0
6	36.0	64.0
7	15.0	85.0
8	3.0	97.0
9	2.8	97.2
10	0.8	99.2

[a] 38 ml of mixture was separated for 9 h at $\Delta T = 90$ °C

The course of the separation of the test-mixture in a batchwise operation is shown in Fig. 2.7.9. The concentration on the top and at the bottom of the column changes initially very rapidly, but then it approaches asymptotically to the values for pure components. In this case a period of 9–12 h is sufficient for the separation.

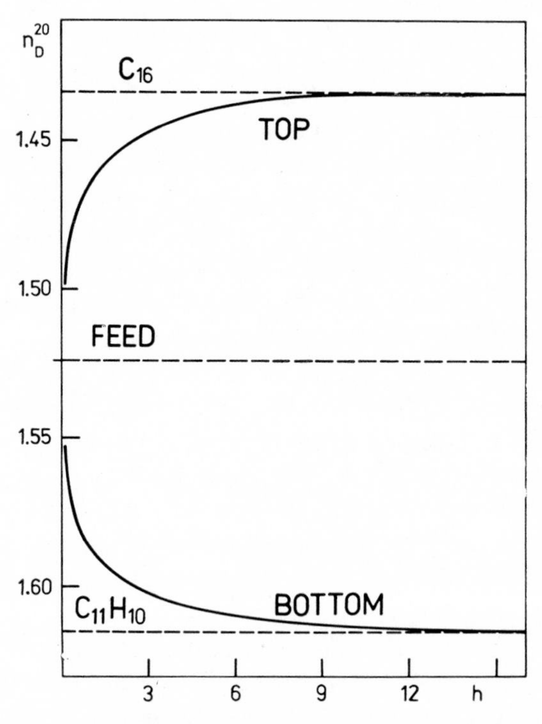

Fig. 2.7.9 Separation of the test mixture *n*-hexadecane—1-methylnaphthalene (50—50 vol. %)
The curves represent the refractive index change at the top and the bottom of the column in dependence on time

In current testing of the columns by means of binary mixtures the observed degree of separation is often expressed as

$$\% \text{ of separation} = \frac{\Delta n_F}{\Delta n_K} \cdot 100$$

where Δn_F is the difference in refractive indices of the upper and the lower thermodiffusion fraction and Δn_K is the difference of refractive indices of pure components of the test mixture. The observed values for the % of separation of the test mixture after 9 h of separation on column E and at various temperature gradients

ΔT	% of separation
60 K	84
90 K	98
120 K	99

show that in accordance with the theory an increase of ΔT above a certain limit has almost no effect on the equilibrium separation.

During the separation of some pairs of isomeric hydrocarbons thermal diffusion displays remarkable efficiency. Jones and Brown[2] found, for example excellent separation of the following pairs: *p*-xylene + *o*-xylene and *m*-xylene + *o*-xylene. In both cases the first mentioned isomer of the pair is the top product. In contrast to this a binary mixture of *m*-xylene and *p*-xylene does not separate.

A good separation effect was also observed in some configurational isomers. For example, during thermal diffusion of 1,2-dimethylcyclohexane the *trans*-isomer is obtained as the top product and the *cis*-isomer as the bottom product, both in a 100 % purity[2]. Also during the separation of decahydronaphthalene the *trans*-isomer concentrates in the upper part and the *cis*-isomer in the lower part of the column. The separation of decahydronaphthalene isomers is often employed for the testing of efficient thermodiffusion columns. Landa and Vaněk[39] when studying the separation of the stereoisomers of various tricycloalkanes found that the stereoisomers with the lowest retention index in gas chromatography, the lowest refractive index, the lowest viscosity and density, tend to accumulate in the upper part of the column. The recorded separation effects were low, however. The given examples show that in the case of separable pairs of stereoisomers of non-polar cyclic hydrocarbons the thermodynamically more stable isomer, with a lower content of energy, is always concentrated in the upper part of the column. This dependence is generally valid because it is connected with the fact that the energetically poorer stereoisomer usually has a larger molecular volume.

Hydrocarbon fractions. With fractions containing hydrocarbons of one structural class (homologous series) poor separation effects are achieved because the components separate on the basis of differences in molecular weights only. Generally, homologues with a higher number of carbon atoms concentrate at the cold wall. Table 2.7.5 gives the result of the separation of the *n*-alkane fraction C_8—C_{14} on the thermodiffusion column *E*. The concentration shift of heavier homologues towards the column bottom is clear, it is true, but is only of little relevance for practical separations.

Complex fractions containing hydrocarbons of various structural types are separated according to the mentioned rules so that individual structural types concentrate in zones from the column top downwards in the following order: light *n*-alkanes, heavy *n*-alkanes, isoalkanes, monocyclanes, dicyclanes, polycyclanes and non-hydrocarbon (high-molecular) fractions.

TABLE 2.7.5

Separation of a mixture of *n*-alkanes by thermal diffusion[a]

Fraction	Composition in weight %						
	C_8	C_9	C_{10}	C_{11}	C_{12}	C_{13}	C_{14}
Feed	1	8	17	22	24	20	8
Top fraction	2	10	20	23	24	16	5
Bottom fraction	0	3	12	20	25	26	14

[a] 38 ml of mixture was separated for 9 h at $\Delta T = 90$ °C

Usually those fractions are submitted to thermal diffusion which are rendered maximally simple by other separation methods. Most commonly *n*-alkanes and aromates are separated beforehand and the thermodiffusion column is charged with the remaining isoalkane-cycloalkane fraction. A concentrate of isoalkanes is then isolated as the top product and a concentrate of cycloalkanes as the bottom product; the latter can be further separated, if necessary, to subfractions with a various number of cycles. When aromatic fractions are separated saturated hydrocarbons are eliminated beforehand; fractions of aromates differing in the number of cycles can then be obtained.

The efficiency of thermal diffusion when applied on more complex fractions can be demonstrated by the results of the separation of light petroleum fractions on column E^{18}: The fraction of paraffinic crude oil (160 – 200 °C) of mean molecular weight 162, freed of aromates by adsorption chromatography on silica gel and of *n*-alkanes by adductive crystallization with urea, was separated at a 90 K temperature gradient for 9 h in two steps as shown in Fig. 2.7.10. The first separation step was repeated

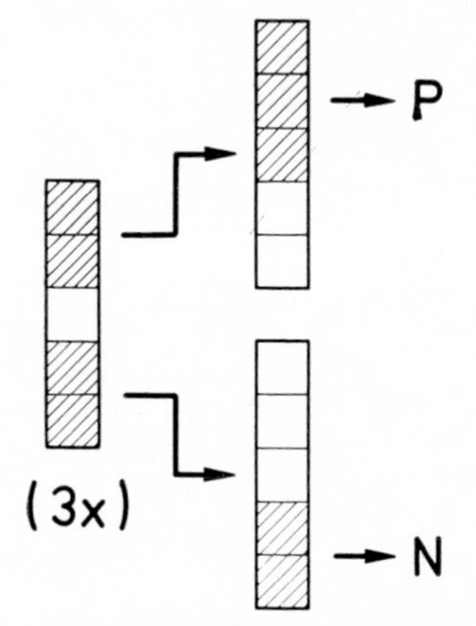

Fig. 2.7.10 Scheme of a two-step separation of isoalkane-cycloalkane petroleum fraction 160—200 °C by thermal diffusion
P — isoalkane concentrate; *N* — cycloalkane concentrate

TABLE 2.7.6

Composition of the products obtained by two-step thermodiffusion separation of isoalkane-cycloalkane fraction of the petroleum fraction 160—200 °C

Fraction	Composition in vol. %[a]		
	Alkanes	Monocycloalkanes	Dicycloalkanes
Feed 160—200 °C	55.5	40.6	3.9
Concentrate of isoalkanes	83.3	16.3	0.4
Concentrate of cycloalkanes	24.7	67.7	7.6

[a] Determined by mass spectrometry according to ASTM D 2789-71

thrice in order to accumulate sufficient material for the separation in the second step. The combined two upper fractions (40 vol. %) and the combined two lower fractions (40 vol. %) were then submitted to thermal diffusion separately in the 2nd step. The compositions of the concentrate of the cycloalkanes from the two lower fractions and of the concentrate of the isoalkanes from the three upper fractions of the second thermodiffusion are compared in Table 2.7.6. When calculated per introduced fraction 160 – 200 °C, the yield of the isoalkane concentrate is 24 vol. % and the yield of the cycloalkane concentrate is 16 vol. %. The degree of concentration of both types can be further increased, but at the cost of a lower yield of products.

TABLE 2.7.7
Composition of the top and the bottom product from a thermodiffusion separation of an alkane-cycloalkane fraction of a lubricant oil (according to Melpolder et al.[47])

Hydrocarbon group	Composition in vol. %[a]	
	Top product (fraction 1)	Bottom product (fraction 43)
Alkanes	70	—
Monocycloalkanes	20	—
Dicycloalkanes	9	—
Tricycloalkanes	1	—
Tetracycloalkanes	—	5
Pentacycloalkanes	—	28
Hexacycloalkanes	—	24
Heptacycloalkanes	—	17
Octacycloalkanes	—	13
Nonacycloalkanes	—	9
Decacycloalkanes	—	4

[a] Determined by mass spectrometry

The best use of thermal diffusion is in the separation of heavy fractions. The separation of isoalkane-cycloalkane fractions from the heart cut (with average molecular weight 474) of the lubricant oil has been described by Melpolder et al.[47]. Applying a multistep separation on 10 thermodiffusion columns they prepared 43 fractions of 5 ml volume each. The results of mass sprectrometric analysis of the top fraction No. 1 and the bottom fraction No. 43 are given in Table 2.7.7. Both fractions have approximately the same molecular weight; the top fraction represents a 70 % concentrate of isoalkanes, while the bottom fraction contains polycycloalkanes with 4 to 19 cycles in the molecule.

In consequence of the low sensitivity of thermal diffusion towards differences in molecular weights an excellent fractionation effect may be achieved with fractions with a wide boiling range. The average molecular weight of individual fractions from a thermodiffusion column does not change considerably in such cases either, but the differences in the properties of the upper and the lower fractions are surprisingly large. For example, according to a study by Jones[2], the furfural extract from the oil distillate 353 – 444 °C with the viscosity index 5.4 separates on a thermodiffusion column with the result that the upper product has viscosity index +176 and the bottom product – 1960.

Thermal diffusion has been applied many times as the terminal step in separation procedures during the investigation of the composition of petroleum fractions and other hydrocarbon materials. In Petrov's[35] group thermal diffusion was applied in the identification of petroleum bicycloalkanes C_8—C_9 in the fraction 125 – 150 °C, in the study of bicycloalkanes and tricycloalkanes in the fraction 150 – 230 °C, and of high-boiling pentamethylene hydrocarbons about C_{24} in the fraction 350 – 420 °C.

In the authors' laboratory thermal diffusion combined with adductive crystallization with thiourea was used for the isolation of adamantane[18] and some diamondoid hydrocarbons[48] from crude oil. A similar procedure was also employed in other investigations[41, 49].

In a series of communications the results have been published of a study of group composition and properties of petroleum fractions separated by thermal diffusion[34, 36, 40, 43, 46, 50]. The separation of complex residual petroleum fraction on a battery of columns interconnected by a system of transversal flow has been described by Jones[51]; The separation of asphalt was also successful[50, 52]; the structures of the components in the obtained fractions could be more closely characterized. This paper documents that the extent of the use of thermal diffusion technique includes even the heaviest petroleum fractions.

Thermal diffusion was also applied for the separation of hydrocarbon groups from shale oil[53]. Cummins and Robinson[54] isolated individual isoprenoid alkanes C_{15}—C_{20} from isoalkane–cycloalkane concentrate from shale oil in a thermal diffusion process lasting 6 weeks.

Landa and Urban[38] also used thermal diffusion for the separation of isoalkane–cycloalkane fraction of low-temperature brown coal tar. Brooks and Smith[37] submitted the tar from low temperature carbonization of bituminous coal to thermal diffusion and achieved a sharp separation of the aliphatic and the aromatic fraction. The aliphatic fraction consisted of straight-chain alkanes and alkenes C_{10}—C_{28}; the aromatic fraction contained alkylated and hydroaromatic hydrocarbons of the naphthalene, anthracene and phenanthrene series.

2.7.3 SPECIAL TECHNIQUES

In addition to various applications in the separation of hydrocarbon groups, thermal diffusion is also applied in the separation of *sulphur compounds* from crude oil[6,55] and as a concentration method for *trace metals* (vanadium, nickel) in crude oil[56]. From the analytical point of view the finding is important that organometallic compounds containing Zn, P, Ca, and Ba, used as lubricating oil *additives* concentrate highly efficiently in the bottom fraction during thermal diffusion[43].

In recent years the use of thermal diffusion is ever more used for the fractionation of *polymers*[57,58]. A systematic study of the separation of dilute solutions of polystyrene has been published by Taylor[59]. Thermal diffusion causes both a separation of the polymer from the solvent, and a fractionation of the polymer according to molecular weights. In conformity with the theory of Ham[60], according to which the coefficient of thermal diffusion of sufficiently large molecules is only little dependent on the molecular weight, it was found that the fractionating effect on a thermogravitational column is caused primarily by differences in the coefficients of regular diffusion. The heavier molecules of the polymer are concentrated at the cold wall because they have a lower tendency to diffuse back into the region of lower concentration near the hot wall.

When thermolabile compounds are separated the ability of the thermodiffusion column to operate even at relatively mild conditions comes into effect. This can be demonstrated by the recent isolation of new phenolic compounds from the urine of pregnant women[61]; the substances were separated on an all-glass column 30 cm high, with a 0.3 mm slit and 5.2 ml capacity, operating under "*physiological conditions*", with temperatures of the walls 0 °C and 37 °C. Columns of this type can be made easily, and in various sizes, from suitably selected glass tubes.

In spite of the extreme simplicity of the thermodiffusion columns used, efforts are continually being made to improve the apparatus. An interesting sample of this development is the new variant of the thermodiffusion column of the Clusius–Dickel type for gaseous mixtures, which has a *laser beam*[62] in the tube axis. In comparison with a classical column with a platinum hot wire of 0.1 mm diameter, the laser thermodiffusion column has a 2 to 3 times higher separating effectiveness with an equal power input.

Thermal diffusion with added component. A very promising way for an improvement of separations on thermodiffusion columns is the addition of another component. This attractive idea has been investigated by several authors: Jones[2, 33] found that a non-separable mixture of *m*-xylene and *p*-xylene can be separated if *o*-xylene is added. Then the order of isomers from the column top is *p*-, *m*-, *o*-. Another non-separable pair, benzene-cyclohexane, separates in the presence of methanol[43].

A substantial improvement in the separation of binary mixtures of an alkane with a dicycloalkane can be achieved by addition of a monocyclic hydrocarbon. The mechanism of the effect of the third component is quite clear in this instance – the added monocyclic hydrocarbon concentrates between the alkane and the dicyclic hydrocarbon in the central part of the column and thus forms the greater part of the intermediate fraction. Ho and Kin[63] improved the separation of *n*-hexadecane from decahydronaphthalene by addition of *o*-xylene. Vámos and Major[7] describe a similar case of an improved separation of *n*-hexadecane from 1-methylnaphthalene in the presence of toluene.

An addition of 60 % of toluene improves the percentage of the separation from the original 42 % (determined by paper thermal diffusion) to 84 %. The addition of an equal amount of *n*-heptane remains without any effect. As the added component can be selected from a very broad range of substances, this technique seems to be very promising even for the separation of complex fractions.

A slightly different effect is exhibited by so-called thermal diffusion *accelerators.* It was found that the separation of oil fractions on a continuous column proceeds better if about 50 % of tetrachloromethane, 1,2-dibromomethane or some other compound[64] are added. The dilution of the oil with an accelerator improves the hydrodynamic situation on the thermodiffusion column to such an extent that the flow through the column can be increased several times without a deterioration in the quality of the separation, or the separation efficiency can be increased substantially at the same flow-rate. The favourable effect of the accelerators was also confirmed for batchwise operations[39]. In the separation of *cis-* and *trans-*isomers of decahydronaphthalene not only the separation achieved was improved after the addition of some of these accelerators, but the time necessary for equilibrium attainment was shortened as well.

The added component can also have the function of an immiscible but *permeable liquid layer* that separates the upper and the lower fraction in the thermodiffusion column and prevents the mixing of the separated components. This new technique, surprising in its simplicity, was demonstrated by Pinheiro and Bott[65] in the separation of the model mixture benzene-tetrachloromethane, when water was used as the permeable layer. The liquid membrane in the column centre improves the separation remarkably. However, in the separation of hydrocarbon fractions the choice of immiscible liquids of suitable density will be a certain problem.

Thermal Field Flow Fractionation (TFFF). TFFF is being developed as a new method for the separation of macromolecules[66, 67] on the theoretical basis postulated by Giddings[68,69]. TFFF combines the *elution technique* of liquid chromatography with the effect of the outer *thermal field.*

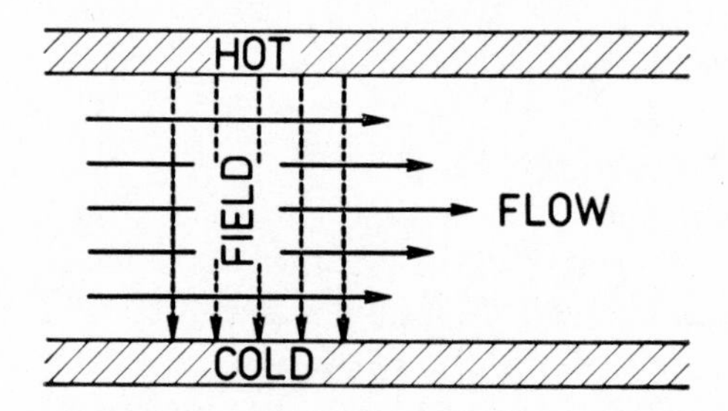

Fig. 2.7.11 Thermal Field Flow Fractionation
Representation of the laminar flow of the eluent in the horizontal slit with a tranverse thermal field

Similarly to thermal diffusion the fractionation takes place in a single homogeneous liquid phase. The TFFF arrangement is illustrated schematically in Fig. 2.7.11. The eluent carrying the sample flows through a horizontal column where it is exposed to a vertical temperature gradient. The construction of the column is similar to the flow-through horizontal plate apparatus for thermal diffusion[8,70]. In the column for TFFF the eluent flows through the slit between two long plates of which the upper one is heated and the lower cooled, as shown in Fig. 2.7.12. The sample

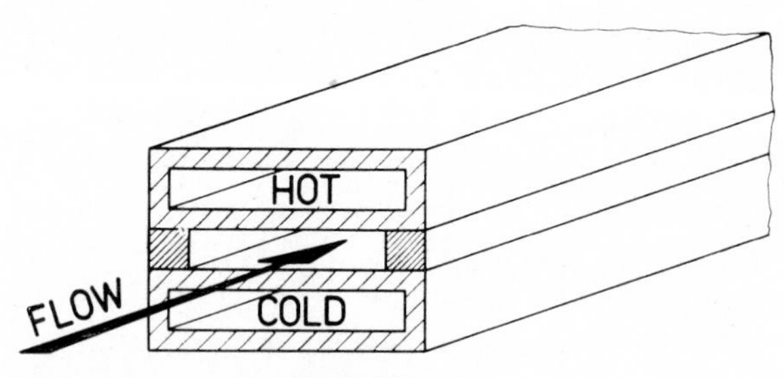

Fig. 2.7.12 Column for TFFF

selected for analysis is injected into the eluent stream in the form of a solution. During the sample movement through the column the separation is caused by two effects:

a) under the effect of a transversal temperature gradient individual components of the sample concentrate at various distances from the cold wall;

b) when the eluent flow through the slit is laminar the components that concentrate in the centre of the slit are carried more rapidly, owing to the non-uniformity of the flow profile, than the components near the walls, where the eluent layers flow slowest. If a suitable detector is connected to the column end the peaks of the separated components can be recorded analogously to liquid chromatography.

TFFF is at the beginning of its development, and the best column dimensions and separation conditions are only now being sought. When the ability of 3 m long TFFF column to fractionate samples was tested[67], toluene was used as eluent (flow 0.4 ml/h) and the temperature gradient was 50 K. The injected material consisted of 0.2 g of sample dissolved in 5 ml of toluene. An improved efficiency was found[71] in a *pressure column* operating with a temperature difference of 180 K. The conditions can be programmed during the analysis[72].

The suitability of TFFF for the fractionation of polymers was compared with gel permeation chromatography[73]. In contrast to GPC, in TFFF the light components are eluted before the heavier ones. According to theoretical assumptions both methods can achieve about the same number of theoretical plates; however, in TFFF it will probably be possible to achieve a several times higher *peak capacity*.

From this review of the applications of thermal diffusion it is evident that the method is not stagnant but is steadily developing. Attention is focussed mainly on the following properties:

a) exceptionally wide applicability of the method – from isotopes of the lightest gases up to metal alloys. It can be applied purposefully in the analysis of petroleum and similar hydrocarbon raw materials for all fractions, from gasoline up to asphalt.

b) relatively high capacity. In the separation of an undiluted sample a common laboratory column gives the separated components in amounts of several grams per hour. Therefore it is suitable mainly for preparative separations.

c) The ability to separate various compounds having identical boiling points, azeotropic mixtures, various compounds of the same molecular weight, thermolabile compounds, biological solutes, polymers and other macromolecules.

d) A thermodiffusion column can be easily controlled and monitored so that it can be left in action overnight without supervision. This speeds up the preparation of large samples and the procedure in multistep thermal diffusion separations enormously.

The manifold advantages of thermal diffusion are obvious. In connection with the general shift of the research directions towards heavier fractions it may be expected that thermal diffusion – either in its classical form or in some newer modification – will be one of the frequently used separation techniques in the future.

REFERENCES (2.7)

1. K. Clusius and G. Dickel, *Naturwissenschaften* **26** (1938) 546.
2. A. L. Jones and G. R. Brown in "Advances in Petroleum Chemistry and Refining", J. J. McKetta (Ed.), Interscience Publishers, New York, Vol. 3, 1960, p. 43.
3. J. E. Powers in "New Engineering Separation Techniques", H. M. Schoen (Ed.), Interscience Publishers, New York, Vol. 1, 1962, p. 1.
4. J. E. Powers, *Ind. Eng. Chem.* **53** (1961) 577.
5. R. Grasseli, G. R. Brown and C. E. Plymale, *Chem. Eng. Progr.* **57,** No. 3 (1961) 59.
6. C. J. Thomson, H. J. Coleman, C. C. Ward and H. T. Rall, *Anal. Chem.* **29** (1957) 1601.
7. E. Vámos and G. Major, *Acta Chim. Hung.* **27** (1961) 193; *J. prakt. Chem.* **14** (1961) 4.
8. B. D. Butler and J. C. R. Turner, *Trans. Faraday Soc.* **62** (1966). 3114, 3121.
9. C. R. Wilke and J. E. Powers, *A. I. Ch. E. Journal* **3** (1957) 213.
10. P. L. Chueh and H. M. Yeh, *A. I. Ch. E. Journal* **13** (1967) 37.
11. H. M. Yeh and H. C. Ward, *Chem. Eng. Sci.* **26** (1971) 937.
12. H. M. Yeh and C. S. Tsai, *Chem. Eng. Sci.* **27** (1972) 2065.
13. H. M. Yeh and S. M. Cheng, *Chem. Eng. Sci.* **28** (1973) 1803.
14. H. M. Yeh and T. Y. Chu, *Chem. Eng. Sci.* **29** (1974) 1421; **30** (1975) 47.
15. D. Frazier, *Ind. Eng. Chem. Proc. Des. Develop.* **1** (1962) 237.
16. R. Grasselli and D. Frazier, *Ind. Eng. Chem. Proc. Des. Develop.* **1** (1962) 241.
17. F. W. Melpolder, R. A. Brown, T. A. Washall, W. Doherty and W. S. Young, *Anal. Chem.* **26** (1954) 1904.
18. S. Hála, M. Kuraš and S. Landa, *Neftekhimiya* **6** (1966) 3; *Ropa a uhlie* **7** (1965) 227.
19. A. M. Henke and H. C. Stauffer, U.S. Pat. 2, 936.889 (1960).
20. R. L. Murphey, M. C. McClure and C. F. Rueping, U.S. Pat. 2, 970.695 (1961).
21. J. H. Ramser, *Ind. Eng. Chem.* **49** (1957) 155.
22. J. W. Beams, U.S. Pat. 2, 521.112 (1950).
23. A. L. Jones, U.S. Pat. 2, 892.544 (1959).
24. A. M. Henke and H. C. Stauffer, U.S. Pat. 2, 791.332 (1957).
25. O. M. Hanson, U.S. Pat. 2, 585.244 (1952).
26. A. L. Jones et al., U.S. Pat. 2, 712.386 (1955); 2, 720.978 (1955).
27. W. E. Scovill, U.S. Pat. 2, 723.759 (1955).
28. J. E. Powers and C. R. Wilke, *A. I. Ch. E. Journal* **3** (1957) 213.
29. L. J. Sullivan, T. C. Ruppel and C. B. Willingham, *Ind. Eng. Chem.* **47** (1955) 208; **49** (1957) 110.
30. M. Lorenz and A. H. Emery, *Chem. Eng. Sci.* **11** (1959) 16.
31. V. Sanchez and L. Estaque, *Chem. Eng. Sci.* **29** (1974) 2283.
32. T. A. Washall and F. W. Melpolder, *Ind. Eng. Chem. Proc. Des. Develop.* **1** (1962) 26.

32a. G. D. Rabinovich, V. P. Ivakhnik, K. I. Zimina, N. G. Sorokina, *Inzhenerno-fizicheskii zhurnal* **35,** No. 2 (1978) 278.

33. A. L. Jones and E. C. Milberger, *Ind. Eng. Chem.* **45** (1953) 2689.

34. A. A. Simeonov, M. V. Tikhomirov and K. I. Zimina, *Neftepererabotka i neftekhimiya* No. **7** (1963) 25.
35. A. A. Petrov et al., *Neftekhimiya* **6** (1966) 165; **7** (1967) 511; **8** (1968) 491; **9** (1969) 487; **13** (1973) 779.
36. B. J. Mair and F. D. Rossini, *Ind. Eng. Chem.* **47** (1955) 1062.
37. J. D. Brooks and J. W. Smith, *Fuel* **43** (1964) 125.
38. S. Landa and M. Urban, *Brennstoff-Chemie* **44** (1963) 377.
39. S. Landa and J. Vaněk, *Coll. Czech. Chem. Commun.* **35** (1970) 3733.
40. S. R. Sergienko, D. N. Ernepesov, Kh. N. Ernepesov, A. G. Korotkii and L. D. Melikadze, *Dokl. Akad. Nauk SSSR* **190** (1970) 1159.
41. L. D. Melikadze, E. A. Usharauli, A. A. Dzamukashvili and M. A. Machabeli, *Bulletin of the Academy of Sciences of the Georgian SSR*, **68** (1972) 81.
42. E. L. Dougherty and H. G. Drickamer, *J. Phys. Chem.* **59** (1959) 443; *J. Chem. Phys.* **23** (1955) 295; **22** (1954) 1157.
43. C. R. Begeman and P. L. Cramer, *Ind. Eng. Chem.* **47** (1955) 202.
44. H. Korsching, *Zeit. Naturforsch.* **18**a (1963) 669.
45. W. J. Korchinsky, *Diss. Abstr.* B-**27,** 1 (1966) 153.
46. F. A. Haak and K. van Nes, *J. Inst. Petroleum* **37** (1951) 245.
47. F. W. Melpolder, R. A. Brown, T. A. Washall, W. Doherty and C. E. Headington, *Anal. Chem.* **28** (1956) 1936.
48. S. Hála, S. Landa and V. Hanuš, *Angew. Chem.* **78** (1966) 1060.
49. V. K. Solodkov, A. A. Mikhnovskaya, B. A. Smirnov and A. A. Petrov, *Neftekhimiya* **9** (1969) 487.
50. G. R. Brown and A. L. Jones, *Petroleum Rafiner* **39,** No. 6 (1960) 156.
51. A. L. Jones, *Petroleum Rafiner* **36,** No. 7 (1957) 153.
52. R. A. Gardner, H. F. Hardman, A. L. Jones and R. B. Williams, *J. Chem. Eng. Data* **4** (1959) 155.
53. H. Nurse, Trudy Tallin. Politekh. Inst. Ser. A, No. 165 (1959) 267; C. A. **56** (1962) 10441h.
54. J. J. Cummins and W. E. Robinson, *J. Chem. Eng. Data* **9** (1964) 304.
55. C. J. Thompson, *J. Chem. Eng. Data* **10** (1965) 279.
56. E. C. Gosset, U.S. Pat. 3, 117.078 (1964).
57. B. Rauch and G. Meyerhoff, *J. Phys. Chem.* **67** (1963) 946.
58. V. B. Grigor'ev, L. A. Grigor'eva, V. O. Reikhsfeld, K. L. Makovetskii and N. I. Smirnov, *Zh. Prikl. Khim.* **38**/**11** (1965) 2592.
59. D. L. Taylor, *J. Polym. Sci., Part A*, **2** (1964) 611.
60. J. S. Ham, *J. Appl. Phys.* **31** (1960) 1853.
61. J. C. Touchstone and M. F. Dobbins, *Separ. Sci.* **10** (1975) 617.
62. F. S. Klein and J. Ross, *J. Chem. Phys.* **63** (1975) 4556.
63. H. L. Ho and L. W. Lin, C. A. 53 (1959) 8601b.
64. J. B. Stothers and J. Walker, U.S. Pat. 3, 180.823 (1965).
65. J. de D. R. S., Pinheiro and T. R. Bott, *Separ. Sci.* **11** (1976) 193.
66. M. N. Myers, K. D. Caldwell and J. C. Giddings, *Separ. Sci.* **9** (1974) 47.
67. G. H. Thompson, M. N. Myers and J. C. Giddings, *Anal. Chem.* **41** (1969) 1219.

68. J. C. Giddings, *Separ. Sci.* **1** (1966) 123; *J. Chem. Educ.* **50** (1973) 667.
69. M. E. Hovingh, G. H. Thompson and J. C. Giddings, *Anal. Chem.* **42** (1970) 195.
70. A. L. Jones, U.S. Pat. 2, 723.034 (1955).
71. J. C. Giddings, *Anal. Chem.* **47** (1975) 2389.
72. J. C. Giddings, L. K. Smith and M. N. Myers, *Anal. Chem.* **48** (1976) 1587.
73. J. C. Giddings, Y. Hee Yoon and M. N. Myers, *Anal. Chem.* **47** (1975) 126.

2.8 Chemical reactions

Although predominantly physical methods are now used for the separation of hydrocarbon mixtures, those based on chemical reactions still retain their importance. Chemical reactions are mainly used for separations of aromates and olefins and for the separation of saturated cyclic hydrocarbons. A simplification of the hydrocarbon mixtures can be achieved either by direct elimination of a certain class of hydrocarbons from the mixture, or by such structural changes as facilitate their elimination from the mixture or their isolation. For raffination chemical methods are used much more frequently than for isolation, because in many cases it is difficult to avoid the possibility of a change of the hydrocarbon skeleton during chemical reactions.

2.8.1 REACTIONS OF UNSATURATED HYDROCARBONS

Aromatic hydrocarbons. The most current method used for the separation of aromates is *sulphonation*. Under optimum conditions sulphonation is a selective method for quantitative separation of aromatic hydrocarbons from saturated hydrocarbons. If olefins are present in the mixture under separation, they are separated together with the aromatic hydrocarbons. The separation by sulphonation is in many cases more efficient than chromatographic separation or extraction[1,2]. A procedure for the separation of aromatic hydrocarbons (and simultaneously of the olefins present) in petroleum fractions is described in the ASTM Method D 1019[3]. The method is applicable for depentanized fractions with the 90 % boiling point of maximum 315 °C.

Separation of aromates plus olefins according to ASTM Method D 1019

The sulphonation mixture is prepared from 70 % (by weight) of sulphuric acid (95—96 %) and 30 wt. % of phosphorus pentoxide. Sulphonation is then carried out in a sulphonation flask with a calibrated neck at the temperature of ice melting. The shape and the dimensions of the flask are described in detail in the method cited. The sulphonation mixture (25 ml) is introduced into the flask and cooled in an ice bath for 10 min. The sample is then pipetted (10 ml) into the flask in such a way as to prevent a mixing of the sample with the sulphonation mixture, and after another 10 min. cooling the mixture is submitted to shaking, mechanical or by hand. The time of shaking necessary for a complete sulphonation is determined with a suitable testing mixture; 10 minutes usually suffice for sulphonation by either mechanical or manual shaking. Both layers are then separated by centrifugation. The flasks are then made up with concentrated sulphuric acid so that the entire unsulphonated part should be in the calibrated part of the neck, and centrifugation is continued until the separation of the layers is complete. The volume of the unsulphonated fraction is measured and the content of the substances separated by sulphonation is calculated from the difference.

With gasolines several hours standing suffices for a complete separation of both layers. With some samples with a high content of easily reacting aromates or olefins (for example gasolines from pyrolysis) a vigorous exothermic reaction may take place at the very beginning, and therefore the shaking should be performed very carefully.

The results from the authors' laboratory shown in Table 2.8.1 indicate that by the described method practically all aromates can be eliminated from gasolines after only 10 minutes, irrespective of their content in the analysed sample.

The course of the sulphonation of aromates and olefins of higher molecular weight with concentrated sulphuric acid was investigated by Kochloefl and Schneider[4]. They found that a complete disappearance of aro-

TABLE 2.8.1

Efficiency of the sulphonation method for the removal of aromates and olefins

Sample	Vol. % of aromates (determined by mass spectrometry)		Bromine number	
	original sample	after sulphonation	original sample	after sulphonation
Gasoline from thermal cracking	8.1	0.4	56	0
Gasoline from catalytic cracking	24.8	1.1	107	1.8

matic hydrocarbons takes place only after 2 hours; after half-an-hour their content dropped to about one half (Fig. 2.8.1). During a 6 hours' sulphonation the saturated hydrocarbons did not change. Although the sulphonation mixture described in the ASTM Method is more effective for the separation of aromates than sulphuric acid alone, it is evident that the separation of aromatic hydrocarbons in higher fractions by sulpho-

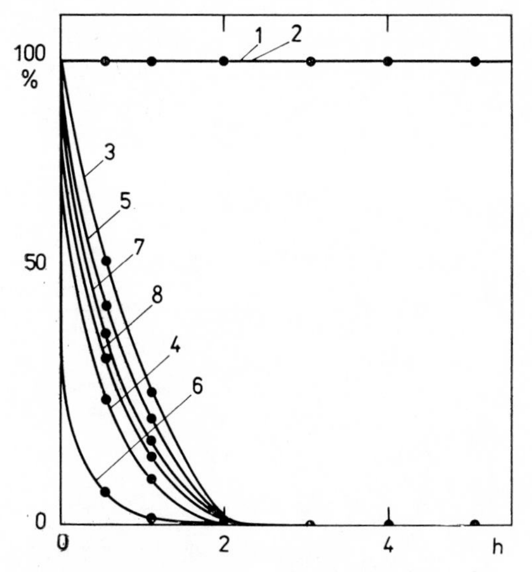

Fig. 2.8.1 Reactivity of hydrocarbons with 100 % sulphuric acid in dependence on the time of shaking
1 — *cis* + *trans*-decalin, 2 — *n*-hexadecane, 3 — 2-ethylnaphthalene, 4 — diphenyloxide, 5 — 1-methyltetralin, 6 — 1,2-dimethylindene, 7 — 3-methyl-3-undecene, 8 — 1,5-dimethylnaphthalene

nation is more difficult and requires a longer reaction time than in the case of aromates in gasoline.

Washall et al.[5] employed a combination of a reaction with acids and adsorption for the separation of saturated from unsaturated hydrocarbons even for their quantitative determination in heavy petroleum fractions (gas oil and high-boiling fractions, including residues). Unsaturated compounds react with fuming sulphuric and nitric acid at room temperature and can be separated from saturated hydrocarbons by adsorption on a column packed with bauxite and silica gel. Saturated hydrocarbons can be obtained from the adsorbent by elution with a suitable solvent. Saturated hydrocarbons are then determined quantitatively by gravimetry, and the unsaturated hydrocarbons from the difference. Higher-molecular saturated hydrocarbons do not react under the conditions of separation.

Lamey and Maloy[6] made use of the different reactivity of hydrocarbons with sulphuric acid for the removal of the remains of anthracene from phenanthrene and 2-phenylnaphthalene. The reaction is carried out in cyclohexane and the required hydrocarbon is obtained in a highly pure state.

In heavy oils the hydrocarbon groups can be separated according to the ASTM Method D 2006[7]. The principle of this method consists in the sample after the removal of asphaltenes with hexane being allowed to react with sulphuric acid in the cold. The fraction that reacts with 85 % sulphuric acid, indicated as "polar substances", contains heterocyclic compounds and very reactive aromatic hydrocarbons. The remaining aromates are further separated to a fraction that reacts with 97–98 % H_2SO_4 and another one that can be separated with fuming sulphuric acid (30 % of free SO_3), or also on silica gel, from saturated hydrocarbons.

Other chemical methods for the separation of aromates are used only exceptionally. The formation of complex compounds with aromates is discussed in Section 2.5.

Olefins. Olefins and aromatic hydrocarbons together can be separated from saturated hydrocarbons by quantitative sulphonation according to ASTM Method D 1019[3] (described above). The results presented in Table 2.8.1 show that after a ten minutes' shaking of gasoline samples the olefins are removed practically quantitatively even from samples with a high content of olefins.

Two methods are used for the separation of olefins from aromatic and saturated hydrocarbons. On the one hand olefins are converted to derivatives from which the original hydrocarbons, are not recovered, similarly to sulphonation. These procedures that are selective and quantitative for olefins, are used mainly for the elimination of olefins from gasoline. On the other hand an isolation method is used in which olefins are converted to derivatives from which the original hydrocarbons can be recovered or at least the structure of the original olefins can be inferred from the products formed. These reactions are also selective for olefins, it is true, but they usually prefer certain types of olefins, and they cannot be used for the isolation of all olefins from petroleum or tar fractions in all instances. These procedures are used mainly for the isolation of olefins from higher-boiling fractions.

A review of some older procedures for the removal of olefins from gasolines has been published by Kurtz et al.[8]. These authors found that *nitrogen dioxide* is a suitable reagent for the removal of monoolefins and conjugated and unconjugated dienes. Lumpkin et al.[9] eliminated olefins from gasolines by addition of *bromine* to the double bond; steam distillation then separates hydrocarbons that do not react with bromine from high-boiling bromo derivatives. Mikkelsen et al.[10] eliminated olefins from cracking gasolines using the reaction with *benzenesulphenyl chloride*. The reaction product formed consists of β-chlorosulphides. This procedure was used for the removal of olefins from gasolines before mass spectrometric analysis. The reaction product has a low vapour tension and does not interfere with hydrocarbons in mass spectrometric analysis.

The removal of olefins from gasoline with benzenesulphenyl chloride

Benzenesulphenyl chloride is added to the sample dropwise at 0 °C until it is present in excess. The excess is then removed by shaking with mercury and the formed precipitate is discarded. The reagent is dark red, while the addition product is colourless. Some substances in gasolines from thermal cracking may react, however, under formation of coloured products. In such instances the reagent should be added very slowly (one drop per 2 minutes); the colour reaction is thus gradually neutralized if the reacting olefins are present. The coloration of the sample that no longer changes and is stable indicates an excess of the reagent and the disappearance of the reacting olefins. The reaction is exothermic and cooling is indispensable. A low temperature decreases the possibility of benzenesulphenyl chloride reacting with other hydrocarbons than olefins.

For the isolation of olefins under simultaneous possibility of their analysis the reaction with *mercuric acetate* and *ozone* is used.

With mercuric acetate olefins give in methanol addition products that are very polar and separate from hydrocarbons on silica gel or alumina columns. After their elution with a polar solvent the original olefins are recovered by decomposition with hydrochloric acid. The reaction was described by Tausz[11] and further elaborated by Martin[12]. Kochleofl and Schneider[4] found with model mixtures that the aliphatic olefins used, 1-dodecene, 4-methyl-4-dodecene, 4-methyl-3-dodecene, 3-methyl-

2-undecene and 3-methyl-3-undecene, react at room temperature to 97 – 99 % within 3 hours. Cyclic olefins (cyclohexene, indene) react more slowly. This procedure was used for the isolation of olefins from brown coal tar fraction.

TABLE 2.8.2

Reactivity of olefins with mercuric acetate[12]

Hydrocarbon	% of substance reacted with mercuric acetate
1-Undecene	96.5
Styrene	98.8
1,9-Decadiene	91.0
1,13-Tetradecadiene	90.7
1,15-Hexadecadiene	93.7
6-Tridecene	43.5
2,4-Dodecadiene	60.0
Dicyclohexen-1-yl	34.3
Cyclohexene	83.9
Norbornene	63.8
2-Methyl-1-octene	90.9
2-Methyl-1-decene	89.9
4-Propyl-3-heptene	40.0
4-Isopropyl-3-heptene	12.5
3-Methyl-2-undecene 3-Methyl-3-undecene	92.7
2-Methyl-3-propyl-2-hexene	4.8
9,10-Octahydronaphthalene	5.0

Hofman et al.[13] in a study of the reactivity of various types of olefins with mercuric acetate found that the reactivity of olefins depends to a considerable extent on their structure. While, for example, 1-undecene, 1-hexadecene and 1,15-hexadecadiene react at 40 °C almost quantitatively within 20 minutes, the reactivity of the olefins substituted at the double bond is lower; tetrasubstituted olefins react to about 5 % only under these conditions (Table 2.8.2). Therefore the reaction with mercuric acetate is suitable especially for the isolation of monosubstituted olefins and asymmetrically disubstituted and trisubstituted olefins. The reaction of mercuric acetate with other olefinic types requires a much longer reaction time. The procedure for the separation of olefins from saturated and aromatic hydrocarbons from the fraction 220 – 280 °C of brown coal tar is described by Schneider et al.[14].

Isolation of olefins from brown coal tar with mercuric acetate

100 g of neutral tar fraction is dissolved in 4 500 ml of methanol and 350 g of mercuric acetate are added. The reaction mixture is kept at 35 °C for 70 h. Methanol and the acetic acid formed are distilled off in a vacuum at 40 °C and the residue is extracted with 1 000 ml of anhydrous benzene. The benzene extract is concentrated in vacuo at 40 °C and the residue separated on a column of silica gel (1 110 g). Hydrocarbons that do not form complexes with mercuric acetate are eluted with 400 ml of dry benzene.

The addition products can be eluted from the column with a suitable polar solvent (methanol, acetic acid solution in methanol) and decomposed with hydrochloric acid[15].

Another method which can be used for the isolation of olefins is the addition of *ozone* to the double bond. Ozonization is selective for double bonds; substitution reactions do not take place. So-called *molozonide* is formed as the primary addition product, the exact structure of which is not known. Molozonide is decomposed under formation of carbonyl compounds which can be isolated and analysed easily, and a bipolar ion which can undergo rearrangement and recombination reactions. Various methods of decomposition of ozonides were investigated by Greiner[16] and Hofman et al.[17]. Griegee and Gunther[18] found that the ozonization of olefins in the presence of tetracyanoethylene leads directly to aldehydes and ketones. The ozonization of olefins was elaborated by Asinger et al.[19] and applied by other authors for preparative and analytical purposes. Smits and Hoefman[20] found that almost all types of aliphatic and cyclic olefins studied by them reacted practically quantitatively. As a rule ozonization is carried out at low temperature (– 60 °C to – 70 °C); the excess of ozone can be indicated by the liberation of iodine from iodine-starch paper. However, an increase in temperature up to + 50 °C affects the yield of the reaction.

In addition to olefins aromatic hydrocarbons and some hetero compounds also react with ozone, but much more slowly[21]. Therefore ozonization at optimum conditions can be considered as an almost selective and quantitative method of separation of olefins. The reactions of ozone are discussed in detail by Bailey[22].

The possibility of the separation of olefins from saturated hydrocarbons by forming their complexes with *silver salts* are mentioned by Quinn[23]. For this purpose both aqueous silver salts solutions ($AgNO_3$, $AgBF_4$), especially suitable for gaseous hydrocarbons, and anhydrous silver salts may be used. Anhydrous silver salts, especially those containing perfluorinated anion, were found advantageous for the separation of linear 1-olefins from *n*-heptane solution; the most perfect separation was achieved with $AgSbF_6$ (Table 2.8.3). The separations of olefins from saturated hydrocarbons by reactions with *cuprous salts* ($CuBF_4$) have been described by Davis and Makin[24].

TABLE 2.8.3

Reactivity of olefins with silver salts[23]

Hydrocarbon	Original concentration wt. %	Final concentration after reaction with	
		$AgBF_4$	$AgSbF_6$
1-Heptene	9.11	0.93	0
1-Decene	9.09	9.05	0.15
1-Tetradecene	8.98	5.56	1.20
1-Octadecene	9.10	0.75	0.44

Acetylenes occur in petroleum products and fractions, with the exception of gaseous, only exceptionally and mostly in low concentration. In Section 3.6 some methods are described for their selective and quantitative determination by chemical methods. On reaction of monoalkylacetylenes with Ag, Cu and Hg ions precipitates are formed which can be filtered off easily[24], and therefore these methods can also be used for the isolation of acetylenes from other hydrocarbons.

2.8.2 DEHYDROGENATION, ISOMERIZATION AND OTHER REACTIONS

The principle of the separation of cycloalkanes from alkanes by chemical methods consists in their conversion to aromatic hydrocarbons which can be separated and analysed easily. Cyclohexane hydrocarbons are aromatized by *dehydrogenation,* while cyclopentane hydrocarbons by isomerization and dehydrogenation. For the dehydrogenation of cycloalkanes sulphur, selenium and metals (Pt) are used as catalysts. Sulphur[26] and selenium[27] are convenient mainly for the determination of the structures of pure substances. For the separation of cycloalkanes as a group, platinum is used. On platinum catalysts dehydroisomerization of cyclopentane hydrocarbons takes place in addition to dehydrogenation of cyclohexane hydrocarbons, so that this method can be used for the separation of both types of hydrocarbons simultaneously. In these reactions of cycloalkanes structural changes can partly take place, so that the aromatic hydrocarbons obtained do not always correspond to the structure of the starting cycloalkanes. For example cycloalkanes with angular or geminal methyl or ethyl groups can be dealkylated. Cycloalkanes with an alkyl group that contains a larger number of methyl groups also afford partly dealkylated aromates. During the separation of cycloalkanes by aromatization work is usually done in the gas phase and in a flow-through system at temperatures of about 300 °C. This method was used successfully for the separation of cycloalkanes from gasoline[28] and from higher-boiling fractions[14].

Landsberg et al.[29, 30] have described a method by which only six-membered (hexamethylene) cycloalkanes can be dehydrogenated, while cyclopentanes and geminal alkylcyclohexanes do not change. In analyses of gasolines they carried out dehydrogenation on a catalyst containing 20 % of Pt and 2 % of Fe. The addition of iron decreases the catalyst's activity which becomes sufficiently selective for the dehydrogenation of six-membered rings. Active charcoal is used as carrier. The selectivity of the catalyst was checked with mixtures of pure cyclopentanes. Under identical conditions aromatization of alkanes does not take place on this catalyst to more than 1 %, while an isomerization of alkanes was not observed. The reaction on Pt-Fe catalyst takes place in a flow-through system at temperatures of about 300 °C and volume rate of $1\ h^{-1}$. Under these conditions dehydrogenation of cyclohexanes is sufficiently selective. If the operation conditions are not adhered to accurately, especially if volume rate is decreased, undesirable side-reactions may also take place in addition to dehydrogenation of the six-membered cycloalkanes. As the dehydrogenation method described is very effective for the separation of cyclopentanes from cyclohexanes, more so than physical methods, we consider it useful to describe the preparation of the dehydrogenation catalyst.

Preparation of the catalyst for the dehydrogenation of cyclohexanes

The preparation of the catalyst from chloroplatinic acid is based on the reaction

$$H_2PtCl_6 + HCHO + H_2O + 6\,KOH = Pt + CO_2 + 6\,KCl + 6\,H_2O$$

The catalyst contains 19.6 % Pt and 2 % Fe and 78.4 % active charcoal. The chloroplatinic acid (containing 37.7 % of Pt) is used in solution that contains 1.33 g of the acid in 1 ml of the solution. Formaldehyde (40 %) is then added to a calculated amount of chloroplatinic acid solution (5 ml formaldehyde per 1 g of Pt), followed by a corresponding amount of $FeCl_3$. The flask with the solution is then shaken until $FeCl_3$ is dissolved completely. This solution is then added dropwise to a corresponding amount of active charcoal in a porcelain dish cooled with ice from outside. A 50 % KOH solution (7 ml per 1 g of Pt) is then added to the reaction mixture. During the addition of KOH solution the active charcoal is stirred with a thermometer and the solution is added at a rate that keeps the temperature within the 30—40 °C interval. After the addition of all the KOH solution the catalyst is heated at 60 °C for 30 min on a water bath and 150 ml of distilled water are then added. After 30 minutes standing the washing with distilled water is continued until the reaction for Cl^- ions with $AgNO_3$ is negative. The catalyst is dried at 105—110 °C and stored in a desiccator.

Dehydrogenation and dehydroisomerization of cycloalkanes to aromates are the bases of catalytic reforming and they are extremely important in the petroleum industry for the production of high-octane gasolines (Section 5.2).

Krasavchenko et al.[31] developed selective *isomerization* of cyclopentanes as a method by which hydrocarbons with five-membered rings can be separated from other cycloalkanes and isoalkanes. The method is based on the ability of aluminium bromide to isomerize the cyclopentane ring slowly to the cyclohexane ring in *n*-hexane solution at 30 °C. The separation of a mixture of cycloalkanes into three structural groups (cyclopentanes – cyclohexanes – bridge-head cycloalkanes) takes place in the following manner:

1. Selective dehydrogenation in the liquid phase[32, 33] converts cyclohexanes to aromatic hydrocarbons which are separated from the mixture.
2. Cyclopentane hydrocarbons are isomerized selectively to cyclohexane hydrocarbons; the hydrocarbons formed are eliminated by repeated dehydrogenation in the form of aromatic hydrocarbons. The remaining fraction forms bridge-head cycloalkanes and less reactive cycloalkanes (cyclopentanes with substituents in angular and geminal positions).

This separation method can also be used for the analysis of heavier petroleum fractions containing polycycloalkanes.

Jakubson et al.[34] used *hydrocracking* for the separation of adamantane hydrocarbons from other tricycloalkanes. On a platinum catalyst (10 % Pt on diatomite) at 400 – 430 °C all tricycloalkanes are hydrocracked quantitatively, with the exception of adamantanes. It was also found that tricycloalkanes in the fractions (200 – 250 °C) of the analysed crude oils consist of 80 to 90 % of hydrocarbons with non-adamantane structure.

REFERENCES (2.8)

1. T. L. Chang and C. Karp, *Anal. Chim. Acta* **26** (1962) 410.
2. K. J. Hunter and M. Vahrman, *J. Appl. Chem.* **10** (1960) 128.
3. ASTM Method D 1019-68, Annual Book of ASTM Standards, Part 23, American Society for Testing and Materials, Philadelphia 1975.
4. K. Kochloefl and P. Schneider, *Brennstoffchemie* **45** (1965) 289.
5. T. A. Washall, W. A. Mameniskis and F. W. Melpolder, *Anal. Chem.* **37** (1965) 748.
6. S. C. Lamey and J. T. Maloy, *Separ. Sci.* **9** (1974) 391.
7. ASTM Method D 2006-70, Annual Book of ASTM Standards, Part 24, American Society for Testing and Materials, Philadelphia 1975.
8. S. S. Kurtz, I. W. Mills, C. C. Martin, W. T. Harwey and M. R. Lipkin, *Anal. Chem.* **19** (1947) 175.
9. H. E. Lumpkin, B. W. Thomas and A. Elliott, *Anal. Chem.* **24** (1952) 1389.
10. L. Mikkelsen, R. L. Hopkins and D. Y. Yee, *Anal. Chem.* **30** (1958) 317.
11. J. Tausz, *Petroleum* **13** (1918) 649.
12. R. W. Martin, *Anal. Chem.* **21** (1949) 121.
13. J. Hofman, J. Podehradská and O. Tománek, Preprints Symposium Alken '73, Mariánské Lázně, June 18—21 1973, Vol. A, p. 246.
14. P. Schneider, K. Kochloefl and V. Bažant, *Coll. Czech. Chem. Commun.* **28** (1963) 3382.
15. D. F. Kuemmel, *Anal. Chem.* **34** (1962), 1003.
16. A. Greiner, *J. Prakt. Chem.* **27** (1965) 69.
17. J. Hofman, J. Šmídová and S. Landa, *Coll. Czech. Chem. Commun.* **35** (1970) 2174.
18. R. Griegee and P. Günther, *Ber.* **96** (1963) 1564.
19. F. Asinger, B. Fell and G. Collin, *Ber.* **96** (1963) 716.
20. M. M. Smits and D. Hoefman, *Z. anal. Chem.* **264** (1973) 297.
21. M. M. Smits and D. Hoefman, *Anal. Chem.* **44** (1972) 1688.
22. P. S. Bailey, *Chem. Rev.* **58** (1958) 925.
23. H. W. Quinn, Progress in Separation and Purification, Vol. 4, E. S. Perry and C. J. Van Oss (Eds.), p. 133, J. Wiley-Interscience, New York 1971.

24. G. D. Davis and E. C. Makin, *Separation and Purification Methods* **1** (1972) 199.
25. S. Siggia, Quantitative Organic Analysis via Functional Groups, 3. Ed., J. Wiley and Sons, New York 1963.
26. L. Ruzicka and J. Meyer, *Helv. Chim. Acta* **4** (1921) 505.
27. V. Jarolím, K. Hejno, F. Hemmert and F. Šorm, *Coll. Czech. Chem. Commun.* **30** (1965) 873.
28. C. I. Areshidze and V. A. Kikvidze, *Dokl. AN SSSR* **121** (1958) 1025.
29. G. S. Landsberg and B. A. Kazanski, *Izv. AN SSSR Khim. Ser.* **1951**, 110.
30. G. S. Landsberg, B. A. Kazanski, P. A. Bazhulin, T. F. Bulanova, A. L. Liberman, E. A. Mikhailova, A. F. Plate, K. E. Sterin, M. M. Sushchinski, G. A. Tarasova and C. A. Ukholin, Opredelenie Individualnogo Uglevodorodnogo Sostava Benzinov, Izdatelstvo AN SSSR, Moscow 1959.
31. M. I. Krasavchenko, A. A. Mikhnovskaya, O. E. Morozova and A. A. Petrov, *Neftekhimiya* **8** (1968) 663.
32. S. R. Sergienko, E. V. Lebedev and A. A. Petrov, *Dokl. AN SSSR,* **123,** (1958) 704.
33. A. A. Petrov, Khimiya naftenov, Izdatelstvo Nauka, Moscow 1971.
34. Z. V. Jakubson, O. A. Arefev and A. A. Petrov, *Neftekhimiya* **13** (1973) 345.

2.9 Other separation and purification methods

In addition to the separation methods mentioned in the preceding sections conventional methods, such as extraction and crystallization, are also used in the analysis of hydrocarbons, but to a lesser extent. We shall mention them briefly because these methods are described in detail in various manuals of laboratory techniques, and in their classical form they are generally known.

In the last paragraph of this section a description of several new separation methods is given, which are interesting in their principle or a new arrangement of the apparatus. Even though these methods are not yet checked in practice, we mention them because of their value as inspiration and as an example of their prospective possibilities in the area of "separation science".

2.9.1 SOLVENT EXTRACTION

Substances with different solubility can be separated by solvent extraction. In the separation of hydrocarbons extraction is used, for example, for the isolation of bitumens from various rocks or in the isolation of hydrocarbon mixtures by liquid-liquid extraction technique. Extractional separation methods are still widely used in the petroleum industry, for example for the isolation of aromates and in refining processes. Extraction has been replaced in many instances on a laboratory scale and in various analytical techniques, by chromatographic methods.

Preference is given, however, to extraction in those cases when a larger amount of material has to be separated and when the difference in solubility of the components are considerable.

Liquid-liquid extraction. The distribution process in liquid-liquid extraction consists in thermodynamic equilibrium between two liquid phases. If a mixture of two components, 1 and 2, is in equilibrium with two solvents, *A* and *B*, so that the two components 1 and 2 form a real solution with each of the solvents, the separation factor of one-step separation can be expressed as

$$\alpha = \frac{(N_1/N_2)^{\mathrm{A}}}{(N_1/N_2)^{\mathrm{B}}} \tag{2.9.1}$$

where N_1^A and N_2^A are the mole fractions of the component A in both solvents and N_1^B, and N_2^B are the mole fractions of the component B in both solvents. For an n-step separation process the equation

$$n - 1 = \frac{1}{\log \alpha} \log \frac{(N_1/N_2)^{\mathrm{A}}}{(N_1/N_2)^{\mathrm{B}}} \tag{2.9.2}$$

applies, from which the number of separation steps, necessary for the achievement of the required separation, between the first and the n-th step can be ascertained.

As a rule a single solvent is used for the separation of hydrocarbons by liquid-liquid extraction, forming one phase, while the hydrocarbon mixture (possibly containing a small amount of solvent) represents the second phase. The requirements of the character of the solvent are described by Rossini et al.[1]: 1. The separated hydrocarbons must have considerably different solubilities in the solvent used. 2. The solubility of the hydrocarbons in the solvent should be within the 3 – 10 % interval. 3. The solvent and the separated mixture of hydrocarbons should have considerably different densities. 4. The solvent should be easily separable from the hydrocarbon mixture. These authors used extraction for the separation of hydrocarbon groups, mainly in higher-boiling petroleum fractions (lubricant oils). On an extraction column 16.8 m long they separated aromatic hydrocarbons with methyl cyanide, and they removed the saturated hydrocarbons in the extract with another solvent (Marcol). By this method they obtained a 92 % concentrate of aromatic hydrocarbons from a narrow distillation cut of the lubricant oil, and they achieved a partial separation of mono- and diaromates. In the case of a narrow distillation cut from petroleum (boiling about 227 °C) they obtained concentrates of alkanes, monocycloalkanes and dicycloalkanes on one-step extraction with formamide.

Recently only few papers have been published[2] on the use of liquid-liquid extraction for the separation of hydrocarbons. Baker described the use of this method of extraction for the concentration of *metalloporphyrins* from petroleum fractions[3]. The principles of liquid-liquid extraction as a separation method are described, for example, by Freiser[4], and some of its applications in industry are mentioned by Francis and King[5].

Extraction is not limited to liquids only. A system of solvent-less extraction also exists, which uses gases as *supercritical fluids*[6]. The still rather frequently used analytical test – *aniline point* – is based on the different solubilities of hydrocarbons of different types in aniline. The aniline point is the lowest temperature at which an equal amount of aniline and the analysed sample are completely soluble; the lowest values of the aniline point are found in hydrocarbons with maximum solubility. For the analytical use of this value see Section 3.7.

Extraction of organic substances from solid material. This method of extraction is used frequently, mainly for the isolation of bitumens in oil shale and other rocks. A single auxiliary phase is used, i. e. the solvent. The extraction can be carried out by shaking the sample with a suitable solvent in a Soxhlet extractor, or by submitting the sample dissolved in the solvent to ultrasonic energy treatment[7]. In comparison with other methods the advantage of the use of ultrasonic energy consists mainly in its considerable rapidity. The extract contains the dissolved organic material separated from the kerogen fraction of the shale. For extraction hexane is often used as solvent. The benzene-methanol mixture (4 : 1, by volume) is more suitable because it dissolves a greater number of types of substances than hexane. The efficiency of the extraction usually increases with temperature. Even when carried out at elevated temperature the extraction does not change the composition of the organic matter from oil shale and similar material. Bitz and Nagy[8] extracted organic material from oil shales in a Soxhlet apparatus before its further processing by ozonization.

When several various solvents are used, soluble high-molecular fractions of *asphalts* can be separated into several groups. For example, from the fraction of asphalt insoluble in gasoline the following fractions can be obtained according to the classical Richardson's separation procedure:
1. *asphalthenes* – substances soluble in CCl_4 and benzene, insoluble in ethyl alcohol,
2. *carbenes* – substances insoluble in CCl_4, soluble in carbon disulphide,
3. *carboids* – substances insoluble in carbon disulphide.

Solvent extraction of coal or peat affords *montan wax*. The choice of suitable solvents, the method of obtaining montan wax and its properties are described by Wollrab and Streibl[9].

Imuta and Ouchi[10] extracted bituminous coal with benzene. From the extract they isolated *n*-alkanes, isoalkanes with the isoprenoid structure, and the tricyclic hydrocarbon adamantane, present in the original coal in a 0.16 ppm concentration.

2.9.2 CRYSTALLIZATION

The separation of substances by crystallization is based – to be the case of extraction – on mass transfer between phases. Usually crystallization is carried out by cooling the mixture to be separated. In some instances hydrocarbons (isomers) of close boiling points can be separated by crystallization. In the separation of hydrocarbons crystallization from solutions and melts can be made use of; so-called adductive (extractive) crystallization with urea or thiourea is described in detail in Section 2.5.

One-step crystallization. The degree of separation which can be achieved by one-step crystallization is dependent on the solid-liquid equilibrium. In practice a true equilibrium is hardly ever achieved. The mass transfer in solids is very slow in view of small diffusion coefficients[11]. Crystallization from solvents usually requires the substance to be separated being present in a high concentration. This concentration can be achieved by a suitable combination of other fractionation procedures. It is used mainly for the isolation of high melting substances that precipitate easily in crystalline form and are easily separated from other components. When hydrocarbons are separated by crystallization from solutions, the mixture of crystals and the liquid can consist of the separated hydrocarbon material only, or this material can be mixed with a suitable solvent. For the elimination of a maximum amount of high melting hydrocarbon from a mixture with other hydrocarbons, it is advantageous to carry out a crystallization without a solvent first. The solvent is then used for further purification of the crystals formed. The separation of several components is possible by fractional crystallization from two solvents[1].

In the case of relatively low melting hydrocarbons, forming *eutectic* systems, one-step crystallization can afford – theoretically – the pure component and the material of eutectic composition. However, the crystals formed are contaminated with the mother liquors that must be removed (by washing with a solution or recrystallization). This method is indicated as normal freezing, but it cannot be employed for the systems forming mixed crystals. Eutectic mixtures are formed, for example, from benzene and aliphatic or aromatic hydrocarbons, xylene isomers, etc.[12]. In this manner Sloan achieved a several-fold concentration of impurities in an enriched residue after the crystallization of benzene; the degree of concentration can be increased by stirring the mixture[13].

However, one-step crystallization, often used for preparative or technological purposes, is of little importance for analytical separations. The required degree of separation is usually not achieved in one step, a part of the crystals remains in the mother liquors and the separation is not quantitative.

Fractional crystallization and fractional melting. For analytical purposes the method of crystallization in a column, described by Baker and Williams[14] is of especial importance. The separation is achieved by placing the mixture to be separated on the top of an inert carrier in a column and exposing it to a temperature and solvent gradient. The linear temperature gradient is maintained along the whole column length and the separation is achieved by elution with a solvent that does not dissolve the separated substances well. At a certain temperature and solvent composition the more soluble components are dissolved and they move down the column with the solvent into the lower temperature region, where they crystallize out. These crystals are then dissolved by the enriched solvent and they proceed down the column in a series of dissolution and crystallization steps. The material leaves the column as a saturated solution. This method is used mainly for polymeric substances. Schulz and Purdy[15] exploited this method for the separation of the isomers of terphenyl and quaterphenyl, and they suceeded in separating *o,p*- and *m,p*-quaterphenyl rather successfully. The rapid method of purification of small amounts of organic substances (milligrams to grams) by multistep crystallization on a microcolumn has been described by Schildknecht et al.[16].

Fractional melting is also based on the crystallization principle and it differs from fractional crystallization in that it does not use solvents and the mixture is separated between the solid phase and the melt. By melting, substances can be separated that melt without decomposition; the method is used with advantage for the separation of substances that are thermally labile and for mixtures with excessively high boiling points.

In the case of mixtures that form series of solid solutions (mixed crystals) the cooling of the original mixture produces a system in equilibrium composed of a crystalline and a liquid phase which contain the separated components in various concentrations. For further separation of these substances the two phases must be separated. Then the first phase is melted and partly cooled, and the liquid phase is submitted to further cooling. Repetition of these operations with subsequent separation of phases at each step gives fractions that are always enriched with one of the separated components. If the number of the separation steps is sufficient, both separated components can be obtained in a pure state (in the case of eutectic mixtures one pure component and a mixture with a constant melting point are obtained). Mixed crystals are formed, for example, by benzene-thiophene and naphthalene-benzothiophene. The most efficient method of separation by fractional melting can be achieved by countercurrent contact of two phases. In practice this is achieved for mixtures forming mixed crystals by transporting the formed crystals from the site of their formation to a zone of higher temperature, where they melt under formation of an equivalent amount of crystals of higher melting point, i.e. richer in the higher-melting component. Mastrangelo and Aston[17] obtained highly pure heptane, iso octane and cyclooctane on gradual cooling and elimination of mother liquors. Fractional melting was used for the preparation of highly pure benzene[18] and some other hydrocarbons[19,20].

Zone melting serves primarily to obtain substances of high purity[21,22].

It is usually carried out so that a small heating furnace is moved over the solid mixture to be separated, placed in a tube or a boat, so that the melted zone permeates the entire substance. Impurities from the substance, forming mostly mixed crystals with it, remain mainly in the melt which thus contains more impurities than the initial material. On further movements of the furnace the melt eventually contains so much impurities that the crystallisate has the same composition as the starting material in front of the melting zone. An equilibrium state is thus attained that changes only when the melting zone reaches the end of the tube. The efficiency of the separation is increased with the number of melting zones. When additional melting zones pass through the substance, the impurities are pushed continually to the end, until a limit state is achieved after many passages. The efficiency of the enrichment of the impurities depends on the distribution constant between the crystals and the melt and the ratio between the length of the column and the melting zone. Zone melting is used for the purification of some crystalline hydrocarbons, such as naphthalene, diphenyl, anthracene, pyrene, chrysene and others[23,24]. The efficiency of purification of some aromatic hydrocarbons by zone melting is shown in Table 2.9.1.

TABLE 2.9.1

Purification of aromatic hydrocarbons by zone melting[24]

Substance	Melting point °C starting	Melting point °C final
Anthracene	217.5—221	219.5—220
Pyrene	150.0—152.0	152.0
Chrysene	248.0—249.5	250.0—250.5

Zone melting, adapted as a chromatographic technique, has been described by Pfann[25]. This technique, called *zonal chromatography*, requires a longer column in which separation takes place and a greater number of passages of zones in current zone melting refinement.

2.9.3 PERMEATION

Permeation is based on differing sorption and diffusion properties of the components of the mixture under separation, in contact with a membrane. The mechanism of permeation consists of three steps, i.e. sorption of the penetrants on the membrane surface, diffusion through the membrane, and the desorption of the penetrants from the other side of the membrane. For permeation mainly membranes of polymer type are used, but in some cases liquid or metallic membranes (for the separation of gases) are also employed.

Both gaseous and liquid mixture can be separated by permeation. The possibilities of separation of hydrocarbons by permeation were investigated mainly by Michaels and Bixler[26].

In the case of gaseous mixtures permeation is used for the preparation of helium from natural gas, for the purification of hydrogen, the separation of oxygen from air, and for some other separations[27,28]. From the point of view of the separation of hydrocarbons the permeability of liquids is more important. Under the effect of organic solvents plastic foils swell and under certain circumstances achieve a state permitting selective diffusion to some type of molecules. The diameter of the small channels between the molecules of the polymer, in which the solvent is present, increases with the time of heating and is dependent on the properties of the solvent used. In consequence of the swelling of polymers with liquids the permeability of the liquids is up to a thousand times greater than the permeability of gases[29].

Michaels and Bixler[26,30] mention, that separation by liquid permeation can be used mainly in the following cases:

1. Separation of mixtures with similar chemical structure and close boiling points, for example benzene–cyclohexane,
2. separation of mixtures of isomers, for example *n*-alkanes and isoalkanes, isomers of xylene,
3. separation of azeotropic mixtures, for example benzene–methanol,
4. separation of mixtures containing instable components.

The methods of separation are divided into *pervaporation* and *perstraction,* according to whether the desorption on the other side of the membrane takes place into a gaseous or a liquid phase. In pervaporation desorption is achieved by decreasing the partial pressure of the permeating components below the saturated vapour pressure. A review of the possibilities of pervaporation for the separation of substances is given in refs.[29,31]. The separation of xylene isomers has been investigated particularly intensively. In perstraction both sides of the membrane are in contact with a liquid phase; on one side the mixture under separation is present the components of which pass through the membrane and dissolve in another liquid – solvent. In principle, perstraction is extraction controlled by a membrane. The solvent should fulfil the following conditions:

1. The permeability for the solvent should be negligible,
2. it must be less volatile than the permeating components,
3. it must dissolve the permeating components well.

Higher boiling hydrocarbons, stable latexes and similar substances are suitable as solvents. The solubility of the permeate is dependent on structure[32]. In the case of *n*-alkanes the permeation rate increases with an increasing number of carbons in the molecule. 1-Hexene permeates three times faster than *n*-hexane. From a mixture containing 90% *o*-, 5 % *m*- and 5 % *p*-xylene, 98 % *p*-xylene was obtained after a six-fold permeation. McCandless[3] found that vinylidene fluoride resin plasticized with 3-methylsulpholene is an efficient membrane for the separation of aromatic hydrocarbons from cycloalkanes by permeation. The resin is almost impermeable for hydrocarbon vapours. The process was further improved by addition of a small amount of substances used for the extraction of aromatic hydrocarbons (dimethylformamide, dimethylsulphoxide)[34]. The possibilities of perstraction for the separation of hydrocarbons were also investigated by Shah and Owens[35]. On an industrial scale permeation still cannot compete with today's separation procedures. Of the permeation procedures perstraction seems most promising for high-capacity separation of liquid mixtures, and its industrial application for the separation of hydrocarbon mixtures is envisaged.

2.9.4 OTHER METHODS

Permeation through liquid membranes. For the separation of substances by permeation liquid membranes are now being investigated which are formed from a surface active substance and water, or a surface active substance and a water-insoluble solvent[36,37]. Liquid membranes are formed as an enveloping layer of droplets of the separated mixture, passing through the solution of the surface active substance or during the emulgation of the mixture under separation in this solvent. The permeate obtained after the removal of the solvent is enriched in the more rapidly permeating substances. This method was tested, for example for the separation of the *n*-hexane – toluene mixture; the possibilities of the separation by liquid membranes are also promising for other hydrocarbon mixtures. Liquid membranes possess certain advantages over polymeric membranes, as for example being less thick, no problems with durability, etc.

Countercurrent extraction. Ito and Bowman[38, 39] have described various arrangements for countercurrent movement of two liquid phases, which can be considered as liquid-liquid chromatography without a solid support. Similarly as in LLC one liquid moves along the other and the separation of the solute takes place on the phase boundary. The sample is constantly in solution and equilibrium is determined by dissolution and distribution isotherms. The authors mentioned developed three different apparatuses enabling mutual movement of the phases, on which almost complete extraction equilibrium can be achieved:

Helix countercurrent chromatography. – The separation takes place inside a horizontal wound narrow tube with a large number of threads. The stationary liquid is fixed in a part of each thread either by a gravitational or centrifugal field. The mobile liquid (the lighter one) is injected into one end of the helically wound separation tube and the effluent is discharged at the other end.

Droplet countercurrent chromatography (DCC). – This method is based on the passage of small droplets of the light phase through a vertical tube filled with the heavier phase. Glass silanized tubes of 20 – 60 cm length and 1.8 mm I.D. are used. The drops of the lighter phase, with low affinity toward the wall surface ascend through the heavier phase, with a visible very active interphase movement. The diameter of the drops should be close to the diameter of the tube and they should follow each other at minimum distance. The content of the tube is thus divided into small segments that prevent longitudinal diffusion in the tube. In this manner a large amount of substances can be separated with a relatively high efficiency. The dimensions of the separation tube should be adjusted for the system of solvents used, especially to the ability of the light phase to form drops. In some systems the drops are formed with difficulty, which limits the usefulness of this procedure. The DCC technique proved practical in the separation of natural substance[40, 41]. The apparatus is already commercially available[42].

Locular countercurrent chromatography. – In this method an oblique teflon column is used, separated by barriers to a large number (over 100) of segments connected with a central opening. Into the column containing the heavy solvent and rotating around its axis the light phase is introduced from below. The solute introduced into the column is submitted to a multistep distribution process during the rotation of the column, and it is eluted eventually from the column top. The compartments that almost completely prevent longitudinal diffusion increase the efficiency of the separation so that the height of the theoretical plate is merely a few mm. This method is very universal and it can also be arranged as *reversed phase* chromatography.

Separation on water surface. The dispersion of the hydrocarbon layer in a thin film on the water surface is accompanied by two effects: 1. the more volatile and soluble components are eliminated preferentially from the original mixture, 2. a concentration gradient is formed in the surface film in the direction from the centre, caused by the fact that individual hydrocarbons disperse with different velocities; the more scattering hydrocarbons are thus separated from the others. Phillips and Groseva[43] found that this separation of hydrocarbons takes place in consequence of their different dispersion coefficients and that it does not depend on their relative volatility and solubility. These coefficients differ considerably for different hydrocarbons, which gives a relatively good separation effect. In a model mixture consisting of toluene, *n*-octane and *n*-decane they found that the molar fraction of toluene with a high scattering coefficient (6.8 dyn/cm) was several times higher even after 5 seconds at a distance of about 10 cm from the centre than in the original mixture. The mole fraction of *n*-octane (scattering coefficient 0.22 dyn/cm) changed but little and in the case of *n*-decane (−2.3 dyn/cm) it was even several times diminished. A somewhat less pronounced separation effect was observed when a model mixture was mixed in a 1 : 1 ratio (by volume) with crude oil.

REFERENCES (2.9)

1. F. D. Rossini, B. J. Mair and A. J. Streiff, Hydrocarbons from Petroleum, Reinhold Publ. Corp., New York 1953.
2. S. N. Aminov, T. A. Muradov, S. A. Zanudtinov and K. S. Akhmedov, Neftepererabotka, Neftekhimiya 1970 (9) 30.
3. E. W. Baker, Organic Geochemistry, G. Eglinton and M. T. J. Murphy (Eds.), p. 470, Springer Verlag Berlin 1969.
4. H. Freiser in B. L. Karger, L. R. Snyder and C. Horvath, An Introduction to Separation Science, p. 247, J. Wiley Interscience, New York 1973.
5. A. W. Francis and W. H. King, The Chemistry of Petroleum Hydrocarbons, B. T. Brooks, S. S. Kurtz, C. E. Boord and L. Schmerling (Eds.), Vol. I, p. 197, Reinhold Publ. Corp. New York 1954.
6. E. Stahl and W. Schilz, *Z. anal. Chem.* **280** (1976) 99.
7. M. T. Murphy, Organic Geochemistry, G. Eglinton and M. T. J. Murphy (Eds.), p. 75, Springer Verlag, Berlin 1969.
8. S. M. C. Bitz and B. Nagy, *Proc. Nat. Acad. Sci. US* **56** (1966) 1383.
9. V. Wollrab and M. Streibl, Organic Geochemistry, G. Eglinton and M. T. J. Murphy (Eds.), p. 576, Springer Verlag, Berlin 1969.
10. K. Imuta and K. Ouchi, *Fuel* (1973) 301.
11. W. R. Wilcox in B. L. Karger, L. R. Snyder and C. Horvath, An Introduction to Separation Science, p. 303, J. Wiley Interscience, New York 1973.
12. W. D. Betts and G. W. Girling, Progress in Separation and Purification, Vol. 4, p. 31, E. S. Perry and C. J. Van Oss (Eds.), J. Wiley, Interscience, New York 1971.
13. G. J. Sloan, *Anal. Chem.* **38** (1966) 1805.
14. C. A. Baker and R. J. P. Williams, *J. Chem. Soc.* (*London*) **1956,** 2352.
15. W. W. Schulz and W. C. Purdy, *Anal. Chem.* **35** (1963) 2044.
16. H. Schildknecht, V. Reimann-Dubbers and K. Maas, *Chem. Zeitung* **94** (1970) 437.
17. S. V. R. Mastrangelo and J. G. D. Aston, *Anal. Chem.* **26** (1954) 764.
18. J. D. Dickinson and E. Eaborn, *Chem. and Ind.* **1956,** 959.
19. T. H. Gouw, *Separ. Sci.* **3** (1968) 313.
20. T. H. Gouw, Progress in Separation and Purification, Vol. 1, p. 143, E. S. Perry (Ed.), J. Wiley Interscience, New York 1968.
21. H. Schildknecht, Zonnenschmelzen, Verlag Chemie, Weinheim 1964.
22. W. G. Pfann, Zone Melting, 2nd Ed., J. Wiley Sons, New York 1966.
23. H. Schildknecht and U. Hopf, *Z. anal. Chem.* **193** (1963) 401.
24. R. Handley and E. F. C. Herington, *Chem. Industries* **1956,** 304.
25. W. G. Pfann, *Anal. Chem.* **36** (1964) 2231.
26. A. S. Michaels, H. S. Bixler and P. N. Rigopoulos, 7th World Petrol. Congr. Proceedings, Vol. 4, p. 21, 1967.
27. S. A. Stern and W. P. Walaswender, *Separ. Sci.* **4** (1969) 129.
28. K. Kammermeyer, Progress in Separation and Purification, Vol. 1, E. S. Perry (Ed.), p. 335, J. Wiley Interscience, New York 1968.
29. R. C. Binning, R. J. Lee, J. F. Jennings and E. C. Martin, *Ind. Eng. Chem.* **53** (1961) 45

30. A. S. Michaels and H. S. Bixler, Progress in Separation and Purification, Vol. 1, p. 143, J. Wiley Interscience, New York 1968.
31. N. N. Li and R. B. Long, Progress in Separation and Purification, Vol. 3, E. S. Perry Ed., Interscience, New York 1970, p. 153.
32. W. Schirmer, *Chem. Tech.* **17** (1965) 259.
33. F. P. Mc Candless, *Ind. Eng. Chem. Prod. Des. Develop.* **12** (1973) 354.
34. F. P. Mc Candless, D. P. Arzheimer and R. B. Hartman, *Ind. Eng. Chem. Prod. Des. Develop.* **13** (1974) 312.
35. N. D. Shah and T. C. Owens, *Ind. Eng. Chem. Prod. Des. Develop.* **11** (1972) 58.
36. N. N. Li, *Ind. Eng. Chem. Prod. Des. Develop.* **10** (1971) 215.
37. M. Sheldrick, *Chem. Eng.* **77** (1970) 52.
38. Y. Ito and R. L. Bowman, *Science* **167** (1970) 281.
39. Y. Ito and R. L. Bowman, *J. Chromatogr. Sci.* **8** (1970) 315.
40. T. Tanimura, J. J. Pisano, Y. Ito and R. L. Bowman, *Science* **169** (1970) 54.
41. Y. Ogihara, O. Inoue, H. Otsuka, K. Kawai, T. Tanimura and S. Shibata, *J. Chromatogr.* **128** (1976) 218.
42. K. Hostettmann, M. Hostettmann-Kaldas and K. Nakanishi, *J. Chromatogr.* **170** (1979) 355.
43. C. R. Phillips and V. M. Groseva, *Separ. Sci.* **10** (1975) 111.

2.10 Combined procedures

In the preceding sections of this chapter various methods of separations of hydrocarbon mixtures have been reviewed and it was shown that for the separation effect even very small differences in some properties of the components suffice, for example in vapour pressure, molecular weight, volume or the shape of the molecule, polarizability, chemical reactivity, etc. In the case of hydrocarbon mixtures with a high number of components it is indispensable to carry out a preliminary separation of the mixture to simpler fractions before analysis. The more complex and the higher boiling the analysed mixture is, the greater number of separation methods must be combined in order to achieve a satisfactory simplification.

With complex unknown samples the analytical procedure should be thoroughly considered beforehand and an operation scheme should be planned according to the available characteristics of the mixture (origin of the sample, boiling range, content of heteroatoms, etc.), in which the separation and the identification methods are rationally connected. Simultaneously the extent of the required information should be envisaged, as well as the duration and the expense of the analysis, and sometimes even the amount of the sample available. A well composed scheme permits the required extent of information to be obtained with a minimum number of steps.

When planning the procedure a problem sometimes arises concerning the choice of the sequence of individual separation methods. Only general principles can be mentioned on this problem, following from the fact that at the beginning of the analysis usually a large amount of sample is available which contains easily separable groups of compounds of different structural classes, while at the terminal phase of the analysis the sample is separated to small fractions containing compounds with only slightly different properties. Therefore it is useful to apply cheap steps at the beginning of the analysis, with a low *fractionation capacity* but with a high *productivity* (for example simple phase separations) and at the terminal steps the methods with a higher fractionation capacity, which – it is true – have a low productivity, but a high *separation efficiency* (chromatographic methods, GPC, etc.).

The great majority of the hydrocarbon mixtures occurring for analysis comes from petroleum, coal and other fossile fuels or from the products of their thermal or chemical processing. As regards their composition and complexity these fractions are similar in many respects and therefore the proved sequence of the separation methods can be applied successfully to samples of various origins. The schemes used are improved from time to time in accordance with the requirements put on the accuracy or the speed of the analysis and the progress of instrumental analytical methods.

As said in the introduction, nowadays three main types of analysis are employed for the evaluation of complex hydrocarbon mixtures: a) the determination of the content of all structural classes present, b) the determination of individual hydrocarbons within one structural class, c) the quantitative determination of all individual hydrocarbons in the fraction.

According to the type of analysis selected, according to the nature of the crude material and the aims set, the combinations of the separation and the analytical methods can be changed in various ways. Some especially efficient and proved methods are often used, however, in wel established sequence. When separation schemes for the analysis of high-boiling petroleum fractions and similar hydrocarbon mixtures are set, the procedures are now preferred which are briefly mentioned below:

If the analysed mixture has an excessively wide boiling range it is separated to several fractions on a *flash-distillation* column (see Section 2.1.4). This prevents changes in the chemical composition of heavy fractions.

Acid and basic non-hydrocarbon components can be eliminated from the fraction by column chromatography on *ion-exchangers*[1], worked out in *API Research Project 60*. The acids are retained on the column packed with an anion exchanging resin (Rohm and Haas, Amberlyst A-29) activated with isopropylamine in methanol. The non-reactive and hydrocarbon material is eluted from the column with pentane and the retained acids are then eluted by a procedure described in greater detail in Section 2.3.4. From the acid-free fraction bases are separated on a column packed with a cation exchanger (Amberlyst A-15) activated with HCl in methanol. After the elution of non-reactive material with pentane, the bases (pyridine, anilines etc.) can be isolated by stepwise elution with benzene, methanol and isopropylamine in methanol.

A shorter procedure for the elimination of acids and bases consists in neutralization washing used, for example, by the group Dooley[2] for the analysis of synthetic petroleum from Synthoil process. The fraction is first extracted with the same volume of 1.5 N NaOH in 1 : 1 methanol-water mixture. The fraction freed from the acids is then extracted with the same volume of 1.5 N HCl in 1 : 1 methanol-water mixture. The acids and bases are isolated from the extract by neutralization and extraction with ether. In the analysis of tar fraction extraction with 10 % NaOH at 50 °C and with dilute sulphuric acid is recommended.

The neutral nitrogen-containing compounds, often present in hydrocarbon fractions, are separated in the form of *complexes* with $FeCl_3$[1]. The lower half of the column is packed with an anion exchanger and the upper part with ferric chloride adsorbed on a mixture of kaolin and Chromosorb W. The fraction from which acids and bases are eliminated is *percolated* through a column and the unreactive oil is eluted with pentane. The complexes of the nitrogen compounds with $FeCl_3$ are then eluted with 1,2-dichloroethane. In contact with the anion exchanger the complexes decompose, the metal salts are retained, and the free nitrogen compounds are obtained from the effluent. The procedure is not quite quantitative and a small amount of nitrogen compounds remains in the hydrocarbon oil.

For further separation of the neutral oil various variants of *liquid chromatography* on silica gel or alumina are used (see Section 2.2.5). According to a procedure used earlier in *API RP 60* for the analysis of heavy ends of petroleum the neutral oil was separated by silica gel chromatography to a class of saturated hydrocarbons and a class of aromatic hydrocarbons (plus neutral sulphur and oxygen compounds). In the case of lighter distillates, containing olefins, three separate groups can be eluted from a silica gel column with pentane: saturated hydrocarbons, aliphatic olefins and aromates[2]. Today suitable combinations of various chromatographic techniques have been elaborated which permit the relatively sharp[3] separation of even the heaviest distillates into several groups. According to a procedure developed by *Bureau of Mines* – API *Research Project 60* the high boiling fractions are separated by gradient elution chromatography on a dual packed column[4] with a fully activated (16 h/400° C) alumina-gel (Alcoa F-20) in the lower half, and a fully activated (16 h/265 °C) silica gel (Davidson grades 12 or 62) in the upper half of the column. The sample of the oil is diluted with a ten-fold amount of pentane and allowed to soak into the column. Pentane elutes first saturated hydrocarbons and then gradient elution is applied, described in Section 2.2.5, by which the groups of aromates, diaromates and eventually polyaromatic hydrocarbons and other polar substances are eluted. The procedure is suitable both for high-boiling petroleum distillates[5, 6] and for heavy fractions of synthetic oils from coal[2].

The isolated class of saturated hydrocarbons can be analysed by *mass spectrometry* directly and the content of aliphatic hydrocarbons and cycloalkanes of various numbers of rings determined as described in Section 3.1. If the saturated hydrocarbons have to be fractionated further, *n*-alkanes are separated usually by means of *molecular sieves* 5A or by *adductive crystallization* with urea. The sharp separation of branched alkanes from cycloalkanes contained in a wide fraction is difficult and not yet solved; best results in the concentration of both types are achieved by *gel permeation chromatography* and *thermodiffusion*. Both methods are also suitable for the separation of cycloalkanes according to the number of rings.

The groups of monoaromates, diaromates and polyaromates obtained by LSC may be analysed for the content of individual structural types by various *spectral methods* (MS, NMR, UV etc.). More accurate results can be obtained, however, if the aromatic groups are first fractionated by gel permeation chromatography.

In special cases, when it is necessary to isolate some group of compounds or even individual hydrocarbons from the separated structural groups, preparative gas chromatography, or high-performance liquid chromatography are most commonly applied as the last separation step.

For illustration two separation schemes are presented. Fig. 2.10.1 shows a scheme as an example of the analysis of individual hydrocarbons in one structural class, used by Gelpi et al.[7] for the isolation of steranes

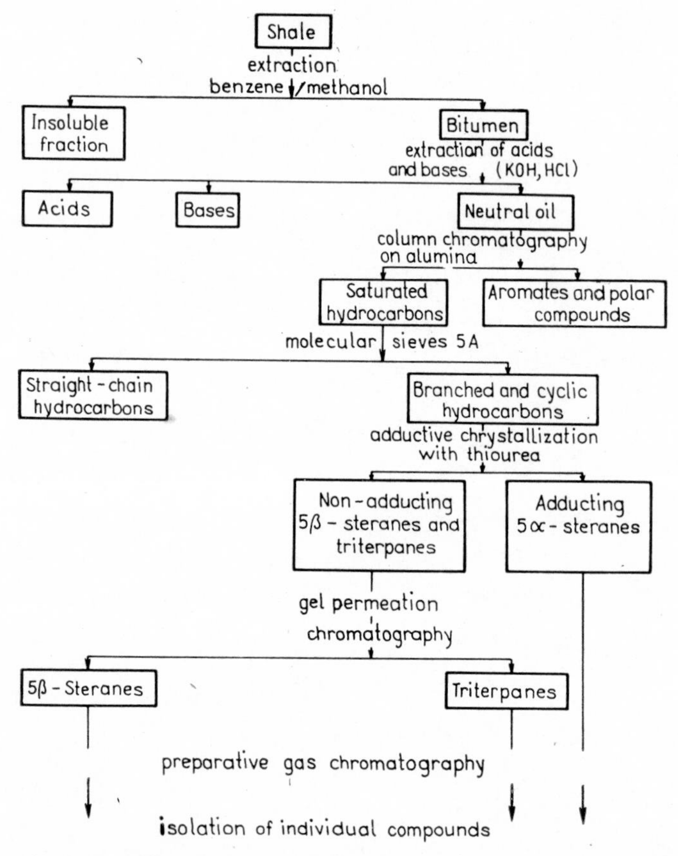

Fig. 2.10.1 The separation scheme used for the isolation of individual steranes and triterpanes from oil shale[7]

and triterpanes from Green River Formation Oil Shale. Fig. 2.10.2 shows a scheme as an example of an analysis aimed at the determination of the content of all structural classes; this scheme was developed for the characterization of high-boiling petroleum distillates[5]. The scheme is relatively universal because it also enables the analysis of acid, basic and other non-hydrocarbon components and it can be applied for the analysis of high-boiling hydrocarbon fractions of various origins.

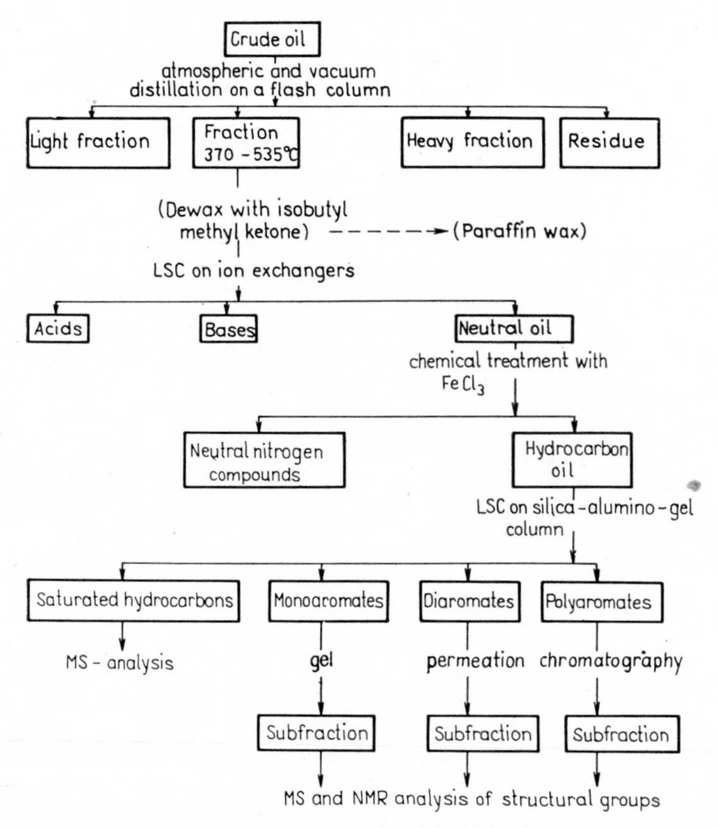

Fig. 2.10.2 The separation scheme for quantitative determination of the content of structural groups in high-boiling hydrocarbon fractions[5]

REFERENCES (2.10)

1. W. E. Haines, C. C. Ward and J. M. Sugihara, Development of Analytical Techniques for Heavy Hydrocarbons and Related Compounds. Preprint No. 24—71 from the Midyear Meeting of the American Petroleum Institute's Division of Refining, San Francisco 1971.
2. J. E. Dooley, G. P. Sturm, P. W. Woodward, J. W. Vogh and C. J. Thompson, Bartlesville Energy Research Center, United States Energy Research a. Development Administration, Reports: BERC/RI-75/7 (August 1975); BERC/RI-75/12 (November 1975); BERC/RI-76/2 (January 1976).
3. D. R. Latham and W. E. Haines, *Prepr., Div. Pet. Chem. Am. Chem. Soc.* **18/3** (1973) 567.
4. D. E. Hirsch, R. L. Hopkins, H. J. Coleman, F. O. Cotton and C. J. Thompson, *Anal. Chem.* **44** (1972) 915.
5. H. J. Coleman, J. E. Dooley, D. E. Hirsch and C. J. Thompson, *Anal. Chem.* **45** (1973) 1724.
6. C. J. Thompson, J. E. Dooley, D. E. Hirsch and C. C. Ward, *Hydrocarbon Processing* **52,** No. 9 (1973) 123; **53,** No. 4 (1974) 93; **53,** No. 7 (1974) 141; **53,** No. 8 (1974) 93; **53,** No. 11 (1974) 187.
7. E. Gelpi, P. C. Wszolek, E. Yang and A. L. Burlingame, *Anal. Chem.* **43** (1971) 864; *J. Chromatogr. Sci.* **9** (1971) 147.

Index

A more detailed index for both parts is given at the end of part B